AF352362

Cell Migration

METHODS IN MOLECULAR BIOLOGY™

John M. Walker, SERIES EDITOR

METHODS IN MOLECULAR BIOLOGY™

Cell Migration

Developmental Methods and Protocols

Edited by

Jun-Lin Guan

*The Department of Molecular Medicine,
College of Veterinary Medicine, Cornell University, Ithaca, NY*

HUMANA PRESS ✳ TOTOWA, NEW JERSEY

Production Editor: C. Tirpak
Cover design by Patricia F. Cleary

Cover Illustration: From Fig. 1, Chapter 7, "Cell-Scatter Assay," by Hong-Chen Chen.

For additional copies, pricing for bulk purchases, and/or information about other Humana titles, contact Humana at the above address or at any of the following numbers: Tel: 973-256-1699; Fax: 973-256-8341; E-mail: humana@humanapr.com; or visit our Website: www.humanapress.com

Photocopy Authorization Policy:
Authorization to photocopy items for internal or personal use, or the internal or personal use of specific clients, is granted by Humana Press Inc., provided that the base fee of US $25.00 per copy is paid directly to the Copyright Clearance Center at 222 Rosewood Drive, Danvers, MA 01923. For those organizations that have been granted a photocopy license from the CCC, a separate system of payment has been arranged and is acceptable to Humana Press Inc. The fee code for users of the Transactional Reporting Service is: [1-58829-382-3/05 $25.00].

Printed in the United States of America. 10 9 8 7 6 5 4 3 2 1

ISSN 1064-3745

E-ISBN 1-59259-860-9

Cell migration : developmental methods and protocols / edited by Jun-Lin Guan.

    p. ; cm. -- (Methods in molecular biology ; 294)

  Includes bibliographical references and index.

  ISBN 1-58829-382-3 (alk. paper)

  1. Cell migration--Laboratory manuals.   I. Guan, Jun-Lin. II. Series:

  Methods in molecular biology (Clifton, N.J.) ; v. 294.

  [DNLM: 1. Cell Movement--physiology--Laboratory Manuals. 2. Biological Assay--methods--Laboratory Manuals. 3. Cell Movement--genetics--Laboratory Manuals.   W1 ME9616J v.294 2005 / QH 647 C3929 2005]

  QH647.C4385 2005

  571.6--dc22

                                        2004016499

Preface

Cell migration plays an essential role in a variety of biological processes, including embryonic development, homeostasis, immune responses, wound healing, and cancer metastasis. Indeed, one of the defining features of life is the ability to move, which occurs at different levels from the macromolecular machines in a cell, to individual or populations of cells, and to the whole organism. The dramatic movements of layers of cells in gastrulation, a critical period of embryonic development of many animals, as well as the swimming of simple unicellular amoeba in search of nutrients, have fascinated biologists for years.

Cell migration is a complex process and has attracted the interest of researchers from diverse disciplines in biology. However, like many other areas of investigation, the recent exciting and explosive growth of cell migration studies are critically linked to the rapid advance of cell and molecular biology and its increasing influence on all the other fields of life sciences and biomedical research. Thanks to rapid progress at the cellular and molecular levels, unprecedented knowledge has been gathered regarding the components and interactions of the macromolecular machinery of cell movement, the signaling mechanisms and pathways involved in cellular responses to chemotactic and haptotactic stimulations, and the regulatory mechanisms that link the signaling pathways to the cell movement machinery.

A second major contributor to our modern understanding of, and research on, cell migration is the recent creation and success in using various model organisms in the study of cell migration. The power of genetics in these systems allowed for the discovery of novel molecules and signaling pathways that are important in the regulation of cell migration. The conservative nature of many of these molecules and pathways has also allowed researchers to move from one system to another when putting together the various puzzle pieces of the regulatory mechanisms, taking advantage of our knowledge in each of these different systems. Last but not least, the use of model organisms has also allowed scientists to study the intrinsically interesting biology of cell migration in the context of development and other biological processes in vivo.

A third emerging trend of modern cell migration research is the increasing use of more sophisticated imaging and other novel genetic approaches. However, it is important to note that many of the simple and classic cell migration assays developed in earlier times are still part of toolboxes of cell migration researchers, and indeed they provide the basis from which the newer variations and novel approaches are developed. The combination of this long history with the recent expansion has ensured the development of a large number of methodologies and experimental protocols in the cell migration field that are derived from diverse biological disciplines, and sometimes as results of interdisciplinary efforts.

The aim of *Cell Migration: Developmental Methods and Protocols* is to bring together a wide range of novel and state-of-the-art methodologies as well as various classic methods in cell migration research in cultured cells, different model organisms, and as applied to specialized cells in normal development and in disease. Each chapter is presented in the user friendly format of the highly successful *Method in Molecular Biology* series by leading experts in their field who have contributed significantly to the development of the methods and/or applied the methods successfully in their research on various aspects of cell migration. This volume begins with an overview of the cell migration field that also discusses the methods presented. Part II (Chapters 2–8) describes various basic assays that apply to all cell migration studies in vitro. Part III (Chapters 9–12) includes cell migration assays for cancer cells, endothelial cells, and neurons both in vitro and in animal models. Part IV (Chapters 13–19) discusses cell migration assays in various model organisms. Part V (Chapters 20–24) combines chapters on biochemical methods to study the mechanisms of cell migration and several new approaches to the study of cell migration.

The overall goal of *Cell Migration: Developmental Methods and Protocols* is to provide comprehensive coverage on all relevant methods in one volume so that anyone who is interested in cell migration research can rely on it as a useful reference. It is intended for both beginning students and active researchers in the field. The primary target audience is cell and developmental biologists. However, it should also be very useful to geneticists, biochemists, and molecular biologists, as well as clinicians and others who are interested in looking at the mechanisms of cell migration or use of cell migration as biological assays in their studies of molecular mechanisms or disease processes.

The publication of *Cell Migration: Developmental Methods and Protocols* would not have been possible without outstanding contributions from all the authors. I am extremely grateful for their generosity in sharing with the readers their expertise and perspective on the various methods and protocols described. My heartfelt thanks also go to Cindy Westmiller, who has provided tremendous assistance in proofreading all the chapters and organizing the manuscripts. I would also like to acknowledge the series editor, Dr. John Walker, for his valuable input and guidance during the editing process.

We all share the hope that this book will find its place on the shelves on many laboratories and be used as a frequent reference by students and experienced researchers alike in the exciting field of cell migration studies.

Jun-Lin Guan

Contents

Contributors

Dominique Alfandari • *Department of Cell Biology, University of Virginia School of Medicine, Charlottesville, VA*

Christoph Ballestrem • *Department of Molecular Cell Biology, The Weizmann Institute of Science, Rehovot, Israel*

Allison L. Berrier • *LSU School of Dentistry, Department of Cell Biology and Anatomy, New Orleans, LA*

Eberhard Bodenschatz • *Department of Physics, Cornell University, Ithaca, NY*

Marianne Bronner-Fraser • *Division of Biology, California Institute of Technology, Pasadena, CA*

Hong-Chen Chen • *Department of Life Science, National Chung Hsing University, Taichung, Taiwan*

Edna Cukierman • *Tumor Cell Biology, Basic Science, Fox Chase Cancer Center, Philadelphia, PA*

Lance Davidson • *Department of Cell Biology, University of Virginia School of Medicine, Charlottesville, VA*

Maria Elena de Bellard • *Division of Biology, California Institute of Technology, Pasadena, CA*

Douglas W. DeSimone • *Department of Cell Biology, University of Virginia School of Medicine, Charlottesville, VA*

Frauke Drees • *Department of Molecular and Cellular Physiology, Beckman Center for Molecular and Genetic Medicine, Stanford University School of Medicine, Stanford, CA*

Andrea Flesken-Nikitin • *Department of Biomedical Sciences, Cornell University, Ithaca, NY*

Benjamin Geiger • *Department of Molecular Cell Biology, The Weizmann Institute of Science, Rehovot, Israel*

Jun-Lin Guan • *Department of Molecular Medicine, Cornell University College of Veterinary Medicine, Ithaca, NY*

Mien V. Hoang • *Division of Cancer Biology and Angiogenesis, Department of Pathology, Beth Israel Deaconess Medical Center and Harvard Medical School, Boston, MA*

Alan F. Horwitz • *Department of Cell Biology, University of Virginia, Charlottesville, VA*

Vidhya V. Iyer • *Department of Cell and Developmental Biology, University of North Carolina-Chapel Hill, Chapel Hill, NC*

ANTONIO JACINTO • *Instituto Gulbenkian de Ciência, Rua da Quinta Grande, Oeiras, Portugal*

JOACHIM KIRCHNER • *Department of Molecular Cell Biology, The Weizmann Institute of Science, Rehovot, Israel*

DENNIS F. KUCIK • *Research Service, Birmingham VA Medical Center, and Department of Pathology, University of Alabama at Birmingham, Center for Biophysical Sciences and Engineering, Pittsburgh, PA*

SUSAN E. LAFLAMME • *Center for Cell Biology and Cancer Research, Albany Medical College, Albany, NY*

SONG LI • *Department of Bioengineering and The Center for Tissue Engineering, University of California, Berkeley, CA*

ERIK A. LUNDQUIST • *Department of Molecular Biosciences, University of Kansas, Lawrence, KS*

MUNGO MARSDEN • *Department of Cell Biology, University of Virginia School of Medicine, Charlottesville, VA*

JOCELYN A. MCDONALD • *Department of Biological Chemistry, The Johns Hopkins University School of Medicine, Baltimore, MD*

DAVID MIKOLON • *Department of Immunology, The Scripps Research Institute, La Jolla, CA*

DENISE J. MONTELL • *Department of Biological Chemistry, The Johns Hopkins University School of Medicine, Baltimore, MD*

SHARVARI M. NADKARNI • *Department of Physics, Cornell University, Ithaca, NY*

W. JAMES NELSON • *Department of Molecular and Cellular Physiology, Beckman Center for Molecular and Genetic Medicine, Stanford University School of Medicine, Stanford, CA*

ALEXANDER YU. NIKITIN • *Department of Biomedical Sciences, Cornell University, Ithaca, NY*

YI RAO • *Department of Anatomy and Neurobiology, Washington University School of Medicine, St. Louis, MO*

AMY REILEIN • *Department of Molecular and Cellular Physiology, Beckman Center for Molecular and Genetic Medicine, Stanford University School of Medicine, Stanford, CA*

DANIEL S. RHOADS • *Department of Molecular Medicine, Cornell University, Ithaca, NY*

ANNE J. RIDLEY • *Department of Biochemistry and Molecular Biology, University College London, London, UK and Ludwig Institute for Cancer Research, Royal Free and University College Medical School Branch, London, UK*

LUIS G. RODRIGUEZ • *Department of Molecular Medicine, Cornell University College of Veterinary Medicine, Ithaca, NY*

MICHAEL D. SCHALLER • *Department of Cell and Developmental Biology, University of North Carolina-Chapel Hill, Chapel Hill, NC*

DONALD R. SENGER • *Division of Cancer Biology and Angiogenesis, Department of Pathology, Beth Israel Deaconess Medical Center and Harvard Medical School, Boston, MA*

DIANE S. SEPICH • *Department of Biological Sciences, Vanderbilt University, Nashville, TN*

M. AFAQ SHAKIR • *Department of Molecular Biosciences, University of Kansas, Lawrence, KS*

LESLIE M. SHAW • *Division of Cancer Biology and Angiogenesis, Department of Pathology, Beth Israel Deaconess Medical Center and Harvard Medical School, Boston, MA*

LILIANNA SOLNICA-KREZEL • *Department of Biological Sciences, Vanderbilt University, Nashville, TN*

LOLING SONG • *Department of Physics, Cornell University, Ithaca, NY*

CHRIS STORGARD • *Department of Immunology, The Scripps Research Institute, La Jolla, CA*

DWAYNE G. STUPACK • *Department of Immunology, The Scripps Research Institute, La Jolla, CA*

CAMILLA VOELTZ • *Department of Physics, Cornell University, Ithaca, NY*

MICHAEL E. WARD • *Department of Anatomy and Neurobiology, Washington University School of Medicine, St. Louis, MO*

DONNA J. WEBB • *Department of Cell Biology, University of Virginia, Charlottesville, VA*

WATT W. WEBB • *Department of Applied and Engineering Physics, Cornell University, Ithaca, NY*

CLAIRE M. WELLS • *Randall Centre, King's College London, London, UK Ludwig Institute for Cancer Research, Royal Free and University College Medical School Branch, London, UK*

REBECCA M. WILLIAMS • *Department of Applied and Engineering Physics, Cornell University, Ithaca, NY*

WILLIAM WOOD • *Instituto Gulbenkian de Ciência, Rua da Quinta Grande, Oeiras, Portugal*

CHUANYUE WU • *Department of Pathology, University of Pittsburgh, Pittsburgh, PA*

XIAOYANG WU • *Department of Molecular Medicine, Cornell University College of Veterinary Medicine, Ithaca, NY*

HUAYE ZHANG • *Department of Cell Biology, University of Virginia, Charlottesville, VA*

WARREN R. ZIPFEL • *Department of Applied and Engineering Physics, Cornell University, Ithaca, NY*

I

INTRODUCTION

Cell Migration

An Overview

Donna J. Webb, Huaye Zhang, and Alan F. Horwitz

Summary

Cell migration is an essential process for normal development and homeostasis that can also contribute to important pathologies. Not surprisingly, there is considerable interest in understanding migration on a molecular level, but this is a difficult task. However, technologies are rapidly emerging to address the major intellectual challenges associated with migration. In this chapter, we outline the basics of cell migration with an emphasis on the diverse systems, methodologies, and techniques described in this book. From the contributions presented, it is apparent that the next few years should produce major advances in our understanding of cell migration.

Key Words: Cell migration; adhesion; signaling; protrusion; development.

1. Introduction

Cell migration is a complex process that is essential for embryonic development and homeostasis *(1,2)*. In gastrulation, migration is particularly robust, where essentially all cells migrate as sheets to form the three layers, including endoderm, ectoderm, and mesoderm that comprise the resulting embryo. Cells within these layers migrate to target locations throughout the developing embryo, where they differentiate and form various tissues and organs. The migration of cells from epithelial layers to their targets is a general phenomenon that occurs throughout development. In the developing cerebellum, neuronal precursor cells migrate from the epithelium to their residences in distinct layers. One special form of migration during development is the extension of neurites. The tip of a developing neurite, the growth cone, shares many similarities with a migrating cell. The precise guidance and target recognition of

From: *Methods in Molecular Biology, vol. 294: Cell Migration: Developmental Methods and Protocols*
Edited by: J-L. Guan © Humana Press Inc., Totowa, NJ

growth cones are central to the establishment of the neuronal network and thus cognitive functions.

Migration is not limited to development, but occurs in the adult, where it is central to both normal and pathological states. For example, the migration of precursor cells from the basal layer to the epidermis functions to continuously renew skin. Other homeostatic processes, including wound repair and mounting an effective immune response, also require migration. Leukocyte migration from the circulation into the surrounding tissue, where they ingest bacteria, is important for mounting an immune response. Migration also can contribute to some pathological processes, such as vascular disease, chronic inflammatory diseases, and tumor formation and metastasis. For example, tumor formation is accompanied by the construction of a new vascular network, which involves migration of the endothelial cells from pre-existing blood vessels into the tumor, where they proliferate and form the new vessels (angiogenesis). Migration also occurs during metastasis when some tumor cells migrate out of the initial tumor into the circulation and move to new locations, where they form a secondary tumor. Because the invasion of tumor cells from the primary site into the surrounding area and angiogenesis is essential for tumor development, assays have been developed to study these processes. In Chapter 9, Shaw outlines an assay for examining the invasion of tumor cells through Matrigel. Assays to dissect the signaling events involved in endothelial cell migration and model systems for angiogenesis are described in Chapters 10, 11, and 19.

Although many studies have examined cell migration in vertebrates, migration is equally important in invertebrates, plants, and some single-cell organisms. For example, during development of *Caenorhabditis elegans*, cells migrate within the embryo along defined trajectories. Each cell stops to divide, and the daughter cells continue to migrate. Because of the small size and the transparency of *C. elegans*, individual cells can be followed as they migrate in living embryos, making it a simple system for studying migration. In Chapter 13, Shakir and Lundquist describe the methods for analyzing migration in *C. elegans*. Another invertebrate model system is the fruit fly, *Drosophila melanogaster*. *Drosophila* has a more complicated body plan and therefore has increased complexity in its migration patterns during development and adult life. One example is primordial germ cells, which migrate through the midgut epithelium and attach to the mesoderm, where they associate with the gonadal precursors and eventually form a gonad on either side of the embryo. Chapters 14 and 15 outline protocols for studying migration of different cell types in *Drosophila* and live imaging of *Drosophila* embryos. The information gained from studying invertebrate cell migration can be very useful for understanding migration in more complex organisms, as confirmed by the high degree of homology between many invertebrate and vertebrate gene products that are involved in migration.

2. The Migratory Cycle

Migration can be thought of as a cyclical process. It begins when a cell responds to an external signal by polarizing and extending a protrusion in the direction of movement. The formation of adhesion complexes functions to stabilize the protrusion by attaching it to the substratum on which the cell is migrating. These adhesions, which serve as traction points for migration, initiate signals that regulate adhesion dynamics and protrusion formation *(3)*. Contraction then moves the cell body forward and release of the attachments at the rear, as the cell retracts, completes the cycle. Slow-moving cells, such as fibroblasts, show these distinct steps of migration, but they are less obvious in other cell types. For example, rapidly migrating cells, such as keratocytes and leukocytes, glide over the substratum by protruding and retracting smoothly without forming obvious attachments.

2.1. Polarization

Many different molecules serve as external agents that initiate and promote migration. For example, some molecules initiate a migratory phenotype (chemokinetic) whereas others reside in soluble (chemotactic)- or substrate (haptotactic)-associated gradients and lead to directed movement. These molecules and their receptors are well studied in leukocytes. These cells can sense the presence of even a shallow gradient, in which they polarize and migrate persistently in one direction *(4)*. Their persistent polarity is apparent when the cells sense changing chemotactic gradients and the entire cell turns rather than extending a new protrusion from another region. In contrast, fibroblasts are more plastic and can extend protrusions from any position in the cell as they change directions. Analysis of directional cell migration using the Boyden chamber and the Dunn chemotaxis chamber are discussed in Chapters 2 and 4. In addition, Chapter 24 outlines a method for generating gradients for studying directional cell migration.

2.2. Adhesion Complexes

Adhesion complexes, which are sites of attachment between the cell and the extracellular matrix (ECM), are composed of a number of proteins, including the integrin family of transmembrane receptors, kinases, adaptor and structural molecules *(3)*. Integrins serve as the functional connection between the ECM and the actin cytoskeleton. The small GTPase, Rac, induces the formation of small adhesions at the leading edge *(5–7)*. These adhesions serve as traction points and transmit strong propulsive forces that move the cell body forward *(8)*. The maturation of these small adhesions into larger, more organized structures actually inhibits migration *(3,9)*. In Chapter 5, Kucik and Wu present protocols for analyzing cell adhesion under static conditions and shear stress, and in Chapter 6, Berrier and LaFlamme provide a method to quantify cell spreading, which is an important aspect of cell adhesion. Finally, in Chapter 21, Drees et al. describe methods for analyzing protein complexes at the cell membrane after cell–cell and cell–ECM adhesion.

Because tyrosine phosphorylation of adhesion components is thought to regulate their dynamics, there is great interest in studying this process. This can be accomplished with a "phosphotyrosine reporter" in which yellow fluorescent protein (YFP) is fused to two phosphotyrosine-binding Src-homology 2-domains derived from c-Src *(10)*. Quantitative fluorescent microscopy with this reporter has been used to study the kinetics of tyrosine phosphorylation in adhesions. Details of this method are provided in Chapter 20.

2.3. Moving Forward and Trailing Behind

Actin–myosin contractility at the front of the cell serves to pull the cell body forward in the direction of movement. Release of adhesions at the cell rear and retraction of the tail are also mediated by myosin. Spatial and temporal regulation of Rho GTPases controls these processes through effectors, such as Rho kinase, that regulate actomyosin contractility. Rho kinase has been implicated in release of adhesions at the cell rear through regulation of myosin II *(11)*. Other molecules implicated in the release of adhesions include the protease calpain and a phosphatase, calcineurin *(12,13)*. Microtubules also function in the regulation of adhesion disassembly, probably through the modulation of Rac activity *(14)*.

3. Modes of Migration

3.1. Single-Cell Migration

As discussed previously, the migration of individual cells requires the asymmetrical organization of cellular activities. In culture, many different cell types can become polarized with a front and rear asymmetry, but this is usually transient and results in random migration. The stabilized, directional movement of cells requires external cues. In tissue, these external cues are typically provided by the surrounding environment. The external cues activate intracellular signaling pathways that control polarization and directed cell movement. The small GTPases, Rac, and Cdc42, play a prominent role in regulating this process *(5,6)*.

3.2. Monolayer Cell Migration

During embryogenesis, many cells do not migrate as single cells but rather as sheets or loosely associated clusters. In vitro, scratching or wounding a cell monolayer induces the synchronized movement of sheets of cells. In Chapter 3, Rodriguez et al. outline a protocol for performing a wound assay to study directional migration. As with single cells, the migrating sheets detect the direction of migration and polarize with protrusive activity constrained to the front. Interactions with their neighbors can provide additional directional cues to cells in a monolayer. Like single cell movement, the migration of cell monolayers is regulated by the Rho GTPases *(15)*. Rac and CDC42 are essential for the polarization and migration of the cell monolayers.

Some tumor cells also adopt this mode of migration. Primary melanoma explants migrate in collagen gels as multicellular clusters that are polarized with a clearly defined leading edge at the front of the clusters and a trailing edge at the rear *(16)*. The invasive movement of the multicellular clusters, like single cell migration, is integrin dependent. Inhibition of $\beta 1$ integrin function produces dramatic phenotypic changes in the migration, such as dispersion of the clusters. The cells respond to this by converting to a single cell mode of migration. The dispersion of clusters of cells can be measured with a scatter assay. In Chapter 7, Chen describes an assay for studying the scatter response of epithelial cells to stimulation with growth factors.

4. Cell Migration In Vivo

Much of our knowledge regarding migration has been obtained from cells growing on flat surfaces, such as cover slips or tissue culture dishes. Only recently, have studies begun to examine migration in environments, which more closely mimic that observed in vivo, such as three-dimensional (3-D) matrices and slice cultures. Migration modes and cell morphologies in 3-D environments can differ significantly from those observed with dissociated cells migrating on planar substrates *(17,18)*. One approach is to put cultured cells between two layers of flexible polyacrylamide substrata, which creates an environment that more closely resembles tissue *(19)*. In another approach, when fibroblasts are placed in 3-D matrices derived from tissues or cell culture, enhanced adhesion and migration are observed compared with cells plated on 2-D substrates *(17)*. In the 3-D matrix, adhesions are very long and slender in contrast to the oval-shaped focal adhesions found in cultured cells. Chapter 8 by Cukierman describes a method for measuring rates and directionality of fibroblasts in 3-D matrices. In slice culture, myogenic precursor cells extend long, highly polarized, persistent protrusions that are not usually seen in fibroblasts migrating on flat, rigid substrates *(18)*. However, like cultured cells, the formation of these exaggerated protrusions is regulated by local activation of the small GTPase, Rac.

Neural crest cells provide another system for studying migration in vivo. Chapter 18 outlines methods for studying neural crest migration in chick embryos. During development, the neural tube forms by rolling up and pinching off from the future epidermis. Along the "crest" where the neural tube pinches off, a number of cells break loose and migrate individually along defined pathways. These neural crest cells eventually form most of the peripheral nervous system as well as other cells and organs. As neural crest cells break away from the neural tube, they undergo an epithelial-mesenchymal transition. This process is similar to what occurs when cancer cells escape from the primary tumor and enter the circulation. Therefore, understanding this process in vivo can provide tremendous insight into the mechanisms of cancer metastasis.

In rodents, the migration of neuronal precursor cells from the subventricular zone (SVZ) to the olfactory bulb is particularly robust and directional. This restricted pathway is known as the rostral migratory stream (RMS). The RMS provides a model system for studying neuronal migration using explants from the forebrain of postnatal rats. Detailed methods for the SVZ migration assay in 3-D gels are provided in Chapter 12 by Ward and Rao. They also provide methods for introducing cDNAs into SVZ neurons and live-cell imaging of neuronal migration. Using this system, Rao and colleagues found that the migration of these neuronal precursors was repelled by the secreted protein Slit and attracted by a novel chemoattractant *(20,21)*. In another study using slice cultures from the forebrain, the migrating neuronal precursors, which were visualized with DiI crystals, were observed to extend a single, long, persistent leading protrusion in the direction of movement and migrate along a defined pathway *(22)*. Both integrins and the netrin-1 receptor, deleted in colorectal carcinoma (DCC), play a role in RMS migration. DCC is involved in directional migration of the neural precursors whereas integrin ligation is essential for any movement of these cells.

Finally, migration can be studied in vivo using zebrafish embryos. During gastrulation, cell migration is more robust than at any other stages of life. The zebrafish is an ideal system for studying this process because of the transparency of the embryos, the availability of the mutants and the ability of the embryos to develop externally. In Chapter 16, Sepich and Solnica-Krezel provide some background and protocols for the analyses of migration during gastrulation. They also present data collected from wild-type and mutants showing morphogenetic defects. Gastrulation cell movements can be similarly studied in *Xenopus* embryos. In Chapter 17, DeSimone et al. describe methods for examining cell migration in *Xenopus* embryos.

5. Imaging Migrating Cells

Cell polarization and migration require the coordinated regulation of signaling and adaptor molecules at specific locations within the cell. This requires the development of probes to measure the activation of these key regulatory molecules in living cells with high temporal and spatial resolution. GFP-tagged sensors that bind to the activated molecules can be used to reveal their locations and monitor changes in their activation over time *(23)*. However, the use of these GFP-tagged sensors is limited because of high background levels and low affinities for the activated molecules. Therefore, probes have been generated to monitor the direct interaction between a regulatory molecule and a binding domain from one of its effectors by fluorescence resonance energy transfer (FRET; **ref.** *24*). In Chapter 22, Ballestrem and Geiger outline the use of FRET for studying dynamic interactions in focal adhesions and provide a protocol for FRET measurements in living cells. To generate an effective FRET signal, the protein levels of the regulatory molecule and its effector need to be comparable. To circumvent this problem, biosensors, in which the fluorescence derivatives of the regulatory molecule and binding domain are encoded on a single cDNA, have been developed.

Recent technical advances have allowed investigators to monitor cell behavior in living animals. Multiphoton microscopy is particularly well-suited for in vivo studies because it provides a way of imaging deeply into living tissues without the contribution of out-of-focus light. In Chapter 23, Flesken-Nikitin et al. describe the use of multiphoton microscopy to study cell migration in living mice. They track the migration of individual cells by labeling them with enhanced GFP. This provides a powerful method that might eventually be used to study migration in vivo on a molecular level.

6. Conclusions

Cell migration is an important process that takes place throughout life. Because of its central importance, migration has become the focus of much research. However, since migration requires the coordinated activity of several individual component processes, the mechanisms by which these component processes are integrated present a major challenge. Fortunately, technologies and methodologies are emerging that allow us to study the spatial and temporal regulation that generates the coordination of these processes some of which are described in this book. For example, FRET biosensors, which allow spatially resolved assays of signaling events in real time in living cells, and the development of 3-D systems that allow imaging of cellular and molecular dynamics under in vivo-like conditions. A rapidly emerging theme is that migration in vivo differs from that studied in vitro, which may reflect different signaling mechanisms or cellular mechanics. With the technologies and methodologies in place the next few years should provide a wealth of information in the area of migration.

References

1. Lauffenburger, D. A. and Horwitz, A. F. (1996) Cell migration: a physically integrated molecular process. *Cell* **14,** 359–369.
2. Ridley, A. J., Schwartz, M. A., Burridge, K., Firtel, R. A., Ginsberg, M. H., Borisy, G., et al. (2003) Cell migration: integrating signals from front to back. *Science* **102,** 1704–1709.
3. Burridge, K. and Chrzanowska-Wodnicka, M. (1996) Focal adhesions, contractility, and signaling. *Annu. Rev. Cell. Dev. Biol.* **12,** 463–518.
4. Zigmond, S. H. (1977) Ability of polymorphonuclear leukocytes to orient in gradients of chemotactic factors. *J. Cell Biol.* **15,** 606–616.
5. Nobes, C. D. and Hall, A. (1995) Rho, rac, and cdc42 GTPases regulate the assembly of multimolecular focal complexes associated with actin stress fibers, lamellipodia, and filopodia. *Cell* **11,** 53–62.
6. Ridley, A. J. and Hall, A. (1992) The small GTP-binding protein rho regulates the assembly of focal adhesions and actin stress fibers in response to growth factors. *Cell* **10,** 389–399.
7. Rottner, K., Hall, A., and Small, J. V. (1999) Interplay between Rac and Rho in the control of substrate contact dynamics. *Curr. Biol.* **9,** 640–648.
8. Beningo, K. A., Dembo, M., Kaverina, I., Small, J. V., and Wang, Y-L. (2001) Nascent focal adhesions are responsible for the generation of strong propulsive forces in migrating fibroblasts. *J. Cell Biol.* **153,** 881–888.
9. Chrzanowska-Wodnicka, M. and Burridge, K. (1996) Rho-stimulated contractility drives the formation of stress fibers and focal adhesions. *J. Cell Biol.* **133,** 1403–1415.
10. Kirchner, J., Kam, Z., Tzur, G., Bershadsky, A. D., and Geiger, B. (2003) Live-cell monitoring of tyrosine phosphorylation in focal adhesions following microtubule disruption. *J. Cell Sci.* **116,** 975–986.
11. Worthylake, R. A., Lemoine, S., Watson, J. M., and Burridge, K. (2001) RhoA is required for monocyte tail retraction during transendothelial migration. *J. Cell Biol.* **154,** 147–160.
12. Huttenlocher, A., Palacek, S. P., Lu, Q., Zhang, W., Mellgren, R. L., Lauffenburger, D. A., et al. (1997) Regulation of cell migration by the calcium-dependent protease calpain. *J. Biol. Chem.* **172,** 32,719–32,722.
13. Lawson, M. A. and Maxfield, F. R. (1995) Ca^{2+}- and calcineurin-dependent recycling of an integrin to the front of migrating neutrophils. *Nature* **177,** 75–79.
14. Kaverina, I., Krylyshkina, O., and Small, J. V. (1999) Microtubule targeting of substrate contacts promotes their relaxation and dissociation. *J. Cell Biol.* **146,** 1033–1043.
15. Etienne-Manneville, S. and Hall, A. (2002) Rho GTPases in cell biology. *Nature* **120,** 629–635.

16. Hegerfeldt, Y., Tusch, M., Brocker, E. B., and Friedl, P. (2002) Collective cell movement in primary melanoma explants: plasticity of cell-cell interaction, beta1-integrin function, and migration strategies. *Cancer Res.* **12,** 2125–2130.
17. Cukierman, E., Pankov, R., Stevens, D. R., and Yamada, K. M. (2001) Taking cell-matrix adhesions to the third dimension. *Science* **194,** 1708–1712.
18. Knight, B., Laukaitis, C., Akhtar, N., Hotchin, N. A., Edlund, M., and Horwitz, A. R. (2000) Visualizing muscle cell migration in situ. *Curr. Biol.* **10,** 576–585.
19. Beningo, K. A., Lo, C. M., and Wang, Y. L. (2002) Flexible polyacrylamide substrata for the analysis of mechanical interactions at cell-substratum adhesions. *Meth. Cell Biol.* **19,** 325–339.
20. Ward, M., McCann, C., DeWulf, M., Wu, J. Y., and Rao, Y. (2003) Distinguishing between directional guidance and motility regulation in neuronal migration. *J. Neurosci.* **13,** 5170–5177.
21. Liu, G. and Rao, Y. (2003) Neuronal migration from the forebrain to the olfactory bulb requires a new attractant persistent in the olfactory bulb. *J. Neurosci.* **13,** 6651–6659.
22. Murase, S. and Horwitz, A. F. (2002) Deleted in colorectal carcinoma and differentially expressed integrins mediate the directional migration of neural precursors in the rostral migratory stream. *J. Neurosci.* **12,** 3568–3579.
23. Parent, C. A., Blacklock, B. J., Froehlich, W. M., Murphy, D. B., and Devreotes, P. N. (1998) G protein signaling events are activated at the leading edge of chemotactic cells. *Cell* **15,** 81–91.
24. Kraynov, V. S., Chamberlain, C., Bokoch, G. M., Schwartz, M. A., Slabaugh, S., and Hahn, K. M. (2000) Localized Rac activation dynamics visualized in living cells. *Science* **190,** 333–337.

II

Basic Cell Migration and Related Assays

2

Boyden Chamber Assay

Hong-Chen Chen

Summary

The Boyden chamber assay, originally introduced by Boyden for the analysis of leukocyte chemotaxis, is based on a chamber of two medium-filled compartments separated by a microporous membrane. In general, cells are placed in the upper compartment and are allowed to migrate through the pores of the membrane into the lower compartment, in which chemotactic agents are present. After an appropriate incubation time, the membrane between the two compartments is fixed and stained, and the number of cells that have migrated to the lower side of the membrane is determined. Therefore, the Boyden chamber-based cell migration assay has also been called filter membrane migration assay, trans-well migration assay, or chemotaxis assay. A number of different Boyden chamber devices are available commercially. The method described in this chapter is intended specifically for measuring the migration of Madin-Darby canine kidney cells using a 48-well chamber from Neuro Probe, Inc.

Key Words: Boyden chamber assay; migration assay; cell motility; cell migration; chemotaxis; hapatotaxis.

1. Introduction

The Boyden chamber, originally introduced by Boyden for the analysis of leukocyte chemotaxis *(1)*, is ideally suited for the quantitative analysis of different migratory responses of cells, including chemotaxis, haptotaxis, and chemokinesis. Random cell motility is generally described as chemokinesis, which is distinguished from directed cell motility toward increasing concentrations of soluble attractants, such as growth factors (chemotaxis), or along a concentration gradient of extracellular matrix (ECM) proteins (for haptotaxis *see* **ref.** *2*). For the induction of chemotactic or haptotactic response of cells, attractants are added to the lower compartment of the chamber. However, for

From: *Methods in Molecular Biology, vol. 294: Cell Migration: Developmental Methods and Protocols*
Edited by: J-L. Guan © Humana Press Inc., Totowa, NJ

the induction of chemokinesis, equal concentrations of the agent are added on both sides of the membrane.

In addition, the use of the Boyden chamber-based motility assay has other advantages. First, it allows one to have versatility in conducting motility experiments. For example, one can examine the effect of an inhibitor that is specific for an intracellular signaling molecule or a functional blocking antibody that is specific to a cell surface protein on cell motility by adding it to the upper chamber, in which cells are loaded. Actually, this type of experiment has been widely used to examine the potential involvement of a particular intracellular signaling pathway or cell surface protein in cell motility in response to various stimuli *(3,4)*. Second, the Boyden chamber assay is relatively time saving and allows for cell-motility analysis on a basis without consideration of the effect from cell proliferation. The time allowed for cells to migrate through a porous membrane in the Boyden chamber is generally within a few hours (4–6 h), which is much shorter than the time required for cells to proceed through a cell cycle *(5)*. Therefore, consideration of the effect of cell proliferation on the results is generally not necessary. Third, the Boyden chamber assay allows for cell motility analysis without consideration of the effect from cell–cell interactions. Cells, in particular epithelial cells, have to release cell–cell contacts for their translocation over a clearly measurable distance. It is already known that cell–cell interactions and cell migration are controlled largely by different mechanisms *(6–9)*. Because some extracellular factors or genes may be able to stimulate cell migration but fail to disrupt cell–cell interactions, it is reasonable to analyze cell–cell interactions and cell motility separately *(10)*. The Boyden chamber assay is conducted on a basis largely independent of cell–cell interactions. Next, the Boyden chamber assay allows one to measure the relative cell migration rates of cell populations transiently transfected with genes of interest. With an appropriate marker, such as β-galactosidase, which allows the migrated cells to be visualized by X-gal staining, the relative cell motility of a transiently transfected cell population can be measured *(11)*. Finally, it is easy for learners to gain acquaintance with the assay and the results are generally very reliable from one experiment to another.

A number of different Boyden chamber devices are available commercially. They basically vary in sample size and quantitation method. In general, cell migration is quantified by simply enumerating migrated cells under a light microscope or by measuring optical density values of stained cell extracts. Once an appropriate chamber has been chosen, the migration experiments will need to be optimized for any given cell type and attractants. Some of the important factors to be considered are the number of cells to load on the chamber, the pore size of the membrane, the type and concentration of the attractant, and the incubation time. Note that the ECM protein can be coated on the mem-

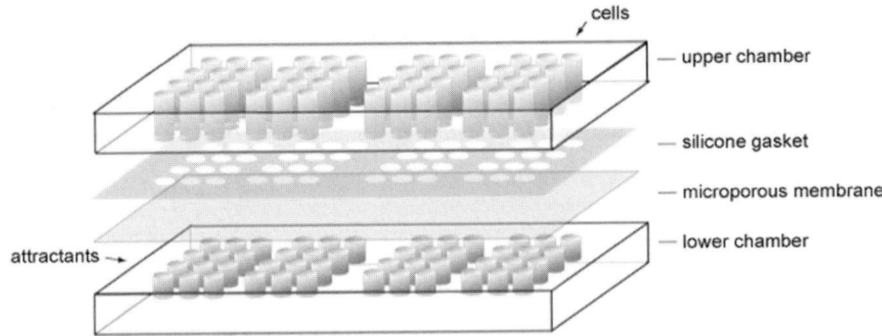

Fig. 1. Components of Neuro Probe standard 48-well chemotaxis chamber.

brane or directly loaded into the lower chamber in its soluble form. In our experience, the results derived from these two loading methods are similar. It is likely that the soluble form of the ECM protein will coat the membrane to form a matrix during the course of the experiment. If a soluble attractant such as growth factor is used, the ECM protein should still be loaded to allow cells to attach. Collagen and fibronectin are usually chosen respectively for epithelial cells and fibroblasts to attach. The following protocol is intended specifically for measuring the migration of Madin–Darby canine kidney (MDCK) epithelial cells using a 48-well chamber from Neuro Probe, Inc., and soluble collagen as an attractant. However, this could easily be adopted to analysis of other cell types in response to other stimuli.

2. Materials

1. Standard 48-well chemotaxis chamber (*see* **Fig. 1**), including: a lower chamber, a silicone gasket, an upper chamber, and thumb nuts (Neuro Probe; Cabin John, MD).
2. Poretics® polycarbonate: polyvinylpyrrolidone-free, 8-μm pore size, 25 × 80 mm (cat. no. K80SH58050; Osmonics; Livermore, CA).
3. MDCK cells (ATCC; cat. no. CCL-34).
4. Dulbecco's modified Eagle's medium (DMEM), high glucose, pH 7.4.
5. Fetal bovine serum (FBS).
6. Versene: 0.2 g of ethylene diamine tetraacetic acid (0.53 mM) and 0.01 g of phenol red in 1 L of phosphate-buffered saline, pH 7.4.
7. 2.5% (w/v) Trypsin (cat. no.15090-046; Gibco Invitrogen): dilute 1:25 in Versene before use.
8. Collagen from calf skin (cat. no. C9791; Sigma-Aldrich): dissolve collagen at 1 mg/mL in 0.1 N acetic acid. Allow to stir at room temperature until dissolved (takes 1–3 h). Keep collagen stock at 4°C.

9. Methanol.
10. Giemsa stain, modified solution (cat. no. GS500; Sigma-Aldrich): dilute 1:10 in distilled H_2O before use.
11. Petri dishes.
12. Glass slide and cover glass (32×24 mm).
13. Nail polish.
14. Terg-A-Zyme® (cat. no. 1304; Alconox; New York, NY): an enzyme active detergent available from Fisher. Dissolve 8 g/L in distilled H_2O.

3. Methods

3.1. Preparing the Cells

1. Seed 5×10^5 MDCK cells per 60 mm-dish in DMEM supplemented with 10% FBS and penicillin-streptomycin and allow them to grow until 50–70% confluency, which usually takes 18 to 24 h.
2. Remove the medium and wash the cells twice with Versene. Add 1 mL of Versene containing 0.05% trypsin and allow the culture to stand at 37°C for 10 to 15 min. Add 3 mL of DMEM with 10% FBS, pipet the cells off the dish, and transfer them to a 15-mL centrifuge tube.
3. Pellet the cells by centrifugation at 150–200g for 5 min. Remove the medium, add 5 mL of DMEM, and centrifuge again.
4. Remove the medium and resuspend the cells in 1.5 mL of DMEM. Count and adjust the cells to 5×10^5 cells/mL in DMEM (*see* **Notes 1** and **2**). If necessary, add the appropriate concentration of an inhibitor or antibody to the cell suspension.

3.2. Loading and Assembling the Chamber

1. Dilute collagen stock in DMEM to 10 µg/mL before loading (*see* **Note 3**). Add 30 µL of the diluted collagen or control reagents to each well of the bottom chamber. The volume should be a slight positive meniscus when the well is filled. If not, re-set the pipetor and load again. It is recommended to use a manual p200 pipetor for loading (*see* **Note 4**).
2. Use forceps to handle polycarbonate membranes. Cut off 1 mm of the corner of a membrane. Lift the membrane by the end using two forceps, and orient it to the chamber so the cut corner corresponds to the Neuro Probe trademark on the lower right corner of the chamber (*see* **Note 5**). Gently place the membrane (shiny side facing up) over the wells of the chamber. Avoid too much movement of the membrane after it has been placed on the wells.
3. Place the silicone gasket over the membrane with cut corner on the lower right.
4. Place the top chamber over the gasket with the Neuro Probe trademark oriented to the lower right corner. Push the top chamber down against the bottom chamber with even pressure on all sides with one hand; with the other hand tighten the thumb nuts.
5. Resuspend the cells by gentle mixing and load 50 µL of cell suspension to each well of the top chamber. Hold the pipetor vertically so that the end of the pipet tip rests against the side of the well just above the membrane. Eject the fluid out

of the pipet tip with a rapid motion to avoid trapping bubbles in the bottom of the well.

6. Wrap the whole chamber in plastic and incubate it at 37°C and 5% CO_2 for 6 h.

3.3. Staining the Membrane

1. Transfer 20 mL of methanol to a Petri dish.
2. Remove the thumb nuts while pressing down the chamber evenly on all sides.
3. Disassemble the chamber, lift the membrane with forceps, and immediately flow it on methanol at room temperature for 10 min to fix cells. The side of the membrane with the migrated cells faces down.
4. Remove the membrane from methanol and allow it to air dry (few min).
5. Dilute 2 mL of Giemsa stain in 20 mL of distilled H_2O in a Petri dish.
6. Flow the membrane (the side with migrated cells facing down) on the diluted Giemsa stain solution for 1 h to stain the cells (*see* **Note 6**).
7. Destain the membrane by briefly (few seconds) rinsing it in distilled H_2O.
8. Drain the excess H_2O from the membrane and place it (the side with migrated cells facing down) on the Petri dish cover. Use a damp cotton swab to wipe the unmigrated cells from the top of the membrane.
9. Cut the membrane into four pieces, each contains 12 wells. Keep track of its orientation by cutting the lower right corner.
10. Attach the membrane with a little amount of nail polish to a glass slide and cover it with a square cover glass.

3.4. Counting the Cells

1. Survey the stained membrane under a light microscope at 50X and 200X magnification and select at least three wells for each experimental group on which the migrated cells are well stained and evenly distributed.
2. Take micrographs using a digital camera (Nikon COOLPIX995) connected to the microscope at 50X magnification. Each micrograph covers a well on the membrane *see* **Note 7**).
3. Remove the camera from the microscope and connect to a computer using the USB cable. Acquire the images from the memory card of the digital camera and view the images with the aid of the software Adobe Photoshop® (Adobe Systems; San Jose, CA). Select the function "view/show" to add grids on the image and count the number of cells on the monitor (*see* **Fig. 2**). Alternative, adjust the size of the image for printing on an A4 size paper and then count the number of cells on the printed image.
4. Calculate the value (mean ± standard error) of the migrated cells from at least three wells for each experimental group.

3.5. Cleaning the Chamber

1. After disassembling the chamber, immediately rinse all parts in running tap H_2O for 30 min. Rinse the chamber components several times with distilled H_2O and let them air dry at room temperature.

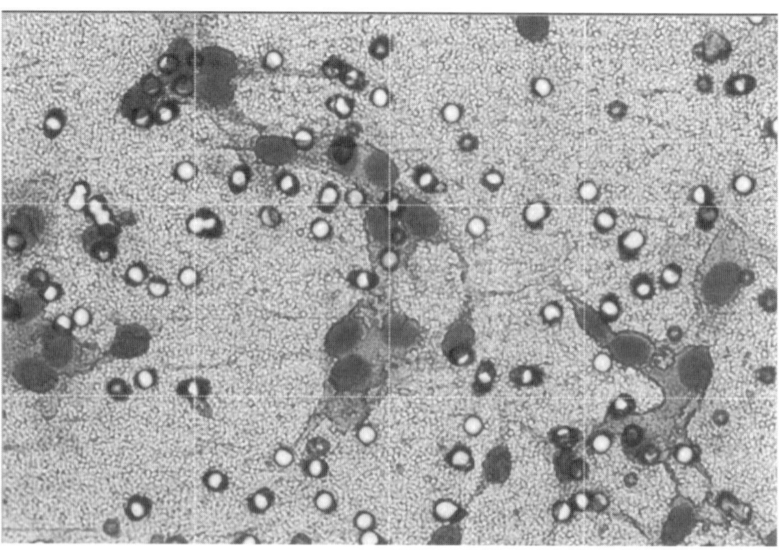

Fig. 2. A representative micrograph of a stained membrane taken by a digital camera under a light microscope.

2. For periodic cleaning, soak all the chamber components in Terg-A-Zyme® solution at 55°C for 1 h. Wash them thoroughly under running tap H_2O for 30 min and rinse them several times with distilled H_2O. Soak them in distilled H_2O overnight and allow them to air dry (*see* **Note 8**).

4. Notes

1. It is recommended that less than six experimental groups are planned for a chamber, which allows at least eight wells for an experimental group. This ensures at least three representative wells that can be selected for quantification for each experimental group. In addition, more experimental groups will take more time for preparation, which causes cells in different experimental groups to stay in suspension with varied time, which may affect cell motility.
2. Because the volume varies from one pipetor to another, it is better to prepare a little more sample for loading. For example, if eight wells per experimental group are planned for loading, prepare enough sample for loading 10 wells. The volume of cell suspension (5×10^5 cells/mL) to load on a well of the upper chamber is 50 µL, that is, 2.5×10^4 cells per well. Make 2.5×10^5 cells in 0.5 mL of DMEM for loading 10 wells.
3. Collagen is less soluble in neutral pH than in low pH. After diluting collagen in DMEM, immediately load it to the lower chamber.
4. To avoid trapping bubbles in the wells of the chamber, the liquid should not be expelled completely from the pipet tip. In addition, for the lower chamber, it is

important to load the correct volume of attractants, which should form a slight positive meniscus when the well is filled. For loading the upper chamber, hold the pipetor vertically so that the end of the pipet tip is against the side of the well just above the membrane and expel the liquid quickly from the pipet tip.

5. Ensure to keep track of the orientation of the membrane by cutting it on the right lower corner.
6. Instead of immersing the membrane in Giemsa stain, we let the membrane flow on the stain solution. This allows only the side of the membrane with the migrated cells to be stained.
7. With the aid of a digital camera and computer, the total number of stained cells in a well can be counted accurately and objectively. Directly counting the cells in selected fields from a well under a microscope at high magnification is not recommended.
8. It is important to immediately immerse all parts of the chamber in water after dissembling the chamber and wash them thoroughly to prevent the accumulation of debris. It is recommended that after approximately five experiments, the chamber components are cleaned with Terg-A-Zyme® (*see* **Subheading 3.5.2.**). Never autoclave the chamber or immerse it in water hotter than 60°C.

Acknowledgments

The author would like to thank Po-Chao Chan and Chun-Chi Liang for technical assistance. Work in the author's laboratory is supported by the National Science Council, Taiwan.

References

1. Boyden, S. V. (1962) The chemotactic effect of mixtures of antibody and antigen on polymorphonuclear leucocytes. *J. Exp. Med.* **115,** 453–466.
2. Carter, S. B. (1967) Heptotaxis and the mechanism of cell motility. *Nature* **213,** 256–260.
3. Reiske, H. R., Kao, S-C., Cary, L.A., Guan, J-L., Lai, J-F., and Chen, H-C. (1999) Requirement of phosphatidylinositol 3-kinase in focal adhesion kinase-promoted cell migration. *J. Biol. Chem.* **274,** 12,361–12,366.
4. Liang, C-C. and Chen, H-C. (2001) Sustained activation of extracellular signal-regulated kinase stimulated by hepatocyte growth factor leads to integrin (α2 expression that is involved in cell scattering. *J. Biol. Chem.* **276,** 21,146–21,152.
5. Cary, L. A., Chang, J. F., and Guan, J-L. (1996) Stimulation of cell migration by overexpression of focal adhesion kinase and its association with Src and Fyn. *J. Cell Sci.* **109,** 1787–1794.
6. Sander E. E., van Delft, S., ten Klooster, J. P., Reid, T., van der Kammen, R. A., Michiels, F., et al. (1998) Matrix-dependent Tiam1/rac signaling in epithelial cells promotes either cell-cell adhesion or cell migration and is regulated by phosphatidylinositol 3-kinase. *J. Cell Biol.* **143,** 1385–1398.
7. Royal, I., Lamarche-Vane, N., Lamorte, L., Kaibuchi, K., and Park, M. (2000) Activation of Cdc42, Rac, PAK, and Rho-kinase in response to hepatocyte growth

factor differentially regulates epithelial cell colony spreading and dissociation. *Mol. Biol. Cell* **11,** 1709–1725.

8. Ren, X. D., Kiosses, W. B., Sieg, D. J., Otey, C. A., Schlaepfer, D. D., and Schwartz, M. A. (2000) Focal adhesion kinase suppresses Rho activity to promote focal adhesion turnover. *J. Cell. Sci.* **113,** 3673–3678.

9. Chen, B-H., Tzen, J. T. C., Bresnick, A. R., and Chen, H-C. (2002) Roles of Rho-associated kinase and myosin light chain kinase in morphological and migratory defects of focal adhesion kinase-null cells. *J. Biol. Chem.* **277,** 33,857–33,863.

10. Lai, J-F., Kao, S-C., Jiang, S-T., Tang, M-J., Chan, P-C., and Chen, H-C. (2000) Involvement of focal adhesion kinase in hepatocyte growth factor-induced scatter of Madin-Darby canine kidney cells. *J. Biol. Chem.* **275,** 7474–7480.

11. Sieg, D. J., Hauck, C. R., and Schlaepfer D. D. (1999) Required role of focal adhesion kinase (FAK) for integrin-stimulated cell migration. *J. Cell Sci.* **112,** 2677–2691.

3

Wound-Healing Assay

Luis G. Rodriguez, Xiaoyang Wu, and Jun-Lin Guan

Summary

The wound-healing assay is simple, inexpensive, and one of the earliest developed methods to study directional cell migration in vitro. This method mimics cell migration during wound healing in vivo. The basic steps involve creating a "wound" in a cell monolayer, capturing the images at the beginning and at regular intervals during cell migration to close the wound, and comparing the images to quantify the migration rate of the cells. It is particularly suitable for studies on the effects of cell–matrix and cell–cell interactions on cell migration. A variation of this method that tracks the migration of individual cells in the leading edge of the wound is also described in this chapter.

Key Words: Cell migration; wound healing; extracellular matrix; transfection; image capture and analysis; time-lapse microscopy.

1. Introduction

The wound-healing assay is one of the earliest developed methods to study cell migration in vitro *(1)*. This method is based on observation of cell migration into a "wound" that is created on a cell monolayer. Although not an exact duplication of cell migration in vivo, this method mimics to some extent migration of cells in wound healing. For example, denuding part of the endothelium in the blood vessels will induce endothelial cell migration into the denuded area to close the wound *(2)*. Depending on the cell type, cells migrate into the wound as loosely connected populations (e.g., fibroblasts) or as sheets of cells (e.g., epithelial and endothelial cells), which also mimic the behavior of these cells during migration in vivo.

In comparison with other popular in vitro methods, such as time-lapse microscopy and Boyden chamber assays, the wound-healing assay is particularly suitable for studies of directional cell migration and its regulation by cell interaction

From: *Methods in Molecular Biology, vol. 294: Cell Migration: Developmental Methods and Protocols*
Edited by: J-L. Guan © Humana Press Inc., Totowa, NJ

with extracellular matrix (ECM) and cell–cell interactions. Migration of the cells is regulated by both the ECM under the cells and soluble factors *(3,4)* as well as intercellular interactions in the case of epithelial and endothelial cells *(5)*. The effects of each of these factors on directional cell migration can be studied using the wound healing method. More recently, this assay also has been combined with microinjection or transfection to assess the effects of expression of exogenous genes on migration of individual cells *(6–8)*. This is probably the simplest method to study cell migration in vitro and only requires common and inexpensive supplies found in most labs that are capable of cell culturing. The basic steps involve creation of a "wound" on monolayer cells, capture of images at the beginning and regular intervals during cell migration to close the wound, and comparison of the images to determine the rate of cell migration.

More recently, a wound-healing assay has been used to track migration of individual cells at the leading edge of the wound to determine the role of particular genes in the regulation of directional migration *(8)*. This requires the aid of time-lapse microscopy and image analysis software and differs from the conventional use of time-lapse microscopy in that it measures directional migration of the cells in the population of migrating cells into the wound instead of random migration of sparsely plated cells. This variation will be described subsequently (*see* **Subheading 3.**).

2. Materials

1. 60-mm or other size tissue culture dishes.
2. Razor or extra fine Sharpie® marker.
3. 1 mg/mL fibronectin stock.
4. 1 mg/mL poly-L-lysine (PLL) stock.
5. Phosphate-buffered saline (PBS).
6. 2 mg/mL Bovine serum albumin (BSA).
7. Versene with trypsin.
8. Dulbecco's modified Eagle's medium cell culture media with supplements (serum, antibiotics).
9. p200 Pipet tips.
10. Phase contrast microscope.
11. Camera.

Optional and for tracking individual cells:

12. Video camera.
13. Image analysis software.
14. Stage incubator.
15. Fluorescence microscope.
16. LipofectAmine and PLUS transfection reagents (Life Technologies).
17. Plasmid-encoding green fluorescent protein (GFP) or other markers.
18. CO_2 independent media.

3. Methods

3.1. Measurement of Migration as Population of Cells

1. Prepare 60-mm dishes with markings on the outer bottom of the dish to be used as reference points during image acquisition. Acquisition of images requires the matching of the first image with the second image acquired in **steps 7** and **9**. Marking the tissue culture dish can be achieved by numerous methods to obtain the same field during the image acquisition. Etching lines lightly with a razor blade on the bottom of the dish provides a good reference under the microscope and can be visualized with the naked eye for creating the wound itself. Keep in mind when using a razor not to separate the lines too much because the goal is to observe two lines in the same field under magnification. Matching the images is essential for quantitative analysis (*see* **Note 1**).

2. Coat the dishes with 3 mL of either 10 µg/mL fibronectin or 50 µg/mL poly-L-lysine as a control by incubating the dishes overnight at 4°C or several hours at 37°C without rotation or shaking. These are for studies of fibroblast migration. For other cell types, other appropriate ECM substrates can be used.

3. When ready to proceed, aspirate off unbound ECM substrate and block the coated dishes with 3 mL of 2 mg/mL bovine serum albumin for 1 h at 37°C.

4. Resuspend subconfluent cells growing in a tissue culture dish by washing cells once with PBS, add trypsin containing versene, resuspend cells with complete media, and obtain cell counts for all cells to be plated. Cells should be dispersed gently by pipetting and/or rocking the dish before incubation so that cells are distributed equally among the plate.

5. Wash the prepared dishes once with PBS and fill the dish with 3–5 mL of media (*see* **Note 2**). Plate 1×10^6 NIH 3T3 cells to create a confluent monolayer onto the prepared 60 mm dish. Allow cells to adhere and spread completely for approx 4 h under proper incubation. If different cell types or different size dishes are used, adjust the number of cells used to create a confluent monolayer.

6. Create a wound by manually scraping the cell monolayer with a p200 pipet tip. Wash the cells once with 1 mL of desired media and replace with 5 mL of the same media. The wound should be created relative to the marking/reference point on the dish. For example, the razor blade method, being the least expensive and most popular, creates a point of reference. The wound is created perpendicular to the marking on the dish.

7. Acquire the first image by using the markings on the culture dish as a reference point. Place the reference mark outside the capture image field but within the eyepiece field of view such that the reference mark can be aligned later for the subsequent image acquisition.

8. Incubate dishes in a tissue culture incubator for 8–18 h (*see* **Note 3**).

9. Match the photographed region acquired in **step 7** and acquire a second image (**Fig. 1**).

10. The images acquired above can be analyzed quantitatively by several methods. The migration of cells can be determined by the number of cells that cross into the wound area from their reference point at time zero. This method provides

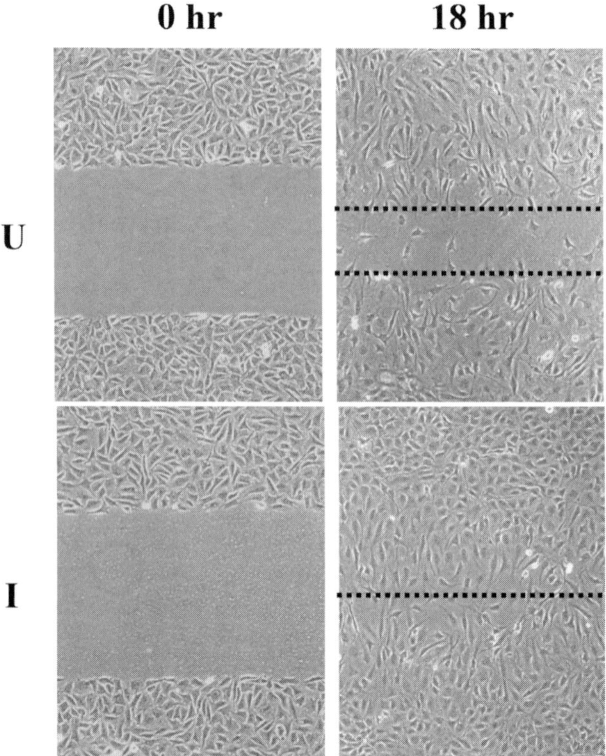

Fig. 1. Example of images acquired at 0 and 18 h in wound-healing assay. Cells shown are uninduced (U) and induced (I), NIH 3T3 cells expressing wild-type 14-3-3β. The dotted lines define the areas lacking cells.

large sample sizes that are easily quantified statistically. Alternatively, various methods can be used to measure the distance or area of the wounded region lacking cells. Free software is available that simplifies this process with convenient tools (Website: http://rsb.info.nih.gov/ij/). Additionally, more elaborate commercial software is available that can automate the measurement process (Media Cybernetics; Carlsbad, CA).

3.2. Tracking Migration of Individual Cells in the Wound-Healing Assay

1. Prepare 60-mm culture dishes as described in **step 1** (**Subheading 3.1.**) to be used to match wounds during image acquisitions. If time-lapse microscopy will be used, this step is not necessary (*see* **Note 4**).
2. Plate growing NIH 3T3 cells at 50–60% confluency 12 to 18 h prior to transfection. Transfect NIH 3T3 cells with plasmid encoding the gene of your interest along with a marker plasmid (i.e., GFP) in a 7:1 ratio. Alternatively, a vector-

encoding GFP fusion protein containing the gene of your interest can be used. Cells are transfected with LipofectAmine and PLUS transfection reagents (Life Technologies) per the manufacturers instructions.

3. When cells reach 100% confluency (usually 24 to 48 h after transfection), create a wound with a p200 pipet tip, as described in **step 6** (**Subheading 3.1.**). Wash the plates once and replace with the desired medium. If time-lapse microscopy will be used, CO_2 independent media will be required.

4. Observe the cells under a fluorescence microscope to ensure that enough cells in the leading edge of the wound are positively transfected (i.e., as marked by GFP; *see* **Note 5**). Acquire both phase contrast and fluorescence images every 2 h by matching the wounded region until the wound has completely closed (usually about 10 h). Return cells to a 37°C cell culture incubator between images unless CO_2-independent media was used.

5. Calculate the rate of migration of the transfected cells by measuring the distance traveled toward the center of the wound after 8 h using a motility program, like OMAware or similar, described previously *(8)*. It is useful to draw an imaginary line in the middle of the wound in the images captured (**Fig. 2**).

4. Notes

1. Alternative to using a razor, marked culture dishes and stickers that attach to the bottom of the dish are available commercially for use as reference markers.

2. The amount of serum in the media needs to be determined for the particular cell types used. It is recommended to use less than the amount in the regular growth media to minimize cell proliferation during the period of assay. However, using too little may lead to apoptosis and/or cell detachment for some cell types. For NIH 3T3 cells, which are normally cultured in 10% calf serum, we find that 1–2% calf serum is optimal for the assay.

3. The amount of incubation time should be determined empirically for the particular cell types under study. However, incubations longer than overnight are not recommended as differences in cell proliferation may start to contribute to the differences in the migration assay. The most desirable length of incubation time is one where the cells in the fastest migrating condition just reach complete closure of the wound. The dishes can be taken out of the incubator and examined periodically and then put back to resume incubation.

4. An automated time-lapse microscope equipped with a temperature control chamber can be used under CO_2-independent media, as the same region would be acquired automatically. However, this usually is not very practical for an 18-h incubation as time constraints on a microscope usually interfere.

5. The cells at the leading edge of the wounded monolayer are most migratory and usually are chosen as the ones that are examined for migration. The rate of their migration reflects the overall rate of wound closure of the migrating monolayer of cells. In addition, these cells are usually not dividing (whereas cells in the back could), thus not presenting the complication of single cells under tracking becoming double. Therefore, the effects of genes on migration are usually assessed using

fluorescence   **phase**

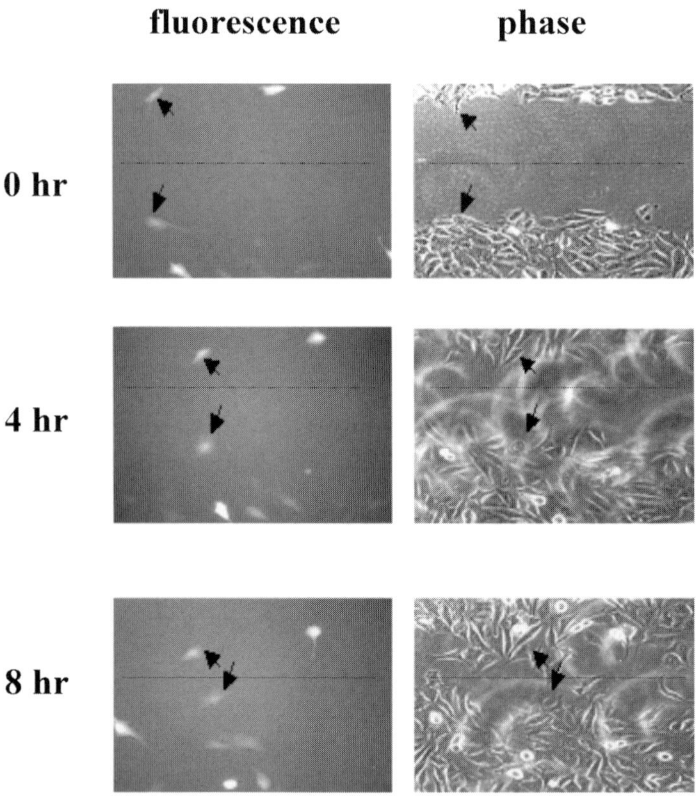

Fig. 2. Measurement of individual cell migration in wound-healing assay. NIH 3T3 cells grown on fibronectin (10 µg/mL) were transfected with a plasmid-encoding GFP. One day after transfection, the cell monolayer was wounded with a pipet tip, incubated at 37°C, and fluorescent and phase contrast images were captured at regular intervals as indicated. The rate of migration was measured by quantifying the total distance that the GFP+ cells (as indicated by arrows) moved from the edge of the wound toward the center of the wound (marked by imaginary dotted lines). This figure is reprinted from **ref. 8** with permission of the American Society for Cell Biology.

the cells at the leading edge, which would be the greatest. In the past, more complex microinjection methods have been used to introduce the genes (and proteins) along with a marker into the cells at the leading edge for this purpose. However, the high-transfection efficiency of the lipofectamine method coupled with a time-lapse microscope now allows such studies in many cell types, such as the NIH 3T3 cells, as outlined in this chapter. This simpler approach, which also avoids the use of sophisticated and expensive equipment, should permit more laboratories to use the method.

Acknowledgments

This work was supported by the NIH grant GM48050 (J. G.) and a predoctoral fellowship GM64086 (L. G. R.).

References

1. Todaro, G. J., Lazar, G. K., and Green, H. (1965) The initiation of cell division in a contact-inhibited mammalian cell line. *J. Cell Physiol.* **66,** 325–333.
2. Haudenschild, C. C. and Schwartz, S. M. (1979) Endothelial regeneration. II. Restitution of endothelial continuity. *Lab Invest.* **41,** 407–418.
3. Han, D. C., Rodriguez, L. G., and Guan, J-L. (2001) Identification of a novel interaction between integrin beta1 and 14,-3-3beta. *Oncogene* **20,** 346–357.
4. Lipton, A., Klinger, I., Paul, D., and Holley, R. W. (1971) Migration of mouse 3T3 fibroblasts in response to a serum factor. *Proc. Natl. Acad. Sci. USA* **68,** 2799–2801.
5. Underwood, P. A., Bean, P. A., and Gamble, J. R. (2002) Rate of endothelial expansion is controlled by cell:cell adhesion. *Int. J. Biochem. Cell Biol.* **34,** 55–69.
6. Etienne-Manneville, S. and Hall, A. (2001) Integrin-mediated activation of Cdc42 controls cell polarity in migrating astrocytes through PKCzeta. *Cell* **106,** 489–498.
7. Fukata, Y., Oshiro, N., Kinoshita, N., Kawano, Y., Matsuoka, Y., Bennett, V., et al. (1999) Phosphorylation of adducin by Rho-kinase plays a crucial role in cell motility. *J. Cell Biol.* **145,** 347–361.
8. Abbi, S., Ueda, H., Zheng, C., Cooper, L. A., Zhao, J., Christopher, R., et al. (2002) Regulation of focal adhesion kinase by a novel protein inhibitor FIP200. *Mol. Biol. Cell* **13,** 3178–3191.

4

Analysis of Cell Migration Using the Dunn Chemotaxis Chamber and Time-Lapse Microscopy

Claire M. Wells and Anne J. Ridley

Summary

The directed migration of cells (chemotaxis) occurs not only during wound healing and inflammatory responses but also during embryonic development. However, the intracellular signaling pathways that enable a cell to detect a chemoattractant and subsequently migrate toward the source are not clearly defined. The Dunn chemotaxis chamber in conjunction with time-lapse microscopy is a powerful tool that enables the user to observe directly the morphological response of cells to a chemoattractant in real time. Here, we describe using the Dunn chemotaxis chamber to study the response of murine bone marrow-derived macrophages to colony stimulating factor-1. This is a particularly useful protocol as it can be adapted to study bone marrow-derived macrophages isolated from genetically modified mice and thus study the requirement for a specific protein in cell migration and chemotaxis.

Key Words: Chemotaxis; Dunn chemotaxis chamber; cell migration; bone marrow-derived macrophages; time-lapse microscopy.

1. Introduction

Directed cell migration (chemotaxis) is important during embryogenesis (*1*), wound healing (*2*), and inflammatory responses (*3*). In addition, cell migration is believed to contribute to the metastatic process (*4*). The technique described in this chapter has been developed to study the intracellular signaling pathways and subsequent changes in cell adhesion and morphology that are required for a cell to initiate a migratory response toward a chemoattractant. Chemotaxis can be studied using a Boyden chamber or transwell assay; however, both these methods are based on scoring cells that have migrated into or through a filter membrane toward a source of putative chemotactic factor. In these assays, the local

From: *Methods in Molecular Biology, vol. 294: Cell Migration: Developmental Methods and Protocols*
Edited by: J-L. Guan © Humana Press Inc., Totowa, NJ

concentration gradients of chemotactic factor in and around the pores of the filter membrane are variable and unknown. Furthermore, the migratory behavior of the cells is unobservable and can only be deduced from the final distribution of the cell population. To overcome these experimental problems, the Dunn chemotaxis chamber (DCC) was developed. In this chamber, the migratory behavior of cells can be directly observed in a gradient of known direction and magnitude. The DCC is a modification of the Zigmond chamber *(5)*, which allows the direct observation of slowly moving cells in a concentration gradient that is stable over longer periods of time *(6,7)*. The DCC consists of two concentric circles ground into one face of a glass slide (referred to as the inner and outer wells). An annular ridge (referred to as the bridge) separates the two wells. The bridge is 20 μm lower than the surrounding glass slide. The outer well contains the chemoattractant whereas the inner well does not (**Fig. 1**). A linear gradient of chemoattractant forms by diffusion across the bridge between these two wells. Cells seeded onto a glass cover slip are inverted over the chamber and it is the cells that lie directly above the bridge that are viewed during the assay. The DCC chamber has been successfully used to characterize the migratory response of human macrophages to colony-stimulating factor-1 (CSF-1; **ref. 8**), fibroblasts to platelet-derived growth factor *(7)* and thrombospondin (TSP1) *(9)*, neutrophils to interleukin-8 *(10)*, and microglia to adenosine triphosphate/adenosine diphosphate *(11)*. In addition, the DCC has been used to study neurite outgrowth *(12)*. We have previously used the DCC to elucidate the role of Rho family proteins in the chemotactic response of BAC1.2F5 macrophages to CSF-1 *(13)*. We describe the use of this chamber to study the response of primary murine bone marrow-derived macrophages (BMMs) to CSF-1, a protocol that can be extended to study the migratory behavior of macrophages and other cell types derived from genetically modified mice.

2. Materials

1. All tissue culture medium and supplements are obtained from Invitrogen unless otherwise stated. Macrophage starve medium consists of RPMI1640 with L-glutamine, 1% essential amino acids, 1% sodium pyruvate, 1% penicillin–streptomyosin, 10% heat-inactivated fetal calf serum, and 0.5% β-mercaptoethanol solution. The β-mercaptoethanol solution (50 mM) consists of 34 μL of 14.3 M β-mercaptoethanol (Sigma; cat. no. M7522) diluted in 10 mL of RPMI containing antibiotics; this can be stored at 4°C. The BMMs are differentiated and maintained in macrophage growth medium, which consists of macrophage starve medium with the addition of 15% conditioned medium from L-929 fibroblasts (L cells) as a source of CSF-1. To provide reproducible and accurate concentration gradients during the chemotaxis assay, recombinant CSF-1 (R and D systems) is used in the DCC. The BMMs are maintained in a humidified incubator at 37°C in the presence of 10% CO_2.

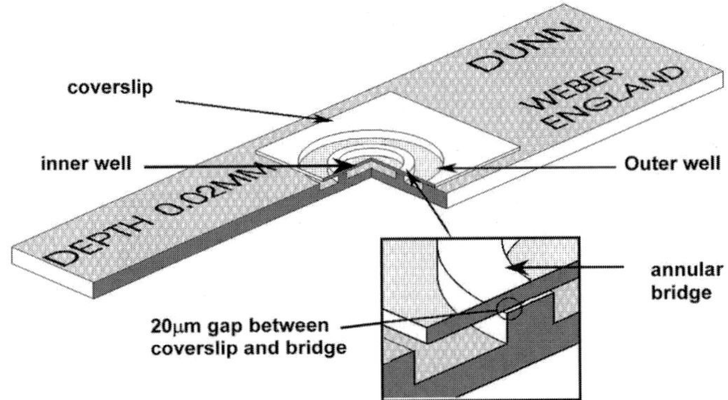

Fig. 1. The Dunn chemotaxis chamber with cover slip in place. The chamber consists of two concentric circles ground into one face of a glass slide (inner and outer wells). An annular ridge (bridge) separates the two wells. Insert, the bridge is 20 µm lower than the surrounding glass slide resulting in a gap between the cover slip and the bridge. Adapted from a figure kindly provided by Professor Gareth Jones.

2. 18 × 18-mm Glass cover slips (no. 3, BDH laboratory supplies) are individually washed with detergent, rinsed six times with distilled water, soaked for 10 min in concentrated HCl, then washed extensively with distilled water and stored in 100% ethanol until use. Before use the ethanol is removed from each cover slip by sweeping it through a naked flame using tweezers until all the ethanol has evaporated. This also serves to sterilize the cover slip.

3. Dunn chemotaxis chambers are supplied by Weber Scientific International. Sealing the cover slips onto the chambers requires dental wax, which can be purchased from Agar Scientific. The wax is applied to the chamber using a small paintbrush. Do not use a paintbrush with plastic bristles as these melt in the hot wax. We use a sable artists paintbrush.

4. It is prudent to clean the chambers immediately after use. The most important principle to observe when cleaning the DCC is to avoid touching the bridge. All the dental wax is removed carefully with a razor blade scraping in the opposite direction to the bridge at all times. Using gloved fingers rub off any residual wax with distilled water and rinse a further six times with distilled water. The chamber is then placed in a glass Coplin jar (or beaker) filled with 100% acetone and sonicated for 10 min using a water bath sonicator (Decon FS Minor, Ultrasonic Ltd). After sonication the chamber is rinsed six times with distilled water. Once rinsed place the chamber in a glass Petri dish grooved side upwards and soak in 30% hydrogen peroxide for 10 min. After acid cleaning the chamber is rinsed a further six times with distilled water. Store the chambers in 100% ethanol until use.

5. To observe live cells in the DCC requires a microscope adapted for time-lapse recording. We use an axiovert 135 microscope (Zeiss; Hertfordshire, UK) equipped with a heated stage (Zeiss), a Uniblitz electronic shutter (Vincent Associates; NY), and a 1CP-M1E CCD camera (Hitachi Denshi; Leeds, UK) coupled to a Meteor Frame Grabber card (Matrox Electronic Systems; Quebec, Canada). The system is controlled and coordinated by Tempus meteor.exe computer software (Kinetic Imaging; Nottinghamshire, UK). Ideally, the microscope should have a heated stage, but a fan heating system can be used *(14)*. Phase-contrast microscopy using an inverted or upright microscope will provide good results. To determine the chemotactic potential of a cell population (using the software described in **Subheading 3.3., step 7**) requires the analysis of a large number of cells. Therefore, a low-magnification objective lens (routinely 10X) is used whereby the full width of the DCC bridge can be visualized.

3. Methods

3.1. Isolation and Culture of Mouse Primary BMMs

1. Isolate and clean the murine femoral bone. Maintain the bones in a small Petri dish containing macrophage starve medium. It is important that all of the tissue surrounding the femur is scraped away to prevent contamination of the BMM preparation.
2. Once the bone has been cleaned use a 19-gage needle to pierce both ends of the bone. Keep the needle in one end of the bone (fatter end). Using a 5-mL syringe, flush the bone marrow out of the bone with 5 mL of macrophage starve medium into a 15-mL Falcon tube. Add an additional 5 mL of macrophage starve medium and resuspend vigorously using a p1000 Gilson pipet.
3. Centrifuge the cell suspension for 5 min at $1000g$ and resuspend the pellet in 5 mL of macrophage starve medium. Count the hematopoietic progenitor cells (do not count the smaller crescent-shaped red blood cells).
4. Seed the cells at 2×10^5 cells/cm^2 on tissue culture plastic (Nunc). We routinely seed 15 mL of cells suspended in growth medium in a 75-cm^2 flask at a density of 1×10^6 cells/mL. Incubate the cells for 3 d.
5. Carefully collect the non-adherent population of cells. The cells attached to the bottom of the flask are a mixed population of differentiated hematopoietic cells and fibroblasts and care should be taken not to dislodge these cells when collecting the non-adherent population. Centrifuge the cell suspension and count all the cells present.
6. The cells are then cryogenically frozen in growth medium containing 10% DMSO at a cell density depending on the subsequent application. We routinely place multiples of 5×10^5 cells in each vial.
7. To recover frozen cells rapidly thaw a cryovial, add 4 mL of warm growth medium dropwise and centrifuge for 5 min at $1000g$.
8. Seed the cells on 6-cm bacterial culture plates (Falcon; *see* **Note 1**) in 5 mL of growth medium at a density of 10^5 cells/mL (i.e., a vial containing 5×10^5 cells

will seed one 6-cm dish). One 6-cm dish will produce enough cells for over 20 Dunn chamber cover slips.

9. Incubate the cells for a further 5 d without refeeding. The cells remain in suspension until approximately d 4. The differentiated BMMs will then become adherent and can be harvested on d 5 (*see* **Note 1**).

10. To harvest the cell carefully remove the medium as the cells are only loosely attached and add 2.5 mL of Versene (Gibco). Incubate for approx 20 min and then collect the cells by vigorous pipetting across the dish. Add 2.5 mL of macrophage starve medium and centrifuge for 5 min at 1000*g*. Resuspend the cells in an appropriate amount of macrophage growth medium.

11. Place a flamed (*see* **Subheading 2, item 2**) acid-washed Dunn chamber cover slip in a 2-cm tissue culture dish and add 2 mL of growth medium containing 2×10^4 cells/mL. Incubate the cells for a further 18 h.

12. To prepare the cells for the DCC, wash twice with starve medium and then incubate in starve medium for 8 h. At the same time place 8 mL of starve medium in the incubator, so that it becomes warm and gassed, to use in the chamber assembly.

3.2. Assembly of the DCC

1. Place some dental wax in a glass beaker and melt slowly using a low setting on a stage heater. Avoid overheating the wax, as a noxious fume will be produced. Ensure that the time-lapse stage is heated to 37°C and that filming can begin immediately after the chamber is assembled.

2. Aliquot 5 mL of the pregassed (*see* **Subheading 3.1., step 12**) starve medium into a 7-mL bijoux tube: this will be used to wash the chambers (wash). Aliquot 1 mL of the starve medium into a 7-mL bijoux tube (control) and 1 mL of starve medium into a 7-mL bijoux tube (test). Add 30 ng/mL recombinant CSF-1 to the 1 mL of starve medium in the test bijoux tube.

3. Place the cover slip with cells attached, wash, control and test bijoux tubes in an incubator close to the DCC assembly site. Alternatively 5 or 10% CO_2 can be pumped from a cylinder across the DCC assembly site. It is important to keep the medium used during assembly of the DCC fully gassed, as once the chamber is sealed by wax there is no opportunity for gas exchange.

4. Place a DCC on tissue paper until the storage ethanol has evaporated. The chamber should be completely dry before proceeding. Do not attempt to wipe the chamber as contact with the bridge can result in damage that may affect the gradient.

5. Initially the chamber is washed to remove any residual traces of ethanol. Using a Gilson, carefully pipet 100–150 µL of wash medium over the center of the chamber without letting the tip touch the bridge. Both wells should fill but not overflow. Carefully remove the medium and repeat six times.

6. Place 150 µL of control medium over the chamber until it fills both wells. Quickly but carefully invert the cover slip over the two wells in the center of the chamber. It is important that inverting the cover slip over the chamber does not incorporate any bubbles into the chamber as these can affect the gradient (*see* **Note 2**).

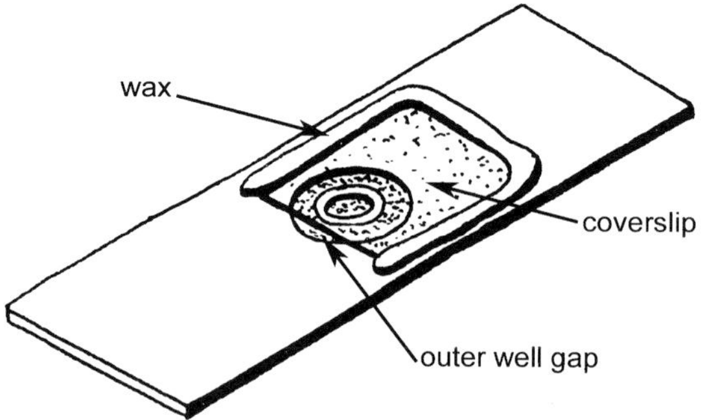

Fig. 2. Part-assembled chamber. Once the cover slip is in place, the chamber is sealed on three sides with wax. The fourth side remains unsealed and the medium is removed from the outer well and replaced with chemoattractant using this outer well gap. Adapted from a figure kindly provided by Dr. Graham Dunn.

The cover slip should be placed slightly off center so that a small gap remains in the outer well (**Fig. 2**). However, the cover slip should completely cover the inner well so that the outer well can be drained without any loss of medium from the inner well.

7. Lightly press down the cover slip around the edges and mop up excess medium with Whatman paper. Using a sable paintbrush wax three sides of the cover slip leaving the side with the outer well gap unwaxed (**Fig. 2**).

8. Tear a piece of Whatman paper and place a small corner of the torn edge just inside the outer well gap until it starts to absorb the medium. Leave the filter paper to absorb all of the medium in the outer well. It is important not to move or lift the paper as this can introduce air into the DCC.

9. Using a Gilson pipet add 100 µL of test solution into the outer well through the gap, making sure that it is bubble-free, until the well is full (*see* **Note 3** for testing the gradient formation). Quickly wax the remaining side ensuring that the gap is completely sealed. Once the wax has set wash the remaining cover slip surface with distilled water and dry with Whatman paper or compressed air.

10. Immediately place the DCC on the microscope stage and start filming. *See* **Note 4** for fixing cells at the end of filming.

3.3. Time-Lapse Microscopy and Migration Analysis

1. Place the assembled chamber on the heated stage.

2. The live image from the microscope is then viewed on the computer monitor using the Acquisition manager (AQM) software (Kinetic Imaging; Liverpool, UK) and a region of the bridge is selected for recording. Ideally, there should be 15–25 cells in the field of view and the cells should be evenly spaced.

outer well

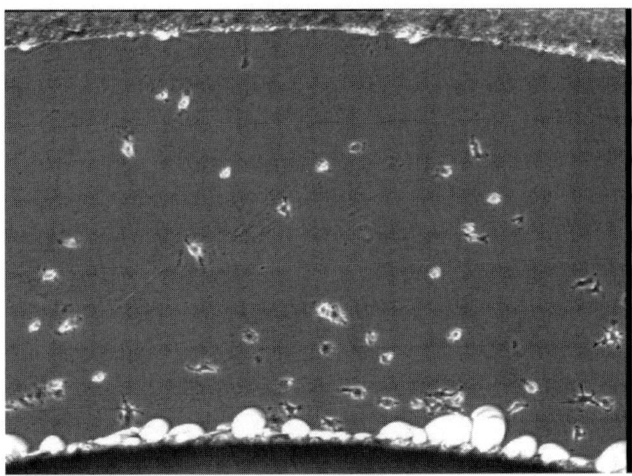

inner well

Fig. 3. BMMs in the DCC as imaged by AQM software. A region of the bridge is selected for recording and the image orientated such that the source of chemoattractant (outer well) is at the top of the screen.

3. The high concentration (outer well) end of the chemotactic gradient is then aligned to the top of the monitor screen (**Fig. 3**). We achieve this by rotating the camera. This is extremely useful for subsequent tracking and mathematical analysis of chemotactic behavior.

4. For analyzing the chemotactic behavior of BMMs in a gradient of recombinant CSF-1, we use a time-lapse interval of 10 min and film the cells for 24 h. BMMs respond slowly to a CSF-1 gradient compared with macrophage cell lines such as Bac1.2F5 *(13)*.

5. Once the assay is finished, the recorded sequence of images can be analyzed using the same AQM software (Kinetic Imaging). Each cell must be individually tracked using the AQM software throughout the sequence. In our laboratory, we adopt the following tracking criteria: only cells present in the first frame are tracked. If any cell present in the first frame divides within the first 60 frames, it is excluded. If a cell present in the first frame divides at a later stage in the film, it is tracked until it ceases to be migratory; for BMMs this is normally one to two frames or 20 min before mitosis occurs. The daughter cells of this division are not tracked.

6. Once all the appropriate cells in the field of view have been tracked the marked positions of each cell in each frame are saved into a file with the extension ".cel." This file records the cell number, frame number, x-coordinate and y-coordinate of every tracked cell in every frame.

7. Finally, we conduct trajectory analysis using a range of software originally designed by Dr. Graham Dunn and Dr. Daniel Zicha at the Randall Centre, Kings College London. This program returns a Rayleigh test for unimodal clustering of directions where the null hypothesis is uniform distribution of cell trajectories. If the null hypothesis is rejected by the Rayleigh test the cells are showing a significant directional response to CSF-1. This software is not commercially available, however Kinetic Imaging has some software analysis to use with the DCC incorporated into the AQM software.

8. The cell trajectories can be plotted so that their starting point is shifted to the origin in Microsoft Excel to give a preliminary indication of chemotaxis. A significant chemotactic response should result in the majority of cell trajectories radiating from the origin in the direction of the highest chemoattractant concentration (**ref. *10*** and **Fig. 4**). To do this the tracking data must also be saved as an .xld file, which can be recognized by Excel. This option is available in the Kinetic Imaging tracking software.

4. Notes

1. The BMMs must be differentiated and cultured on bacterial plates because they adhere too strongly to tissue culture plastic and are impossible to passage. BMMs change in morphology and migratory behavior during passaging. They gradually become more elongated and less motile thus reducing their chemotactic potential. In our motility assays, we only use cells for 7 d after the 5-d differentiation period.

2. Successful assembly of the DCC is a specialized technique that requires patience and practice. It is advisable for a new user to practice assembling the chamber initially using a wet cover slip with no cells attached. It is essential that the user can assemble a chamber that contains no bubbles after a 24-h incubation. The presence of bubbles particularly in the inner well will seriously interfere with the formation of the gradient and produce unreliable results. Bubbles can result from incorrect sealing of the chamber but are most commonly the result of the cover slip being lifted away from the chamber during the draining of the outer well. Once the chamber can be assembled correctly the user should be able to proceed to assembling the chamber with cover slips seeded with cells. If the cells are still viable 24 h later then the user can confidently begin to conduct chemotaxis assays.

3. The time taken for a gradient to form between the two wells of the DCC depends on the molecular weight of the chemotactic factor. Factors with a molecular weight between 350 and 370 Da will form a gradient in 10 min, whereas factors with a molecular weight between 10 and 20,000 Da will take 30 min *(7)*. Similarly, the molecular weight of the chemoattractant will determine the stability of

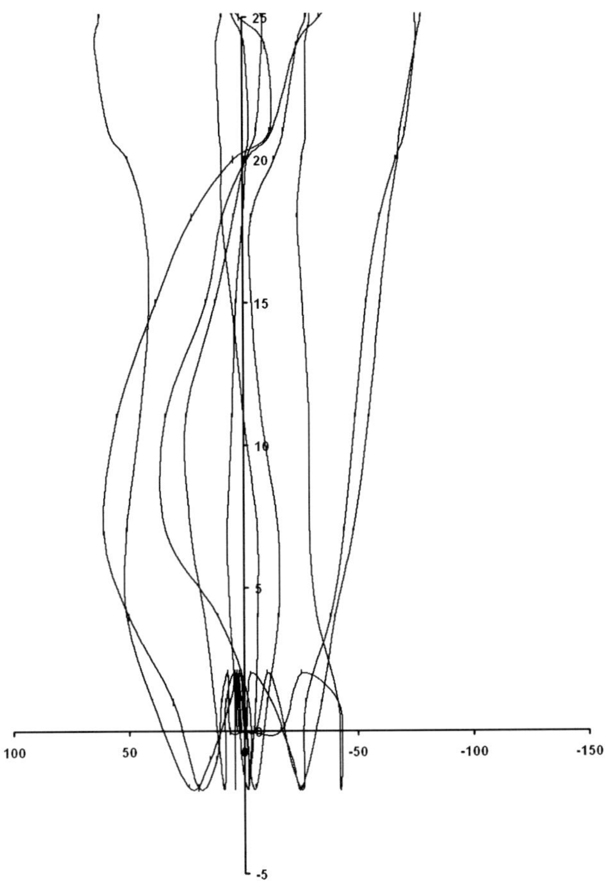

Fig. 4. Trajectory plot of cells. A plot of 10 BMM cell trajectories (pixel coordinates) where their starting point is shifted to the origin in Microsoft Excel. The source of chemoattractant is at the top of the plot.

the gradient—generally the greater the molecular weight the longer the gradient will be maintained. It is advisable for a new user to test the formation and stability of the gradient in the DCC using a fluorescein isothiocyanate-conjugated dextran of comparable molecular weight to the intended chemoattractant. Assemble the chamber as normal using the fluorescein isothiocyanate–dextran in the outer well. The formation of the gradient can then be monitored using a standard epi-fluorescence microscope. A similar method was used to establish that a gradient of CSF-1 forms within 25 min and is stable for at least 17 h in the DCC *(15)*.

4. In addition to studying the response of BMMs isolated from knock-out mice, the DCC can be used to characterize the chemotactic response of cells expressing tagged proteins introduced either by microinjection of recombinant protein or deoxyribonucleic acid *(16)* or by transfection. To identify expressing cells the cover slip is removed from the chamber and the cells fixed immediately after the filming period ends. To fix the cells, remove the wax using a razor blade and carefully lift the cover slip away from the chamber using watchmaker forceps. Immediately place the cover slip in fixative (e.g., 4% paraformaldehyde). The cover slip can then be stained for expressed protein following normal protocols. If a GFP tag is used in conjunction with time-lapse fluorescence microscopy the cells can be directly identified and tracked from the resulting film.

References

1. Dormann, D. and Weijer, C. J. (2003) Chemotactic cell movement during development. *Curr. Opin. Genet. Dev.* **13,** 358–364.
2. Seppa, H., Grotendorst, G., Seppa, S., Schiffmann, E., and Martin, G. R. (1982) Platelet-derived growth factor is chemotactic for fibroblasts. *J. Cell Biol.* **92,** 584–588.
3. Jones, G. E. (2000) Cellular signaling in macrophage migration and chemotaxis. *J. Leukoc. Biol.* **68,** 593–602.
4. Kassis, J., Lauffenburger, D. A., Turner, T., and Wells, A. (2001) Tumor invasion as dysregulated cell motility. *Semin. Cancer Biol.* **11,** 105–117.
5. Zigmond, S. H. and Hirsch, J. G. (1973) Leukocyte locomotion and chemotaxis. New methods for evaluation, and demonstration of a cell-derived chemotactic factor. *J. Exp. Med.* **137,** 387–410.
6. Dunn, G. A. and Zicha, D. (1993) Long-term chemotaxis of neutrophils in stable gradients: preliminary evidence of periodic behavior. *Blood Cells* **19,** 25–39; discussion 39–41.
7. Zicha, D., Dunn, G. A., and Brown, A. F. (1991) A new direct-viewing chemotaxis chamber. *J. Cell Sci.* **99,** 769–775.
8. Jones, G. E., Zicha, D., Dunn, G. A., Blundell, M., and Thrasher, A. (2002) Restoration of podosomes and chemotaxis in Wiskott-Aldrich syndrome macrophages following induced expression of WASp. *Int. J. Biochem. Cell Biol.* **34,** 806–815.
9. Orr, A. W., Elzie, C. A., Kucik, D. F., and Murphy-Ullrich, J. E. (2003) Thrombospondin signaling through the calreticulin/LDL receptor-related protein co-complex stimulates random and directed cell migration. *J. Cell Sci.* **116,** 2917–2927.
10. Zicha, D., Allen, W. E., Brickell, P. M., Kinnon, C., Dunn, G. A., Jones, G. E., et al. (1998) Chemotaxis of macrophages is abolished in the Wiskott-Aldrich syndrome. *Br. J. Haematol.* **101,** 659–665.

11. Honda, S., Sasaki, Y., Oshawa, K., Imai, Y., Nakamura, Y., Inoue, K., and Kohsaka, S. (2001) Extracellular ATP or ADP induce chemotaxis of cultured microglia through Gi/o-coupled P2Y receptors. *J. Neurosci.* **21,** 1975–1982.
12. Maden, M., Keen, G., and Jones, G. E. (1998) Retinoic acid as a chemotactic molecule in neuronal development. *Int. J. Dev. Neurosci.* **16,** 317–322.
13. Allen, W. E., Zicha, D., Ridley, A. J., and Jones, G. E. (1998) A role for Cdc42 in macrophage chemotaxis. *J. Cell Biol.* **141,** 1147–1157.
14. Jones, G. E., Ridley, A. J., and Zicha, D. (2000) Rho GTPases and cell migration: measurement of macrophage chemotaxis. *Methods Enzymol.* **325,** 449–462.
15. Webb, S. E., Pollard, J. W., and Jones, G. E. (1996) Direct observation and quantification of macrophage chemoattraction to the growth factor CSF-1. *J. Cell Sci.* **109,** 793–803.
16. Ridley, A. J. (1998) Mammalian cell microinjection assay to study the function of Rac and Rho. *Methods Mol. Biol.* **84,** 153–160.

5

Cell-Adhesion Assays

Dennis F. Kucik and Chuanyue Wu

Summary

One of the most important properties of cells that are derived from multicellular organisms is their ability to adhere to extracellular matrix proteins or other cells. Analysis of cell–extracellular matrix and/or cell–cell adhesion, therefore, is of important value to experimental biologists as well as clinical investigators. Over the past several decades, many different cell-adhesion assays have been developed. Based on the experimental conditions, most of the cell-adhesion assays fall into two categories, namely static adhesion assays and flow adhesion assays. Static assays are widely used to assess the adhesion of many types of cells (e.g., epithelial cells and fibroblasts) to the extracellular matrix. The flow adhesion assays are more appropriate for analysis of blood cell (e.g., leukocyte) adhesion to endothelial cells, to each other, or extracellular matrix proteins. This chapter describes two basic protocols, one for analysis of cell adhesion under static conditions and the other for measurement of cell adhesion under shear stress. In addition, variations to the basic protocols and areas where special attention is required for successful application of these methods are discussed.

Key Words: Cell adhesion; integrins; extracellular matrix; fibronectin; N-acetyl-β-D-glucosaminidase; enzyme-linked immunosorbent assay; flow; shear stress.

1. Introduction

Cell adhesion is a fundamental process that is critically involved in embryonic development and diseases. On the basis of the experimental conditions under which adhesion is measured, methods for measurement of cell adhesion can be in general divided into two types. In the first type of the methods, cell adhesion is analyzed under static conditions. Static assays are widely used to assess the adhesion of different types of cells, including epithelial cells and fibroblasts. They are relatively simple to perform and provide a valuable assessment of the adhesiveness of cells to a defined extracellular matrix sub-

From: *Methods in Molecular Biology, vol. 294: Cell Migration: Developmental Methods and Protocols*
Edited by: J-L. Guan © Humana Press Inc., Totowa, NJ

strate (e.g., fibronectin). However, static assay methods poorly simulate adhesion that occurs in blood, or even lymph vessels, under shear stress. Therefore, a second method is provided to measure cell adhesion under shear stress, which can be achieved by using flow chambers. The use of a flow chamber enables the researcher to simulate blood flow to reconstruct cell systems in the presence of shear. Some adhesive events occur only under shear and, thus, cannot be characterized under static conditions *(1,2)*. Flow chambers are widely available commercially and allow the researcher to introduce cells between two flat surfaces under conditions of laminar flow while the process of adhesion is visualized using a microscope *(3)*. These assays are most commonly used to study leukocyte adhesion, either with endothelial cells or to substrates of purified ligands. They are also useful to study bacterial adhesion *(4)*. An additional advantage of using flow chambers for adhesion assays is that they produce well-defined forces, in contrast to some of the wash methods of static assays. Flow cell assays can also detect rapid events, so that adhesion events, such as remodeling of contacts during rolling, can be studied on a time scale as small as a fraction of a second *(5,6)*. Furthermore, the entire process of adhesion can be observed, as a leukocyte progresses from rolling to firm adhesion and finally migration through the endothelium *(7)*. In this chapter, we first describe a basic protocol of static adhesion assay. Next, we provide a protocol for measuring cell adhesion under shear stress. Finally, we discuss variations to the basic protocols and areas where special attentions are required for successful application of the methods.

2. Materials

2.1. Static Adhesion Assays

1. 96-Well enzyme-linked immunosorbent assay (ELISA) plates (Corning).
2. Extracellular matrix proteins (fibronectin, laminin, collagens, or other extracellular matrix proteins, either alone or in combination).
3. PBS: 2.7 mM KCl, 137 mM NaCl, 1 mM KH$_2$PO$_4$, 10 mM Na$_2$HPO$_4$, pH 7.4.
4. Bovine serum albumin (BSA; 10 mg/mL in PBS; denatured at 85°C for 10 min prior to use).
5. α Minimum essential medium (α-MEM; Life Technologies).
6. Substrate solution: 3.75 mM p-nitropheno-N-acetyl-β-ᴅ-glucosaminide (Sigma) in 50 mM citrate buffer, pH 5.0 (aliquoted and stored at –20°C).
7. Stop solution: 5 mM ethylene diamine tetraacetic acid (EDTA), 50 mM glycine buffer, pH 10.4.

2.2. Measurement of Adhesion in Shear Stress

1. Cultured endothelial cells or substrate coating materials (fibronectin, gelatin, or other extracellular matrix proteins, either alone or in combination).

2. Adhesion media (primary = Hanks Balanced Salt Solution [HBSS], alternate = serum-free culture media with 10 m*M* HEPES buffer).
3. BSA.
4. In vitro flow chamber.
5. Inverted phase-contrast microscope.
6. Low power (approx 10–20X) objective.
7. Programmable syringe pump.
8. 25-mL Syringes.
9. Video camera.
10. VCR.
11. TV monitor.
12. Video cables.

    Recommended:
13. Fluorescence optics.
14. Stage incubator.

    Optional:
15. Computerized image digitization system.
16. Motion analysis software.

3. Methods

3.1. Static Adhesion Assay

The following protocol described has been successfully used for analyses of Chinese ovary hamster cell adhesion to fibronectin *(8)*. With minor modifications, it can be used to analyze static adhesion of almost any cell types.

1. Coat wells of 96-well ELISA plates with 10 µg/mL fibronectin in PBS at 37°C for at least 1 h (*see* **Note 1**).
2. Incubate each well with 200 µL of heat-denatured 10 mg/mL BSA in PBS at 37°C for at least 1 h.
3. Rinse the wells twice with α-MEM.
4. Harvest Chinese hamster ovary cells with 0.3 m*M* EDTA in PBS, rinse the cells three times with α-MEM, and suspend them to a final density of 3×10^5 cells/mL (*see* **Note 2**).
5. Add 100 µL of the cell suspension to each well of the fibronectin-coated ELISA plates and incubate for 60 min in a 37°C incubator under a 5% CO_2–95% air atmosphere (*see* **Note 3**).
6. Rinse the wells three times with PBS (*see* **Note 4**).
7. Add 60 µL of the substrate solution to each well (*see* **Note 5**).
8. In parallel experiments, add 1 mL of the cell suspension to a 1.5-mL microfuge tube that is precoated with heat denatured 10 mg/mL BSA in PBS. Pellet the cells immediately by centrifugation in a microfuge at 3800*g* for 15 min and then keep the cell pellets in –20°C freezer. Simultaneous to **step 7**, add 600 µL of the substrate solution to the microtube, vortex, and then transfer the cell/substrate mixture to new wells of the 96-well ELISA plates (60 µL/well; *see* **Note 6**).

9. Incubate the plates at 37°C in 100% humidity until color develops.
10. Add 90 µL of the stop solution to each well of the 96-well ELISA plates.
11. Measure absorbance at 405 nm (A405 nm) using a microplate reader.
12. Determine cell adhesion using the formula: cell adhesion (%) = A405 nm of the adhered cells (**steps 1–7**) divided by A405 nm of the "total" cells (**step 8**) ×100% (*see* **Note 6**).

3.2. Measurement of Adhesion in Shear Stress

The subsequent methods described outline 1) preparation of adhesive substrates, 2) loading of cells with fluorescent marker, 3) the flow assay, and 4) data analysis.

3.2.1. Preparation of Adhesive Substrates

Commercial flow chambers incorporate either the surface of a culture dish or a cover slip as one of the two parallel plates between which laminar flow occurs. This surface should be coated with either endothelial cells or extracellular matrix proteins as an adhesive substrate. This is described in three alternate procedures (cell monolayer preparation, **steps 1–4**; purified substrate on small part of dishes, **steps 5–8**; and purified substrate on cover slip, **steps 9–18**).

3.2.1.1. Cell Monolayer Preparation

1. Incubate plastic tissue culture dishes or glass cover slips with 1% gelatin in PBS for 30 min at room temperature.
2. Remove gelatin solution.
3. Add endothelial cells to result in a cell density of approx 5000 cells/cm^2. Transfer to 37°C incubator for culture.
4. Feed cells every 2–3 d until confluent. Depending on the cell type, cells are usually suitable for assays for 2–3 d after they reach confluency (*see* **Note 7**).

3.2.1.2. Preparation of Purified Substrate on Plastic Dishes

Purified ligands can be used instead of endothelial cell monolayers for a well-defined adhesive substrate. Because leukocytes and endothelial cells use several sets of adhesion molecules simultaneously to establish adhesion, purified ligand substrates may be necessary to isolate a particular adhesion receptor-ligand interaction. Purified substrates also allow the researcher to vary properties such as ligand density, which is difficult to do with whole cells. The following procedure is designed to coat only a small portion of the plastic dish to conserve reagents.

1. Outline an area to coat with a marker or diamond stylus. If a diamond stylus is used, keep the scratches shallow so as not to interfere with sealing of flow chamber gaskets.

2. Add 25 µL of coating solution (10 µg/mL matrix protein, such as fibronectin or collagen, in PBS, pH 7.4) and incubate at room temperature for 1 h.
3. Remove coating solution and add 1 mL of 1% w/v BSA in PBS. Incubate at room temperature for 1 h.
4. Remove BSA solution. Wash twice with PBS.

3.2.1.3. PREPARATION OF PURIFIED SUBSTRATE ON GLASS COVER SLIPS

Covalently linking matrix molecules to adsorbed poly-L-lysine (PLL), as in the following protocol, results in tighter, more uniform binding to cover slips than direct adsorption. If, however, you prefer to avoid PLL, direct adsorption of matrix materials, as in the plastic dish protocol, may provide an adequate substrate. Cover slips precoated with PLL are also commercially available.

1. Sterilize the cover slips by autoclaving, ultraviolet light, or flaming with ethanol.
2. Place cover slips in 35-mm tissue culture dishes or multi-well plates.
3. Outline an area to coat with a marker or diamond stylus. If a diamond stylus is used, keep the scratches shallow so as not to interfere with sealing of flow chamber gaskets.
4. Add enough PLL (100 µg/mL in PBS, pH 7.4) to coat the outlined area. We use PLL MW 70K-150K (Sigma Chemical; St. Louis, MO). Incubate at room temperature for 10 min.
5. Wash cover slips three times with PBS.
6. Add 1% glutaraldehyde and incubate at room temperature for 30 min.
7. Wash cover slips three times with PBS.
8. Add 25–100 µL of coating solution (depending on the size of the cover slip) at 10 µg/mL and incubate at room temperature for 1 h.
9. Add 1% BSA (enough to completely cover the cover slip) to block. Incubate at room temperature for 30 min.
10. Wash cover slips once with PBS. Add fresh PBS to keep moist until used for flow assay.

3.2.2. Loading Cells With Fluorescent Dye

Although rolling and adherent cells can be visualized without fluorescence, the increase in contrast afforded by labeling the perfused cells greatly facilitates observation and counting, and is sometimes a requirement for object tracking software.

The following protocol uses BCECF-AM as a fluorescent marker. We have used this dye extensively and find it to have no adverse effects on adhesion. Several other cell markers will also work, but concentrations and incubation times may have to be adjusted.

1. Wash cells in HBSS, then resuspend at $1-2 \times 10^6$/mL in 25 mL of HBSS.
2. Add BCECF-AM (Molecular Probes; Eugene, OR) to make 1 µM (from 2 mM stock in DMSO).

3. Incubate for 30 min at room temperature in the dark.
4. Wash the cells three times with 35 mL of HBSS.
5. Resuspend in HBSS to a concentration of 0.5×10^6 cells/mL.

3.2.3. Flow Assay

In the past, investigators constructed their own laminar flow chambers, but this is no longer worthwhile unless special characteristics are required. Several flow chambers are now available commercially, and, in general, they all work well. The same basic methods are used to conduct flow cell adhesion experiments regardless of the particular flow chamber used. The primary difference among apparatuses is how the substrate (either a cell monolayer or purified ligand) is incorporated. The following method usually assumes use of a Glycotech flow chamber (Glycotech; Rockville, MD), in which cells are grown in a tissue culture dish that becomes part of the flow chamber upon assembly. However, this protocol can be easily adapted for use with other chambers. Simply modify the assembly procedure according to manufacturer's instructions, and use cell- or ligand-coated glass coverslips instead of coated dishes if required.

1. Turn on the stage incubator and warm adhesion buffer prior to experiment (if experiments are to be done at 37°C; *see* **Note 8**).
2. Load the cells to be perfused with a fluorescent dye (optional; *see* **Subheading 3.2.2.** dye-loading protocol).
3. Load leukocytes or other cells to be perfused into a syringe, if using the "pushing" method, or a tube or other reservoir, if using the "pulling" method (*see* **Note 9**).
4. Attach the tubing to the syringe and remove any air bubbles from the system. Assemble the flow chamber, according to manufacturer's instructions, incorporating an adhesive substrate of either cells or purified ligand. (Cells or ligand will be pre-coated on either a tissue culture dish or a cover slip, depending on the manufacturer of the flow chamber.) Work out any bubbles in the system that may have formed during the assembly process.
5. Program the syringe pump to give the desired flow according to manufacturer's instructions (*see* **Note 10**).
6. Mount the flow chamber on the microscope, focus the system, and adjust the optics, camera, and any recording equipment (*see* **Note 11**).
7. Choose an area to observe (*see* **Note 12**), start the recording equipment, and begin the flow. Record 3–6 min of data.
8. Stop the flow (to conserve cells) and move to a new area of observation to collect more data. In general, several areas of observation are required to collect enough data for statistical significance. It is a good idea to resuspend the cells at this point. If using the "pushing" method, with the cells in the syringe, rotate the syringe 180°. If using the "pulling" method, drawing the cells from a tube or other reservoir, resuspend by gently shaking or swirling the tube or reservoir.

3.2.4. Data Analysis

3.2.4.1. QUANTIFYING ADHERENT CELLS

The raw data from flow cell adhesion experiments will be movies of cells as they interact with an adhesive substrate, either other cells or purified matrix materials. These interactions can be quantified in a number of ways. It is important to keep in mind that the method of analysis should be tailored to the scientific questions to be answered. For example, if you are interested only in the number of adherent cells, a simple count of adherent cells/min can be obtained by visual review of videotapes. Even when using simple, low-tech methods of analysis, though, it is important to have precise definitions of the quantities measured. For example, under shear stress, cells are rarely completely stopped. Therefore, a reasonable definition of firmly adherent cells might be, "cells that moved less than 1 cell diameter in 5 s." There is some variability in the literature as to the actual numbers used to define firm adhesion, and this often reflects differences in experimental conditions (higher shear stresses and less adhesive substrates lead to briefer adhesion). To enable comparison of studies, however, a quantitative criterion should be chosen and clearly specified in materials and methods.

The number of adherent cells per minute will depend on the size of the field observed, so this number must be normalized for area of observation. Area can be calculated using a stage micrometer, which is a glass slide with accurate distance scales etched into the surface. These can be purchased from several scientific supply companies (e.g., Fisher Scientific; Pittsburgh, PA). Simply image this known length standard on your system, and use it to calibrate distances (in microns) on the monitor.

The following protocol can be performed with nothing other than a VCR and monitor. It can be made far less tedious, however, if the sequences are digitized into AVIs with a video capture card. Many inexpensive cards are available for this purpose. This will give the investigator greater control over functions such as rewind and freeze frame. Some loss of both temporal and spatial resolution usually occurs in this process, however, so be sure that you choose a card that produces movies of sufficient quality to reliably distinguish adhesion events.

1. Choose a video segment of defined length (e.g., 1 min) to analyze.
2. Visually review the segment, counting the number of cells that stop and marking their locations with a lab marker. If the number is large, it may help to use a marker to divide the screen into sections and count each separately.
3. Review the segment again to be sure that each cell that seemed to stop meets your quantitative criterion for firm adhesion.

3.2.4.2. QUANTIFYING ROLLING CELLS

More care must be taken if rolling cells are to be identified visually. This is because of the properties of laminar flow, fluid in a flow chamber moves more slowly near the chamber walls. Because the adhesive substrate constitutes one wall of the chamber, cells carried by hydrodynamic flow near the substrate, without any adhesive interactions, may appear to be rolling *(13)*. An objective measure developed to identify rolling cells is the critical velocity *(14)*. This is based on the fact that adhesive interactions will substantially decrease cell velocity, distinguishing it from a cell that is moving in the slowest hydrodynamic flow. Knowing the size of the cell, the viscosity of the fluid (near that of water for simple buffers), and the shear stress, an upper limit of velocity compatible with rolling can be calculated. Therefore, even if your experimental question deals only with the number of rolling cells, and not their velocity, it is a good idea to calculate a critical velocity for a few cells to verify that cells counted as rolling do indeed have adhesive interactions with the substrate. Once you know what to look for, it may then be feasible to identify rolling cells visually.

A simple calculation for critical velocity is:

$$V_{critical} = \beta \cdot r \cdot \gamma$$

where r = the radius of the cell, γ = the shear rate, and β is a drag factor. A reasonable estimate of β that can be used for cells in a flow chamber is 0.5. Cells moving slower than the critical velocity can be confirmed as being rolling cells.

3.2.4.3. ROLLING VELOCITIES

A simple calculation of rolling velocity can be performed by marking a known length (in microns) on the monitor (again, calibrated using a stage micrometer) and measuring the time required for a cell to move this distance. The temporal resolution of this method will be limited, since times less than a few seconds may be difficult to measure. Greater time resolution can be achieved using professional-quality videotape machines that can accurately count frames, or by digitizing the sequences. Alternatively, more accurate velocity determinations can be made using more sophisticated image processing software that can accurately localize objects in each image and convert pixels to distances. A variety of commercial software packages are available to track cell motion, and most of these will work for tracking rolling cells. NIH Image, available online for free, will also track object motion (Website: http://rsb.info.nih.gov/nih-image/about.html).

4. Notes

1. Different extracellular matrix proteins (laminin, collagen I) or their fragments can be used to coat 96-well plates. Although 10 µg/mL is often a good starting point, the concentrations of the coating solutions should be optimized for different matrix proteins and cells under study.

2. If cells are not effectively detached by treatment with EDTA alone, solutions containing both EDTA and trypsin (e.g., solution containing 0.05% trypsin and 0.53 m*M* EDTA from Life Technologies) could be used. In this case, cells should be washed at least once with serum-containing medium after harvesting with the trypsin-EDTA solution, followed by washing with serum-free medium (*see* **step 4**). The number of cells added to each well should be adjusted for different cell types and it should not exceed the maximal number of cells that can adhere to the coated extracellular matrix protein in a well. The maximal number of adhered cells can be estimated by adding a different number of the cells to each well. After incubating the cells in a 37°C incubator under a 5% CO_2–95% air atmosphere for a prolonged period of time (e.g., 2 h), the wells are examined under a microscopy and the number of cells that form monolayer reaching 100% confluence represents the maximal number of cells that can adhere to the coated extracellular matrix protein in a well.

3. Depending on the purpose of the experiments, cells can be treated with various agents (e.g., inhibitory or activating anti-integrin antibodies) before the addition of the cells to the wells. The length of incubation should also be optimized based on the purpose of the experiments, the cell type and the extracellular matrix protein. Samples should be analyzed in duplicate or triplicate under each experimental condition.

4. The simplest way to remove unattached cells is by washing the wells with a buffer as described in the basic protocol. Even, gentle force should be applied during the wash to avoid washing away adhered cells. Alternatively, unattached cells can be removed by centrifugation *(9)*.

5. In the basic protocol, the number of cells is quantified by measuring the activity of *N*-acetyl-β-D-glucosaminidase, a ubiquitous lysosomal enzyme, using *p*-nitropheno-*N*-acetyl-β-D-glucosaminide as a substrate. This method is sensitive and produces a linear relationship between $A_{405\ nm}$ and the cell number for many different types of cells over a wide range of cell numbers *(10)*. Alternatively, the number of cells can be quantified by staining the cells with dyes such as Crystal Violet *(11)*.

6. Alternatively, the numbers of the total cells and adhered cells in a well can be estimated by photographing multiple (>3) randomly selected microscopic fields before and after the washes and counting the cell number. Cell adhesion (%) can be presented as: the number of the adhered cells (after wash) divided by the number of the "total" cells (before wash).

7. The quality of the monolayer will depend on the cell type and the initial coating density. While a monolayer can be established more quickly by coating at higher density, coating at a lower density and allowing more time may result in a more orderly monolayer. After the cells are confluent, they may change properties such as shape and adhesion molecule expression. It is a good idea to always plate cells at the same density and do the experiments on the same day post-confluency. Optimal conditions for each cell type should be worked out empirically. The choice of plating cells on plastic vs glass will largely depend on the flow chamber selected. Some flow chambers incorporate glass cover slips as one wall of the cham-

ber, while others use a plastic tissue culture dish. Growing cells in tissue culture plastic generally results in superior adhesion as compared to glass cover slips, even if both are coated with extracellular matrix materials. The advantage of glass cover slips, however, is superior optical properties. Birefringence of plastic can be a problem if Nomarski imaging is used. Plastic also tends to have background fluorescence. Analysis of adhesion under flow does not generally require optimal optical conditions, though, so visualization through plastic is usually adequate for this purpose.

Endothelial cells generally need to be stimulated to express the adhesion molecules necessary for leukocyte adhesion. Whether, and how, you choose to stimulate your endothelial cells, however, will depend on the physiologic conditions you are mimicking. Interleukin-1, tumor necrosis factor-α, and lipopolysaccharide have been considered good paradigms for proinflammatory mediators and are often used experimentally on cultured endothelial cells to simulate inflammation *(12)*.

8. Many investigators perform experiments at room temperature, although adhesion may be stronger at 37°C. The choice of temperature will depend, among other things, on the experimental question, and how important metabolic processes are to the aspect of adhesion being studied.

9. The optimal cell concentration depends on the experiment. Cells should be dense enough that sufficient events are observed in the few minutes of recording. This will depend on the adhesiveness of the cells being studied. If the cell concentration is too high, however, cell behavior could be dominated by collisions with other rolling or adherent cells. We find that 0.5×10^6 cells/mL works well for many types of leukocytes. Remember, for meaningful comparison of experiments, the number of cells perfused per minute must be the same under all experimental conditions. This requires that cells be loaded into the syringe or reservoir at a consistent density and not be allowed to settle out of suspension.

A syringe pump can be used either to push or pull cells through a flow chamber. To push, the syringe itself can be filled with cells. However, many investigators prefer to attach the syringe to the outlet side of the chamber to pull fluid and cells through the system. An advantage of the pulling method is that the cells can be kept in a tube and warmed in a water bath. Also, the tube can be shaken or swirled periodically to keep the cells in uniform suspension. Both warming and mixing can be much more problematic if the cells are in a syringe mounted on a syringe pump. For most experiments, a syringe size of 10–25 mL works well. Larger syringes enable longer periods of uninterrupted flow, but smaller syringes can produce smoother flow, especially at low flow rates. The number of cells required to do an experiment will depend on the rate of flow and the length of observation needed to gather sufficient data to answer the particular experimental question.

10. The desired flow will depend on the experiment to be performed, but generally should result in a physiologic shear stress. Simulation of arterial flow will require higher shear rates than simulation of venous or post-capillary venule conditions. Typically, physiologic shear stresses are roughly 0.5–5.0 dynes/cm^2.

Only shear rates and shear stresses can be compared for flow chambers or blood vessels of different geometry; flow rates are meaningless for comparison, because it is the combination of the flow rate and geometry that determine the forces experienced by the cells. Flow chamber manufacturers often provide a chart or table to make this conversion. If necessary, shear stress can be calculated from the dimensions of the flow chamber and the rate of fluid flow *(13)*. We recommend doing a range of shears to determine the shear dependence of any phenomena observed.

11. Most investigators currently record on a VHS videotape for later digitization. The technology to record directly to computer hard drive at high frame rates now exists, though, and this may become the standard soon. For some cameras, this may be the only option. DVD recorders are also available now at reasonable cost, and represent another option. Whatever the recording medium, we recommend making a sample recording of a minute or so before starting the experiment to make sure that all of the cables are connected correctly and that the equipment is working properly.

12. When using endothelial cells as an adhesive substrate, it is important to assess the integrity of the monolayer. This can be performed by simply visually examining the endothelial cells using bright-field or phase contrast optics. Rips or other defects in the monolayer will result in interaction of flowing cells with not only endothelial cells, but also with bare plastic or glass, which may make the data uninterpretable.

Acknowledgments

This work was supported by NIH grants GM65188 and DK54639 (to C. Wu) and by a VA Merit Award and NIH grant R21 AI 54552-01 (to D. F. Kucik).

References

1. Lawrence, M. B., Kansas, G. S., Kunkel, E. J., and Ley, K. (1997) Threshold levels of fluid shear promote leukocyte adhesion through selectins (CD62L,P,E). *J. Cell Biol.* **136,** 717–727.

2. Finger, E. B., Puri, K. D., Alon, R., Lawrence, M. B., von Andrian, U. H., and Springer, T. A. (1996) Adhesion through L-selectin requires a threshold hydrodynamic shear. *Nature* **379,** 266–269.

3. Lawrence, M. B. and Springer, T. A. (1991) Leukocytes roll on a selectin at physiologic flow rates: distinction from and prerequisite for adhesion through integrins. *Cell* **65,** 859–873.

4. Poelstra, K. A., van der Mei, H. C., Gottenbos, B., Grainger, D. W., van Horn, J. R., and Busscher, H. J. (2000) Pooled human immunoglobulins reduce adhesion of Pseudomonas as aeruginosa in a parallel plate flow chamber. *J. Biomed. Mater. Res.* **51,** 224–2321.

5. Alon, R., Hammer, D. A., and Springer, T. A. (1995) Lifetime of the P-selectin-carbohydrate bond and its response to tensile force in hydrodynamic flow. *Nature* **374,** 539–542.

6. Smith, M. J., Berg, E. L., and Lawrence, M. B. (1999) A direct comparison of selectin-mediated transient, adhesive events using high temporal resolution. *Biophys. J.* **77,** 3371–3383.

7. Cinamon, G., Grabovsky, V., Winter, E., Franitza, S., Feigelson, S., Shamri, R., et al. (2001) Novel chemokine functions in lymphocyte migration through vascular endothelium under shear flow. *J. Leukoc. Biol.* **69,** 860–866.

8. Wu, C., Chung, A. E., and McDonald, J. A. (1995) A novel role for alpha 3 beta 1 integrins in extracellular matrix assembly. *J. Cell Sci.* **108,** 2511–2523.

9. Lotz, M. M., Burdsal, C. A., Erickson, H. P., and McClay, D. R. (1989) Cell adhesion to fibronectin and tenascin: quantitative measurements of initial binding and subsequent strengthening response. *J. Cell Biol.* **109,** 1795–1805.

10. Landegren, U. (1984) Measurement of cell numbers by means of the endogenous enzyme hexosaminidase. Applications to detection of lymphokines and cell surface antigens. *J. Immunol. Methods* **67,** 379–388.

11. Corbett, S. A. and Schwarzbauer, J. E. (1999) Beta3 integrin activation improves alphavbeta3-mediated retraction of fibrin matrices. *J. Surg. Res.* **83,** 27–31.

12. Cines, D. B., Pollak, E. S., Buck, C. A., Loscalzo, J., Zimmerman, G. A., McEver, R. P., et al. (1998) Endothelial cells in physiology and in the pathophysiology of vascular disorders. *Blood* **91,** 3527–3561.

13. Lawrence, M. B., McIntire, L. V., and Eskin, S. G. (1987) Effect of flow on polymorphonuclear leukocyte/endothelial cell adhesion. *Blood* **70,** 1284–1290.

14. Goldman, A. J., Cox, R. G., and Brenner, H. (1967) Slow viscous motion of a sphere parallel to a plane wall-II couette flow. *Chem. Eng. Sci.* **22,** 653–660.

15. Ley, K. and Gaehtgens, P. (1991) Endothelial, not hemodynamic, differences are responsible for preferential leukocyte rolling in rat mesenteric venules. *Circ. Res.* **69,** 1034–1041.

6

Cell-Spreading Assays

Allison L. Berrier and Susan E. LaFlamme

Summary

A method is provided to quantitate the extent of cell spreading as a function of the expression level of transfected recombinant proteins. This chapter contains protocols for 1) replating and staining transfected cells for immunofluorescence microscopy, 2) optimizing image acquisition so that fluorescence intensity can be measured independent of cell morphology, and 3) quantitating cell area and expression levels of recombinant proteins for individual transfected cells using ImagePro-Plus software. This method can be used to further our understanding of intracellular signals and protein interactions that regulate cell spreading.

Key Words: Cell spreading; cell morphology; signaling proteins; integrins; adhesion signaling; integrin β cytoplasmic domain.

1. Introduction

Cell adhesion to extracellular matrix proteins (ECM) mediated by cell surface integrin receptors is necessary for many cellular processes, such as cell migration, wound healing, differentiation, proliferation and survival *(1–4)*. Cell spreading is an important aspect of cell adhesion for many cell types and is regulated by signaling pathways that become activated upon integrin-mediated cell attachment. Integrin engagement with an ECM ligand triggers the formation of multiprotein complexes at integrin receptor cytoplasmic domains that contain cytoskeletal-associated proteins, adaptor molecules and signaling proteins *(5)*. These cytoplasmic complexes regulate the activation of protein kinases, phosphoinositide kinases, and small guanosine trisphosphate-binding proteins that promote cell spreading *(5–13)*.

Characterization of the structural–functional properties of the heterodimeric α/β integrin receptor revealed that the integrin β subunit cytoplasmic domain

From: *Methods in Molecular Biology, vol. 294: Cell Migration: Developmental Methods and Protocols*
Edited by: J-L. Guan © Humana Press Inc., Totowa, NJ

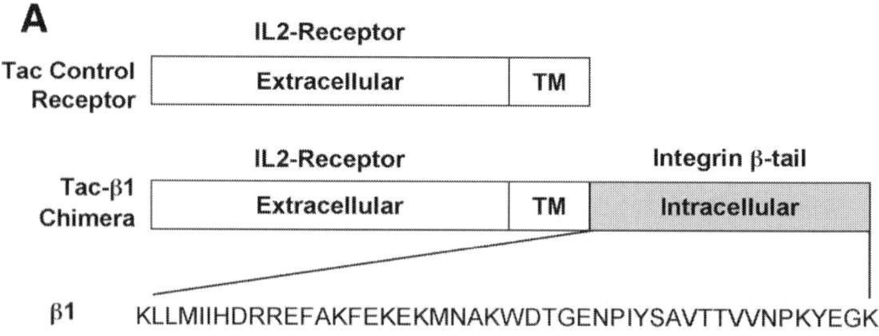

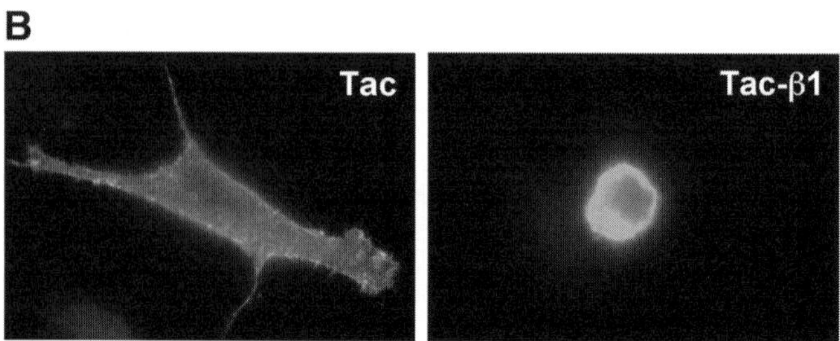

Fig. 1. Expression of tac-β1 inhibits the spreading of HFSFs replated onto collagen I. (**A**) A schematic diagram of the control tac receptor and the tac-β1 chimeric receptor is shown. The control tac receptor is comprised of the extracellular and transmembrane domains of the interleukin (IL-2) receptor small non-signaling subunit. Tac-β1 contains the same domains of the IL-2 receptor fused to the intracellular domain of the integrin β1 subunit. (**B**) HFSFs were transiently transfected with expression vectors for the control tac receptor or tac-β1. The transfected cells were replated onto collagen I and allowed to adhere for 1 h. The adherent cells were stained with FITC-conjugated antibodies to CD25 to detect the tac and tac-β1 expressing cells. The morphology of a representative tac and tac-β1 expressing cell is shown.

(β-tail) is required for the adhering integrin receptor to mediate cell spreading *(14)*. The integrin β-tail may coordinate cell spreading by anchoring the integrin β-tail to the actin cytoskeleton and/or by triggering signaling pathways that promote cell spreading *(5,15,16)*. The role of the β-tail in cell spreading has been studied using a "trans" dominant inhibitor of integrin function *(17–21)*. Expression of an isolated integrin β1-tail connected to the small non-signaling tac subunit of the interleukin-2 receptor (tac-β1) inhibits human foreskin fibroblast (HFSF) cell spreading on several ECM proteins (**Fig. 1; refs. *18,21***). Presumably, tac-β1 functions as a "trans" dominant inhibitor of integrin function in cell spreading by titrating proteins from the endogenous integrin β-tail, such as signaling, adaptor or cytoskeletal-associated proteins.

In this chapter, we describe a method to identify signaling proteins that are important for integrin β-tail function in HFSF cell spreading *(22)*. Cell spreading is assayed by replating cells onto an ECM protein. Initially, replated cells have round morphologies with small cell areas. After cell attachment, membrane protrusions and new cell-matrix adhesions are formed that increase cell area resulting in cell spreading. HFSFs expressing moderate levels of tac-β1 attach to the ECM yet fail to undergo the transition to a spread-cell phenotype. To test whether the expression of constitutively active signaling proteins that regulate cell morphology can rescue cell spreading inhibited by tac-β1, we assayed the extent of cell spreading by measuring the cell area as a function of the expression levels of tac-β1 and the recombinant signaling protein. Using this approach, we found that activated forms of R-Ras, Rac1, phosphatidylinositol 3-kinase, and protein kinase Cε were each capable of rescuing the tac-β1 inhibition of cell spreading by mechanisms requiring Rac1 activity and an intact β-cytoplasmic domain on the adhering integrin receptor *(22)*.

This spreading assay could be adapted to further our understanding of intracellular events that regulate cell spreading. For instance, the protein interactions at the integrin β-tail that are required to promote cell spreading are not yet defined. To identify candidate integrin β-tail binding proteins that are required for cell spreading, the ability of recombinant β-tail binding proteins to restore cell spreading inhibited by tac-β1 could be assayed *(16,23–25)*. Additional mutagenesis studies would then be needed to demonstrate that the binding of these proteins to the β-tail is required for the rescue of cell spreading. Our assay could also be used to identify downstream effectors of signaling proteins that are essential for cell spreading. For example, dominant-negative Rac1 could be used to inhibit cell spreading, and various Rac1 downstream effectors could be assayed for their ability to restore cell spreading. Similarly, the downstream signaling effectors of protein kinase C that promote cell spreading could be identified in a rescue of cell spreading assay performed with cells treated with inhibitors of protein kinase C. As novel signaling pathways that regulate cell morphology are identified, our assay provides an important tool for understanding how signals from these pathways are integrated to control cell morphology.

2. Materials

1. Primary human foreskin fibroblasts.
2. Electroporation buffer: 20 m*M* HEPES, pH 7.05, 137 m*M* NaCl, 5 m*M* KCl, 0.7 m*M* Na$_2$HPO$_4$, 6 m*M* dextrose, 1 mg/mL bovine serum albumin (BSA, Fraction V).
3. Electroporator: Gene Pulser (Bio-Rad) with a Capacitance Extender.
4. Electroporation cuvets: Gene-Pulser 4 mm (Bio-Rad).

5. Serum-free medium: DMEM supplemented with 100 U/mL penicillin, 100 µg/mL streptomycin sulfate, and 2 m*M* L-glutamine.
6. Complete medium: DMEM supplemented with 10% fetal bovine serum (heat-inactivated at 56°C for 30 min), 100 U/mL penicillin and 100 µg/mL streptomycin sulfate, and 2 m*M* L-glutamine.
7. 5 m*M* Sodium butyrate, pH 7.2.
8. 20 µg/mL Collagen I [Vitrogen 100 (Cohesion)] in 1X PBS.
9. Glass cover slips (Fishers finest 4.8 cm^2).
10. 1X PBS: 138 m*M* NaCl, 2.7 m*M* KCl, 8.1 m*M* Na_2HPO_4, 1.5 m*M* KH_2PO_4, pH 7.2.
11. Trypsin: 0.05% trypsin in 0.53 m*M* EDTA, store at 4°C.
12. Trypsin inhibitor: 0.5 mg/mL soybean trypsin inhibitor in DMEM, store at 4°C.
13. Fixative: 1X PBS containing 5% sucrose and 4% formaldehyde.
14. Blocking solution: 150 m*M* glycine containing 3% BSA (Fraction V), pH 7.2.
15. Mounting medium: 1X PBS containing 1 mg/mL phenylenediamine (EM Science) and 10% glycerol; aliquot and store at –20°C.
16. Antibody to the Tac epitope: fluorescein isothiocyanate (FITC)-conjugated mouse monoclonal antibody to human CD25 (Becton Dickinson) used at a 1:100 dilution.
17. Antibody to human integrin β1: FITC-conjugated mouse monoclonal antibody to human CD29 (Coulter-Immunotech) used at a 1:100 dilution.
18. Permeabilization buffer: 0.4% Triton X-100 in 1X PBS.
19. Goat Ig: 10 mg/mL purified goat Ig reconstituted in water.
20. Primary antibodies to the epitope tag of the signaling protein: rabbit polyclonal antibodies to the myc epitope and Flag epitope.
21. Secondary antibodies: rhodamine-conjugated goat anti-rabbit antibodies.
22. ImagePro-Plus Software Version 3 (Media Cybernetics).
23. SlideWrite Version 4 (Advanced Graphics Software).
24. Micrometer (Applied Image).
25. Camera: HRD060-NIK video coupler 0.60X magnification cooled 1CCD Spot Camera Model 1.0.0 and Spot 3.0.5 software (Diagnostic Instruments).
26. Microscope: Olympus BX60 upright microscope with three Olympus objectives:
    a. 100X Magnification oil immersion, 1.35 numerical aperture, flat field optical correction and apochromatic aberration correction, adjustable numerical aperture with iris diaphragm, infinity corrected;
    b. 40X Magnification oil immersion, 1.00 numerical aperture, flat field optical correction and apochromatic aberration correction, adjustable numerical aperture with iris diaphragm, infinity corrected; and
    c. 10X Magnification, 0.30 numerical aperture, flat field optical correction, fluorite aberration correction, infinity corrected.
27. Humidified chamber: disposable 150-mm Petri dish with Whatmann 3-mm paper saturated in water and covered with parafilm.
28. Microscope slides: precleaned 25 × 75 × 1 mm.
29. Nail polish: quick-drying clear nail polish.

3. Methods

3.1. Transient Transfections of Normal HFSFs

Perform two transfections for each experimental condition. This provides a sufficient number of cells to seed three matrix-coated cover slips (*see* **Note 1**).

1. For each transfection, resuspend approx 1.5×10^6 HFSFs in 0.5 mL of electroporation buffer and transfer the cells to an electroporation cuvet.
2. Add 20 μg of expression vector tac or tac-β1 encoding to the cuvet containing the cells for one transfection (*see* **Note 2**).
3. Add 40 μg of the signaling protein expression vector to the cells in one cuvet for each co-transfection.
4. Electroporate the HFSFs at 170 Volts, 960 μFarads (*see* **Note 2**) and then allow the cells to recover for 10 min at room temperature.
5. Collect the cells from each experimental condition (i.e., two transfections) into a 15-mL conical tube containing 5 mL of complete medium supplemented with sodium butyrate.
6. Centrifuge the cells (*see* **Note 11**) and aspirate the medium to remove the electroporation buffer.
7. Resuspend these cells in complete medium supplemented with sodium butyrate, seed in a 100-mm tissue culture dish and incubate overnight in a tissue culture incubator.

3.2. Coating the Glass Cover Slips With Collagen I

1. Coat individual glass cover slips with 250 μL of 20 μg/mL collagen I for 1 h at 37°C in a humidified chamber *(26)*.
2. Remove the excess collagen I from the cover slips by dipping the cover slips once in 200 mL of PBS. Then, transfer the individual collagen-coated cover slips to individual wells of a six-well tissue culture plate and wash two times by addition and aspiration of 2 mL of PBS per well.
3. Store the coated cover slips in PBS until use.

3.3. Cell Spreading

1. Harvest the transfected HFSFs 16 h after electroporation by trypsinization (0.5 mL of trypsin per 100-mm dish) at room temperature. As soon as the cells lift from the plate, add chilled trypsin inhibitor (0.5 mL of trypsin inhibitor per 100-mm dish) to prevent excess trypsin digestion of cell surface proteins.
2. Collect the cells in a 15-mL conical tube and centrifuge (*see* **Note 11**).
3. To remove any remaining trypsin and trypsin inhibitor, resuspend the cell pellet in 10 mL of chilled PBS, centrifuge, and repeat the PBS wash.
4. Resuspend the cell pellet in 6 mL of cold serum-free medium and remove an aliquot to count the cells.
5. Centrifuge the cells and then resuspend the cell pellet in cold serum-free medium at a density of $1–2 \times 10^5$ cells/mL.

6. Allow the cells to recover from trypsinization by incubating them in suspension for 40 min in the tissue culture incubator (37°C, 5% CO_2, humidified atmosphere) with the cap of the conical tube loose to maintain the pH of the medium (*see* **Note 3**).

7. Seed the cells from each transfection condition (two cuvets that were seeded onto a single 100-mm dish) onto three collagen I-coated cover slips in six-well tissue culture plates at a density of $2–4 \times 10^5$ cells per 2 mL per well and incubate for 1 h in the tissue culture incubator (*see* **Note 4**).

8. Wash the cover slips twice with PBS as described in **Subheading 3.2.** to remove unattached cells.

9. Fix the adherent cells with 2 mL of fixative per well for 30 min at room temperature.

10. Aspirate the fixative, and wash the cells in the six-well dishes three times with 2 mL of PBS.

11. Store the fixed cells in PBS at 4°C until ready for use.

3.4. Immunofluorescence

1. Transfer the cover slips containing the fixed cells from the six-well dishes to humidified chambers with the cells facing up.

2. Dilute the anti-tac antibody in PBS (*see* **Note 5**), and centrifuge to precipitate aggregates (*see* **Note 11**).

3. Add 250 μL of the supernatant to each cover slip and then incubate for 40 min at room temperature in a humidified chamber.

4. After staining, remove excess antibody by dipping the cover slips in 200 mL of PBS. Place the cover slips into the original six-well dishes, and then wash twice with 2 mL of PBS. To enhance washing, place the six-well plate on a horizontal shaker (30 rpm) for 15 min for each PBS wash.

5. To detect co-transfected cells, initially stain for the tac epitope at the cell surface as described in **steps 1–4**. Next, permeabilize the cells by incubating the cover slips in the six-well dishes with permeabilization buffer (2 mL per well) for 5 min at room temperature.

6. Remove the permeabilization buffer by three quick washes with 2 mL of PBS.

7. To reduce nonspecific staining, incubate the permeabilized cells in blocking solution (2 mL per well) for 30 min at room temperature and then wash three times with 2 mL of PBS.

8. Transfer the cover slips to humidified chambers, and incubate with primary antibodies that recognize the epitope tag of the recombinant protein. Dilute the antibodies in PBS and then centrifuge. Add 250 μL of this supernatant to each coverslip and incubate for 40 min at room temperature.

9. Remove the excess primary antibody by washing with PBS as described for the anti-tac antibody staining in **step 4**. Transfer the cover slips to humidified chambers.

10. Dilute the secondary antibodies in PBS containing 100 μg/mL goat Ig and then centrifuge to clarify. Add 250 μL of this supernatant to each cover slip and incubate for 40 min at room temperature.
11. Remove the excess secondary antibody by washing the cover slips with PBS as described for anti-tac staining in **step 4** above.
12. Store the cover slips in the dark at 4°C in PBS in six-well dishes.
13. To mount, place each coverslip cell side down onto a drop (5 μL) of mounting medium on a microscope slide. Seal the edges of the cover slip to the microscope slide with nail polish to prevent dehydration.

3.5. Optimizing Conditions for Image Acquisition

3.5.1. Choice of Objective

The sensitivity of the camera and the quality of the objectives will influence the choice of objective. When choosing the objective, make sure the field depth is greater than the height of round cells, and that the objective will detect the fluorescence intensities required for your experiment within the linear range (*see* **Note 6**).

3.5.2. Choice of Exposure Time and Gain Setting

Adjust the exposure time such that the fluorescence of round cells is below saturation for the selected objective. Optimize the gain setting to minimize effects of cell shape on fluorescence intensity. To accomplish this select a cell surface protein that can be stained with a fluorescent antibody to yield a fluorescence intensity similar to the expected fluorescence intensity in the experiment (*see* **Note 7**). Next, generate populations of round and spread cells by varying the length of times that cells are allowed to adhere to a cover slip coated with an ECM protein. Fix the adhered cells and detect the expression of the selected marker protein. Use ImagePro-Plus software to analyze the cell area and fluorescence intensity of the spread and round cells (*see* **Subheading 3.6.**). For different gain settings, generate scatter plots to analyze the cell area as a function of fluorescence intensity for randomly chosen cells (*see* **Note 7, Fig. 2**). Choose a setting that results in similar fluorescence intensities for the spread and round cells. The sensitivity of the camera and the quality of the objectives will influence the camera settings; thus, it is critical to adjust the camera settings to the available imaging system.

3.6. Data Analysis

1. Calibrate the ImagePro-Plus software to the magnification of the selected objective and camera. Acquire an image of a micrometer with this objective. Open the image file in ImagePro-Plus and perform the spatial calibration function to define the length of 100 microns using the image of the micrometer.

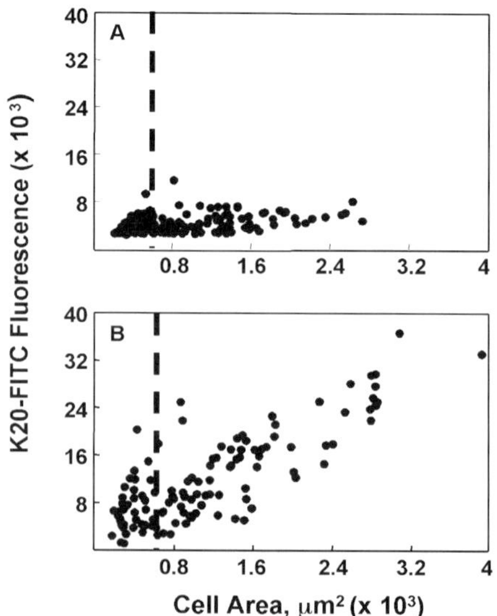

Fig. 2. Detection of similar levels of fluorescence from a surface marker on spread and round cells. HFSFs were seeded onto collagen I for 10 and 60 min. The adherent cells were washed, fixed and then stained for levels of $\beta 1$ integrins with anti-CD29 FITC-conjugated antibody. The Spot camera settings used were (**A**) an exposure time of 2 s and a gain 4 and (**B**) an exposure time of 2 s and a gain 8. The cell area and fluorescence intensity for randomly chosen cells are shown on the scatter plot. The *x*-axis is cell area from 0 to 4000 μm^2, the *y*-axis is FITC fluorescence intensity from 0 to 4.0×10^4 arbitrary units defined by ImagePro-Plus. The vertical line indicates the separation of spread cells to the right, and not spread cells to the left. Note that similar levels of FITC fluorescence were observed for round and spread cells with the camera exposure settings of time 2 s and gain 4.

2. To analyze the transfected cell morphology and expression of the recombinant protein, first outline the perimeter of transfected cells using the define object function. Then quantitate the area and fluorescence intensity for each cell using the object area function and integrated optical density function to measure the total fluorescence signal in a given area (*see* **Notes 8** and **9**). The fluorescence units are arbitrarily defined by ImagePro-Plus.

3. Compile the data in Excel spread sheets.

4. Report the data on SlideWrite scatter plots of cell area as a function of fluorescence intensity for 100 randomly chosen transfected cells. The extent of cell spreading is variable in short-term spreading assays and scatter plots capture the ranges in cell spreading and recombinent protein expression that are observed.

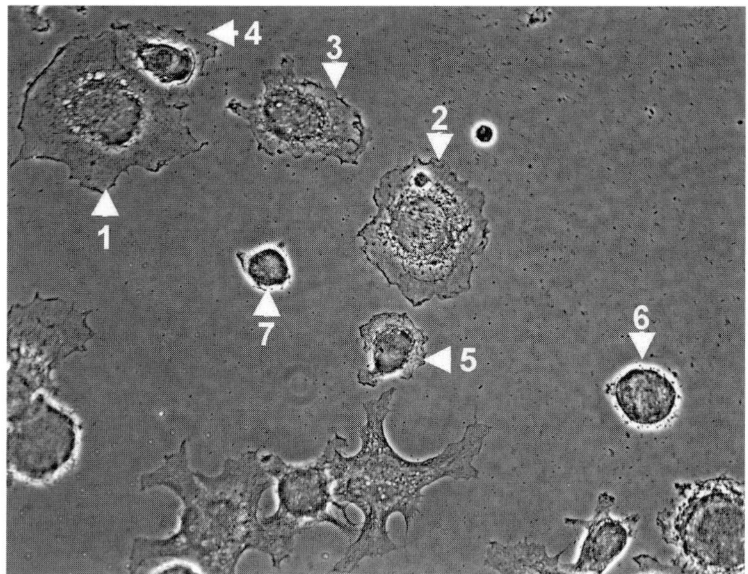

Fig. 3. The cell area and morphology of adherent HFSFs 1 h after replating onto collagen I. Shown is a phase micrograph containing a field of spread and not spread cells. Individual cells are numbered and the corresponding cell areas (μm^2) are as follows: (1) 3212, (2) 2310, (3) 1534, (4) 826, (5) 615, (6) 548, and (7) 278.

The morphologies and cell areas of a representative field of HFSFs replated onto collagen I and allowed to adhere for 1 h is shown (**Fig. 3**). We defined a spread cell as a cell with an area greater than 600 μm^2 (*see* **Note 10**). The ability of signaling proteins to rescue the tac-β1 inhibition of cell spreading can be quantitated by calculating the percentage of transfected cells that are spread (*see* **Note 12**). In the example shown, the scatter plots indicate the morphologies of transfected fibroblasts replated onto collagen I that express either tac, tac-β1, or tac-β1 and L61Rac1 from a representative experiment (**Fig. 4**). Expression of tac-β1 decreases the percentage of cells that are spread when compared to cells expressing tac alone. Cells coexpressing tac-β1 and L61Rac1 have an increase in the percentage of spread cells when compared to cells expressing tac-β1 alone. The percentages of transfected cells that are spread for the experiment shown are tac 91%, tac-β1 11%, tac-β1, and L61Rac1 80%.

5. Statistical analysis of the data from at least three trials can be used to determine whether the percentage of spread cells expressing tac-β1 alone is significantly different from the percentage of spread cells coexpressing tac-β1 and a recombinant signaling protein.

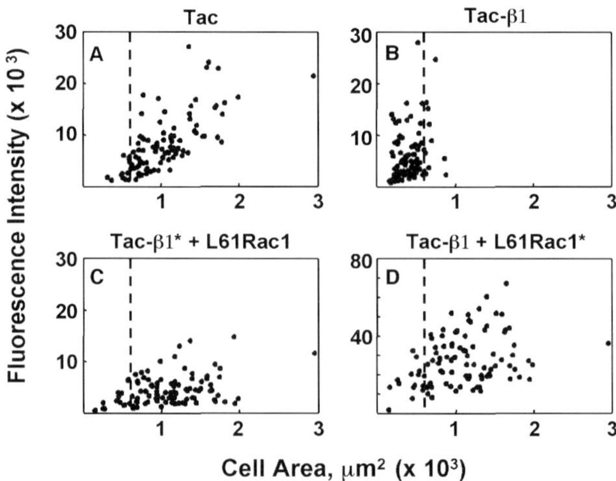

Fig. 4. Morphologies of cells expressing either tac alone, tac-β1 alone, or tac-β1 and L61Rac1. HFSFs were transiently transfected with tac alone (**A**), tac-β1 alone (**B**), and tac-β1 and myc-L61Rac1 (**C and D**). Sixteen hours after transfections, HFSFs were replated onto collagen I for 1 h. The cells were then fixed and stained for tac expression, subsequently permeabilized, and stained for the myc-epitope tag. The area and fluorescence intensity of cells were measured using ImagePro-Plus as described in **Subheading 3**. The scatter plots report the cell area and expression level of the recombinant proteins for one hundred transfected cells. The cell areas and expression levels of tac (**A–C**) or expression levels of myc (**D**) for the same transfected cells in (**C**) are shown. The x-axis is cell area from 0 to 3000 μm², the y-axis is FITC fluorescence from 0 to 30,000 arbitrary units ImagePro-Plus (**A–C**), and rhodamine fluorescence from 0 to 80,000 arbitrary units ImagePro-Plus (**D**). The dotted line at 600 μm² provides the separation of spread cells to the right from not spread cells to the left.

4. Notes

1. The extent of cell spreading varies on a cover slip and between cover slips potentially because of variations in collagen coating and the spreading efficiency of individual cells. To control for these variations, images are acquired from three different cover slips for each parameter in a single trial. We also include tac expressing cells as an internal control in each experiment to account for variations in the extent of cell spreading and immunofluorescence staining that can occur between experiments.

2. Details of the electroporation protocol including synchronizing the cells prior to electroporation have been previously described (*26*). The amount (μg) of plasmid used per transfection needs to be optimized for each expression construct, cell type and transfection protocol. If the cells are transfected by electroporation, the voltage should be optimized for different cell types.

3. The length of time necessary to recover from trypsinization and the coating concentration of the matrix protein will need to be determined empirically for different cell types and matrix preparations.

4. The density of the cells when seeding is important. We performed our studies in 6-well tissue culture dishes using primary HFSFs and observed that at too low a cell plating density ($<1 \times 10^4$ cells per well) there is very poor cell spreading. In contrast, at too high a cell plating density ($>1 \times 10^6$ cells per well), there are substantial numbers of cells that have overlapping projections. Efficient spreading of the nontransfected cells with relatively few overlapping transfected cell projections is observed at a cell plating density of 2×10^5 cells per well. However, when overlapping cells are observed, they are excluded from the analysis. The optimal seeding density also needs to be determined empirically for different cell types and dishes.

5. The optimal dilution for each antibody needs to be determined empirically.

6. The selection of a particular objective will be influenced by the available imaging system. At the time of our initial studies, we had 10X, 40X oil, and 100X oil objectives on an Olympus BX60 upright microscope with a CCD Spot camera as described in **Subheading 2**. The 100X objective was problematic because the field depth is small compared to the height of a round cell; thus, it was difficult to quantitate the fluorescence intensity of round cells. The 10X objective has a larger field depth; however, our Spot camera was not sensitive enough to acquire fluorescence images with the 10X objective for the range of fluorescence intensities in these studies because of the color filter inside our Spot camera. We chose the 40X objective because its field depth was sufficient to quantitate the fluorescence of a round cell in a single image file. The sensitivity of the 40X oil objective allowed us to quantitate the range of fluorescence intensities necessary for our experiments. It may be possible to use a lower magnification objective if a more sensitive black and white camera is used.

7. Previous flow cytometry experiments revealed that cells stained for the β1 integrin with a FITC-conjugated K20 antibody had fluorescence intensities similar to cells transiently transfected with tac-β1 and stained with the anti-tac antibody *(18)*. To optimize the gain setting, HFSFs were seeded onto collagen I for 10 and 60 min to generate populations of round and spread cells respectively. Integrin β1 was detected at the surface of the fixed cells. ImagePro-Plus software was used to analyze the cell area and FITC fluorescence intensity of the cells from both time points. For different gain settings, scatter plots were generated to analyze cell area as a function of FITC fluorescence from at least one hundred randomly chosen cells. The setting was chosen that resulted in similar integrin β1 expression levels for the spread and round cell populations (**Fig. 2A**). We observed that with higher gain settings the FITC fluorescence intensity of spread cells was higher than round cells (**Fig. 2B**).

8. Fluorescent images were acquired for each experiment over a 24-h period to minimize fading of the fluorescent antibody. Additionally, we never collected data with mercury bulbs that had been used for over 300 h, because the fluores-

cence signal generated by these bulbs fluctuates and decreases in intensity. Fluorescent images were acquired in parallel lines across the cover slip in order to avoid acquiring images of the same cell twice. The FITC-conjugated anti-tac antibody faded more noticeably than the rhodamine-conjugated antibody. Thus, the FITC fluorescence was acquired first. To avoid including data from photobleached cells, only the first image captured for the FITC fluorescence was used for analysis.

9. A weakness of quantitating the fluorescence intensity on a per cell basis is the time required for image analysis. Once the fluorescence thresholds for a transfected cell are set, the ImagePro-Plus software outlines the cell perimeter. In many cases, there are transfected cells with membrane projections with fluorescence intensities below the threshold setting. The analysis is labor intensive because the demarcated cell perimeter needs to be adjusted manually to reflect the actual cell perimeter. It was problematic to adjust the fluorescence threshold to detect these membrane projections because their fluorescence was close to the background.

   There are other software packages available that measure cell areas and fluorescence intensity including more recent versions of ImagePro-Plus, Slidebook (Intelligent Imaging Innovations) and MetaVue or MetaMorph (Universal Imaging Corporation, a division of Molecular Devices). Some of these software packages can acquire 3-D images of cells, and thus can be used to calculate the fluorescence intensity per cell volume in addition to cell area.

10. For our studies, we needed to determine the area of round cells. Round cells lack a discernible cytoplasm and are inhibited in cell spreading. To generate a population of round cells, fibroblasts were seeded onto collagen I for 10 min, which is sufficient for cell attachment but not cell spreading. The cells were fixed and then fluorescently labeled at the cell surface with a FITC-conjugated antibody to $\beta1$ integrins. Fluorescent images were captured using the Spot camera with the 40X oil objective. Cell area was determined using ImagePro-Plus software. We found that round HFSFs had cell areas < 600 μm^2. Hence, we defined cells with areas <600 μm^2 as inhibited in cell spreading. Cell size varies with cell type; therefore, the area of a round cell needs to be determined for each cell type.

11. Cells were centrifuged in a 15-mL conical tube at $300g$ for 10 min. Dilutions of antibodies for immunofluorescence were clarified at $1000g$ for 5 min.

12. The expression level of the recombinant protein sufficient to rescue the inhibition of cell spreading by tac-$\beta1$ is informative. Constitutively active signaling proteins such as L61Rac1 and V38R-Ras restored cell spreading largely independent of expression level, indicating that these signaling proteins trigger pathways that promote cell spreading at low, moderate and high levels of expression. The rescue of cell spreading by recombinant β-tail binding proteins may only occur above a threshold sufficient to saturate the tac-$\beta1$ tail. Thus, the rescue may only occur at either moderate or high levels of recombinant protein expression. With pharmacological inhibitors or dominant negative signaling proteins the rescue of cell spreading is predicted to occur with low, moderate or high expression levels of constitutively active signaling proteins. Nonetheless, the

expression level of the recombinant signaling protein sufficient to restore cell spreading will be influenced by its steady state level and the abundance of downstream activators and inhibitors in the signaling pathway.

Acknowledgments

We would like to thank Dr. Alan Hall for the L61Rac1 construct, Dr. Joseph Depasquale for assistance with quantitation of fluorescence intensity on a per cell basis, and Drs. Gang Liu and Carlos Reverte for helpful comments during the preparation of this chapter. We also thank Debbie Moran for her secretarial assistance in the preparation of this chapter. This work was supported by NIH grants GM51540 (SEL) and T32 HL07529 (ALB) and an American Heart Association Postdoctoral Fellowship 0020180T (ALB).

References

1. Chen, C. S., Mrksich, M., Huang, S., Whitesides, G. M., and Ingber, D. E. (1997) Geometric control of cell life and death. *Science* **276,** 1425–1428.
2. Hynes, R. O. (1992) Integrins: Versatility, modulation, and signaling in cell adhesion. *Cell* **69,** 11–25.
3. Giancotti, F. G. and Ruoslahti, E. (1999) Integrin signaling. *Science* **285,** 1028–1032.
4. Schwartz, M. A. and Assoian, R. K. (2001) Integrins and cell proliferation: regulation of cyclin-dependent kinases via cytoplasmic signaling pathways. *J. Cell Sci.* **114,** 2553–2560.
5. Miyamoto, S., Teramoto, H., Coso, O. A., Gutkind, J. S., Burbelo, P. D., Akiyama, S. K., and Yamada, K. M. (1995) Integrin function: molecular hierarchies of cytoskeletal and signaling molecules. *J. Cell Biol.* **131,** 791–805.
6. King, W. G., Mattaliano, M. D., Chan, T. O., Tsichlis, P. N., and Brugge, J. S. (1997) Phosphatidylinositol 3-kinase is required for integrin-stimulated AKT and Raf-1/mitogen-activated protein kinase pathway activation. *Mol. Cell Biol.* **17,** 4406–4418.
7. Chun, J-S., Ha, M-J., and Jacobson, B. S. (1996) Differential translocation of protein kinase Cε during HeLa cell adhesion to a gelatin substratum. *J. Biol. Chem.* **271,** 13,008–13,012.
8. Vuori, K. and Ruoslahti, E. (1993) Activation of protein kinase C precedes α5β1 integrin-mediated cell spreading on fibronectin. *J. Biol. Chem.* **268,** 21,459–21,462.
9. Del Pozo, M. A., Price, L. S., Alderson, N. B., Ren, X-D., and Schwartz, M. A. (2000) Adhesion to the extracellular matrix regulates the coupling of the small GTPase Rac to its effector PAK. *EMBO J.* **19,** 2008–2014.
10. Berrier, A. L., Martinez, R., Bokoch, G. M., and LaFlamme, S. E. (2002) The integrin β tail is required and sufficient to regulate adhesion signaling to Rac1. *J. Cell Sci.* **115,** 4285–4291.
11. Guan, J.-L. (1997) Role of focal adhesion kinase in integrin signaling. *Int. J. Biochem. Cell Biol.* **29,** 1085–1096.

12. Miranti, C. K. and Brugge, J. S. (2002) Sensing the environment: A historical perspective on integrin signal transduction. *Nat. Cell Biol.* **4,** 83–90.
13. Arthur, W. T. and Burridge, K. (2001) RhoA inactivation by p190RhoGAP regulates cell spreading and migration by promoting membrane protrusion and polarity. *Mol. Biol. Cell* **12,** 2711–2720.
14. Ylanne, J., Chen, Y., O'Toole, T. E., Loftus, J. C., Takada, Y., and Ginsberg, M. H. (1993) Distinct functions of integrin α and β subunit cytoplasmic domains in cell spreading and formation of focal adhesions. *J. Cell Biol.* **122,** 223–233.
15. Brakebusch, C. and Fassler, R. (2003) The integrin-actin connection, an eternal love affair. *EMBO J.* **22,** 2324–2333.
16. Liu, S., Calderwood, D. A. and Ginsberg, M. H. (2000) Integrin cytoplasmic domain-binding proteins. *J. Cell Sci.* **113,** 3563–3571.
17. Chen, Y. P., O'Toole, T. E., Shipley, T., Forsyth, J., LaFlamme, S. E., Yamada, K. M., et al. (1994) "Inside-out" signal transduction inhibited by isolated integrin cytoplasmic domains. *J. Biol. Chem.* **269,** 18,307–18,310.
18. LaFlamme, S. E., Thomas, L. A., Yamada, S. S., and Yamada, K. M. (1994) Single subunit chimeric integrins as mimics and inhibitors of endogenous integrin functions in receptor localization, cell spreading and migration, and matrix assembly. *J. Cell Biol.* **126,** 1287–1298.
19. Lukashev, M. E., Sheppard, D., and Pytela, R. (1994) Disruption of integrin function and induction of tyrosine phosphorylation by the autonomously expressed β1 integrin cytoplasmic domain. *J. Biol. Chem.* **269,** 18,311–18,314.
20. Smilenov, L., Briesewitz, R., and Marcantonio, E. E. (1994) Integrin β1 cytoplasmic domain dominant negative effects revealed by lysophosphatidic acid treatment. *Mol. Biol. Cell* **5,** 1215–1223.
21. Bodeau, A. L., Berrier, A. L., Mastrangelo, A. M., Martinez, R., and LaFlamme, S. E. (2001) A functional comparison of mutations in integrin β cytoplasmic domains: Effects on the regulation of tyrosine phosphorylation, cell spreading, cell attachment and β1 integrin conformation. *J. Cell Sci.* **114,** 2795–2807.
22. Berrier, A. L., Mastrangelo, A. M., Downward, J., Ginsberg, M., and LaFlamme, S. E. (2000) Activated R-Ras, Rac1, PI 3-kinase and PKCϵ can each restore cell spreading inhibited by isolated integrin β1 cytoplasmic domains. *J. Cell Biol.* **151,** 1549–1560.
23. Dedhar, S. and Hannigan, G. E. (1996) Integrin cytoplasmic interactions and bidirectional transmembrane signalling. *Curr. Opin. Cell Biol.* **8,** 657–669.
24. Hemler, M. E. (1998) Integrin associated proteins. *Curr. Opin. Cell Biol.* **10,** 578–585.
25. Calderwood, D. A., Fujioka, Y., de Pereda, J. M., Garcia-Alvarez, B., Nakamoto, T., Margolis, B., et al. (2003) Integrin β cytoplasmic domain interactions with phosphotyrosine-binding domains: a structural prototype for diversity in integrin signaling. *Proc. Natl. Acad. Sci. USA* **100,** 2272–2277.
26. Mastrangelo, A. M., Bodeau, A. L., Homan, S. M., Berrier, A. L., and LaFlamme, S. E. (1999) The use of chimeric receptors in the study of integrin signaling, in *Signaling Through Cell Adhesion Molecules* (Guan, J.-L., ed), CRC Press, New York, NY, pp. 3–17.

7

Cell-Scatter Assay

Hong-Chen Chen

Summary

Cell scattering is used to describe the dispersion of compact colonies of epithelial cells induced by certain soluble factors such as growth factors, cytokines, and phorbol esters. The dispersal of epithelial colonies is a dynamic process usually initiated by membrane ruffling and centrifugal spreading of cell colonies. Subsequently, some cells within the colony begin to detach from their neighboring cells and exhibit a shape resembling that of motile fibroblasts. These cells continue to migrate, finally leading to a "scatter" phenomenon. Because the scattering of epithelial colonies possesses characteristics of epithelial–mesenchymal transition, such as the loss of epithelial cell–cell junctions and the acquisition of a motile mesenchymal cell phenotype, the scatter assay has been used for studying epithelial–mesenchymal transition and for detecting factors able to induce migratory behavior of cells. The method described in this chapter is intended specifically for measuring the scatter response of Madin–Darby canine kidney cells to hepatocyte growth factor stimulation.

Key Words: Scatter assay; cell scattering; epithelial–mesenchymal transition; cell dissociation; hepatocyte growth factor; Madin–Darby canine kidney cells.

1. Introduction

The cell-scatter assay was originally described for the detection of a migration-inducing (scattering) factor present in the culture medium of MRC-5 human embryo fibroblasts, able to induce scattering activity of Madin–Darby canine kidney (MDCK) cells (1–3). The scatter factor was later identified as being identical to hepatocyte growth factor (HGF; refs. 4,5). HGF/SF is a heparin binding protein that is processed from its single-chain secreted precursor into an active 69/34-kDa heterodimer (6), which stimulates mitogenic, motogenic, and morphogenic activities in various cell types (7). The diverse bio-

From: *Methods in Molecular Biology, vol. 294: Cell Migration: Developmental Methods and Protocols*
Edited by: J-L. Guan © Humana Press Inc., Totowa, NJ

logical effects of HGF/SF are transmitted through activation of its transmembrane receptor encoded by the c-*met* proto-oncogene *(8,9)*. Although a number of growth factors are known to modulate cell motility, HGF is unique because of the intensity with which it stimulates motility and induces epithelial–mesenchymal (E-M) transition in vitro *(10,11)* and in vivo *(12)*. In addition to MDCK cells, a number of other cells, including epidermal keratinocytes *(13)*, BSLC monkey kidney cells *(14)*, endothelial cells *(15,16)*, rat liver epithelial cells *(17)*, and bladder carcinoma cells *(17)* can also be induced to migrate in response to HGF stimulation.

The scatter response of MDCK cells to HGF stimulation has been used extensively as a model to study E-M transition, characterized by the loss of epithelial polarity, the disruption of E-cadherin-mediated cell–cell adhesions, and the acquisition of a migratory mesenchymal cell phenotype *(18–20)*. Upon HGF stimulation, the scatter of MDCK cells can be visualized first as membrane ruffling and centrifugal spreading of cell colonies (after 1–3 h) followed by cell–cell dissociation (after 3–6 h) and subsequent cell migration (from 6 h; **refs.** *18,21,22*; *see* **Fig. 1**). E-M transitions and epithelial dispersal are tightly regulated and require the coordinated activation and targeting of structural and signaling complexes that modulate the remodeling of cytoskeleton essential for cell–cell dissociation and cell migration. Several intracellular signaling pathways have been implicated to act downstream of the HGF receptor to mediate scatter response. For example, phosphatidylinositol 3-kinase and small GTPase *Ras* have been shown to be essential for cell dissociation and migration after stimulation of MDCK cells with HGF *(18,19,23)*. In addition, the activation of the Rho family GTPase Rac is required for HGF-induced spreading and motility of MDCK cells, whereas an increase in active Rho inhibits it *(18)*.

Because the compactness of the MDCK cell colonies is essential for a sensitive assay, it is strongly recommended to use a recloned subline of MDCK cells for the assay because the cells available from the American Type Culture Collection (Rockville, MD) show a high background of single migratory cells in the absence of HGF *(18,21,22)*. The compact colonies of MDCK cells are dispersed by HGF and the extent of cell scattering is easily evaluated microscopically after fixation and staining. Because HGF-induced scatter of MDCK cells can be roughly divided into three phases, that is, cell spreading, cell dissociation, and cell migration *(18,21,22)*, methods to analyze each of these phases are described in this chapter.

2. Materials

2.1. Cell-Scatter Assay

1. MDCK cells (strain II, clone 3B5; **refs.** *24,25*; *see* **Note 1**).

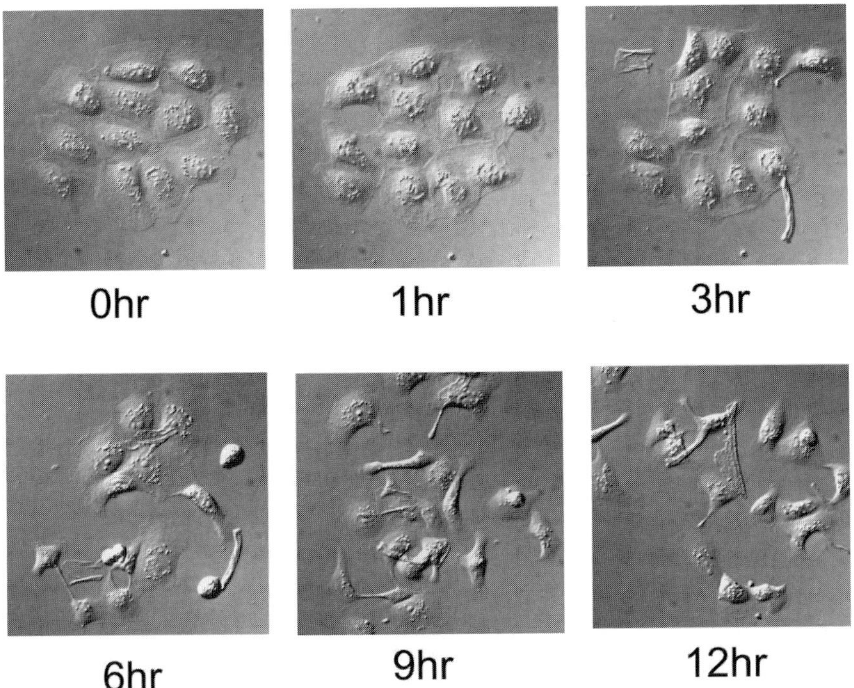

Fig. 1. The scatter response of MDCK cells to HGF. MDCK cells were sparsely seeded on collagen-coated glass and allowed to grow as colonies. The time-lapse micrographs were taken by a cooled CCD under a differential interference contrast microscope every hour for 12 h to record the process of HGF-induced scattering. Upon HGF (20 ng/mL) stimulation, the scatter of MDCK cells can be visualized first as membrane ruffling and centrifugal spreading of cell colonies (after 1–3 h) followed by cell–cell dissociation (after 3–6 h) and subsequent cell migration (from 6 h).

2. Hepatocyte growth factor (cat. no. H1404; Sigma-Aldrich): reconstitute lyophilized HGF in phosphate-buffered saline (PBS) to a concentration of 10 ng/µL and store it at −80°C.
3. Dulbecco's modified Eagle's medium (DMEM).
4. Fetal bovine serum (FBS).
5. Methanol.
6. Giemsa stain, modified solution (cat. no. GS500; Sigma-Aldrich): dilute 1:10 in distilled H_2O before use.

2.2. Cell-Spreading Assay

1. Versene: 0.2 g ethylene diamine tetraacetic acid (0.53 mM) and 0.01 g phenol red in 1 L of PBS, pH 7.4.

2. 2.5% (w/v) Trypsin (cat. no. 15090-046; Gibco Invitrogen): dilute 1:25 in Versene before use.
3. Collagen from calf skin (cat. no. C9791; Sigma-Aldrich): dissolve collagen at 1 mg/mL in 0.1 *N* acetic acid. Allow to stir at room temperature until dissolved (takes 1–3 h). Store collagen stock at 4°C.
4. Bovine serum albumin (BSA): 2 mg/mL in PBS, pH 7.4. Store it at 4°C.

2.3. Cell-Dissociation Assay

1. Cell scraper.
2. Hematocytometer.

2.4. Cell Migration Assay

See Chapter 2; Boyden chamber assay.

3. Methods

3.1. Cell-Scatter Assay

1. Seed 2×10^2 MDCK 3B5 cells on a 60-mm dish in 3 mL of DMEM supplemented with 10% FBS and allow them to grow as discrete colonies at 37°C in a humidified atmosphere of 5% CO_2 and 95% air atmosphere (*see* **Note 2**).
2. 60 to 72 h later, when the majority of colonies contain 30–40 cells, replace the medium with 1.5 mL of fresh DMEM containing 5% FBS and 20 ng/mL HGF (*see* **Notes 3** and **4**).
3. 12 h after HGF stimulation, remove the medium and wash the cells twice with PBS.
4. Remove PBS and immediately add 4 mL of methanol at room temperature for 10 min to fix the cells. Remove the methanol and allow the dish to air dry.
5. Add 4 mL of diluted (1:10 in distilled H_2O) Giemsa stain to the dish at room temperature for 1 h.
6. Remove the stain solution, wash the dish several times with distilled H_2O, and allow it to air dry.
7. Draw 25 grids on the bottom of the dish using a marker pen.
8. Take one micrograph per grid using a digital camera (Nikon COOLPIX995) connected to an inverted phase-contrast microscope at 50X magnification. Take at least twenty micrographs for a dish. Each micrograph shows two to three colonies.
9. Remove the camera from the microscope and connect it to a computer. Acquire the images from the memory card of the digital camera and view the images with the aid of the software Adobe Photoshop® (Adobe Systems; San Jose, CA).
10. Measure the percentage of scattered colonies in the total 50 colonies (*see* **Note 5**). When half of the cells in a given colony have lost contact with their neighbors and exhibits a fibroblast-like phenotype, it is judged as a "scattered" colony (*see* **Fig. 2**).

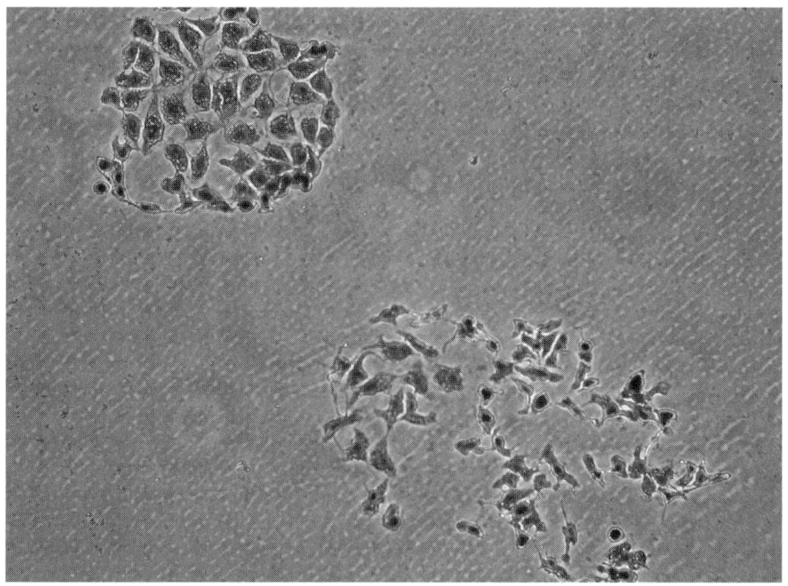

Fig. 2. Cell-scatter assay. Twelve hours after HGF stimulation, MDCK cell colonies were stained and the percentage of scattered colonies in total counted colonies was determined. A colony is judged as a "scattered" one when half of the cells in it have lost contact with their neighbors and exhibits a fibroblast-like phenotype. Two representative colonies are shown. The lower colony is judged as a "scattered" one.

3.2. Cell-Spreading Assay

3.2.1. Coating Dishes With Collagen

1. Add 2 mL of diluted collagen solution (10 µg/mL) to a 60-mm culture dish and allow it to stand at 4°C overnight.
2. Remove the collagen solution, add 2 mL of BSA (2 mg/mL) and incubate at 37°C for 1 h to block the dish.
3. Remove BSA solution and wash the dish twice with DMEM before use.

3.2.2. Preparing Cells

1. Allow MDCK cells to grow as discrete colonies and treat with HGF (*see* **Subheading 3.1., steps 1** and **2**).
2. Two hours after HGF stimulation, remove the medium and wash the cells twice with Versene. Add 1 mL of Versene containing 0.05% trypsin and allow the culture to stand at 37°C for 10 to 15 min. Add 3 mL of DMEM with 10% FBS, pipet the cells off the dish, and transfer them to a 15-mL centrifuge tube.

3. Pellet the cells by centrifugation at 150–200g for 5 min. Remove the medium, add 5 mL of DMEM, and centrifuge again.
4. Remove the medium and resuspend the cells in 1.5 mL of DMEM. Count and adjust the cells to 5×10^4 cells per milliliter in DMEM.

3.2.3. Replating Cells and Counting Spread Cells

1. Transfer 2 mL of cell suspension to a collagen-coated dish and allow cells to spread at 37°C for 20 to 30 min.
2. Take micrographs using a digital camera connected to an inverted phase-contrast microscope at 50X magnification.
3. Remove the camera from the microscope and connect to a computer. Acquire the images from the memory card of the digital camera and view the images on the monitor.
4. Measure the percentage of spread cells in total counted (200–300) cells. Cells with extend processes and not phase bright are defined as spread cells.

3.3. Cell-Dissociation Assay

1. Allow MDCK cells to grow as discrete colonies and treat with HGF (*see* **Subheading 3.1., steps 1** and **2**).
2. Four hours after HGF stimulation, remove the medium and wash the cells twice with PBS at room temperature.
3. Add 1 mL of PBS, gently scrape the cells off the dish, and transfer them to a 15-mL centrifuge tube.
4. Pipet the cells 30 to 50 times with a p1000 pipetor or under a constant force.
5. Measure the number of cell clusters (particles; Np) containing more than three cells under a microscope with a hematocytometer.
6. Pellet the cell particles by centrifugation at 150–200g for 5 min. Remove PBS and resuspend the cells in 1 mL of Versene containing 0.05% trypsin to disrupt cell particles.
7. Add 3 mL of DMEM with 10% FBS and pellet the cells by centrifugation at 150–200g for 5 min.
8. Resuspend the cells in 1 mL of DMEM and measure the cell number (Nc) under a microscope with a hematocytometer. The cell dissociation index is expressed as $Np/Nc \times 100\%$.

3.4. Cell-Migration Assay

1. Allow MDCK cells to grow as discrete colonies and treat with HGF (*see* **Subheading 3.1., steps 1** and **2**).
2. Twelve hours after HGF stimulation, collect cells by trypsinization and subject them to the Boyden chamber assay using collagen as an attractant (*see* Chapter 2).

4. Notes

1. The use of the strain II clone 3B5 of MDCK cells for the scatter assay is strongly recommended. Alternatively, one could establish his/her own subclones from the

MDCK cells available from the American Type Culture Collection. Single MDCK cell colonies can be selected by a cloning cylinder or limiting dilution. The ability of selected cell clones to form compact cell colonies and their response to HGF stimulation should be verified.

2. For cell scatter assay, MDCK cells should be sparsely seeded on a culture dish, which allows them to grow as discrete colonies and have enough space for subsequent scattering.

3. The potency of HGF to induce cell scattering is likely to be different among suppliers. Therefore, it is important to optimize the concentration of HGF for the assay. The concentration of HGF between 10 and 50 ng/mL is in an acceptable range for the assay. Note that too much HGF does not facilitate cell scattering, instead it causes cells to round up.

4. Scatter assay using MDCK cells should be carried out in DMEM supplemented with HGF and 5% FBS. In serum-free medium, HGF fails to induce MDCK cell scattering.

5. Because the larger colonies take more time to reach the criteria for being judged as a "scattered" colony after HGF stimulation, it is important to choose colonies with a similar number of cells for judgment. The number of cells in a colony between 30 and 40 is in an acceptable range.

Acknowledgments

The author would like to thank Chien-Lin Chen, Chi-Hui Huang, and Chun-Chi Liang for technical assistance. Work in the author's laboratory is supported by the National Science Council, Taiwan.

References

1. Stoker, M. and Perryman, M. (1985) An epithelial scatter factor released by embryo fibroblasts. *J. Cell Sci.* **77,** 209–223.

2. Stoker, M., Gherardi, E., Perryman, M., and Gray, J. (1987) Scatter factor is a fibroblast-derived modulator of epithelial cell mobility. *Nature* **327,** 239–242.

3. Gherardi, E., Gray, J., Stoker, M., Perryman, M., and Furlong, R. (1989) Purification of scatter factor, a fibroblast-derived basic protein that modulates epithelial interactions and movement. *Proc. Natl. Acad. Sci. USA* **86,** 5844–5888.

4. Naldini, L., Weidner, K. M., Vigna, E., Gaudino, G., Bardelli, A., Ponzetto, C., et al (1991) Scatter factor and hepatocyte growth factor are indistinguishable ligands for the MET receptor. *EMBO J.* **10,** 2867–2878.

5. Weidner, K. M., Arakaki, N., Hartmann, G., Vandekerckhove, J., Weingart, S., Rieder, H., et al. (1991) Evidence for the identity of human scatter factor and human hepatocyte growth factor. *Proc. Natl. Acad. Sci. USA* **88,** 7001–7005.

6. Furlong, R. A. (1992) The biology of hepatocyte growth factor/scatter factor. *BioEssays* **14,** 613–617.

7. Zarnegar, R. and Michalopoulos, G. K. (1995) The many faces of hepatocyte growth factor: from hepatopoiesis to hematopoiesis. *J. Cell Biol.* **129,** 1177–1180.

8. Bottaro, D. P., Rubin, J. S., Faletto, D. L., Chan, A. M-L., Kmiecik, T. E., Vande Woulde, G. F., et al. (1991) Identification of the hepatocyte growth factor receptor as the c-met proto-oncogene product. *Science* **251,** 802–804.

9. Naldini, L., Vigna, E., Ferracini, R., Longati, P., Gandino, L., Prat, M., et al. (1991) The tyrosine kinase encoded by the MET proto-oncogene is activated by autophosphorylation. *Mol. Cell. Biol.* **11,** 1793–1803.

10. Weidber, K. M., Sachs, M., and Birchmeier, W. (1993) The Met receptor tyrosine kinase transduces motility, proliferation, and morphogenic signals of scatter factor/hepatocyte growth factor in epithelial cells. *J. Cell Biol.* **121,** 145–154.

11. Zhu, H., Naujokas, M. A., and Park, M. (1994) Receptor chimeras indicate that the Met tyrosine kinase mediates the motility and morphogenic responses of hepatocyte growth factor/scatter factor. *Cell Growth Differ.* **5,** 359–366.

12. Birchmeier, C. and Gherardi, E. (1998) Developmental roles of HGF/SF and its receptor, the c-Met tyrosine kinase. *Trends Cell Biol.* **8,** 404–410.

13. Matsumoto, K., Hashimoto, K., Yoshikawa, K., and Nakamura, T. (1991) Marked stimulation of growth and motility of human keratinocytes by hepatocyte growth factor. *Exp. Cell Res.* **196,** 114–120.

14. Stoker, M. (1989) Effect of scatter factor on motility of epithelial cells and fibroblasts. *J. Cell. Physiol.* **139,** 565–569.

15. Rosen, E. M., Jaken, S., Carley, W., Luckett, P. M., Setter, E., Bhargava, M., et al. (1991) Regulation of motility in bovine brain endothelial cells. *J. Cell Physiol.* **146,** 325–335.

16. Bussolino, F., Di Renzo, M. F., Ziche, M., Bocchietto, E., Olivero, M., Naldini, L., et al. (1992) Hepatocyte growth factor is a potent angiogenic factor which stimulates endothelial cell motility and growth. *J. Cell Biol.* **119,** 629–641.

17. Feindler, F., Schramke, H., Seebacher, T., and Bade, E. G. (1993) Migration-correlated expression of EIP-1/PAI-1 and p60/stromelysin by NBT II rat bladder carcinoma cells. *Eur. J. Cell Biol.* **60,** 49.

18. Ridley, A. J., Comoglio, P. M., and Hall, A. (1995) Regulation of scatter factor/hepatocyte growth factor responses by Ras, Rac, and Rho in MDCK cells. *Mol. Cell. Biol.* **15,** 1110–1122.

19. Royal, I. and Park, M. (1995) Hepatocyte growth factor-induced scatter of Madin-Darby canine kidney cells requires phosphatidylinositol 3-kinase. *J. Biol. Chem.* **270,** 27,780–27,787.

20. Sander E. E., van Delft, S., ten Klooster, J. P., Reid, T., van der Kammen, R. A., Michiels, F., et al. (1998) Matrix-dependent Tiam1/Rac signaling in epithelial cells promotes either cell-cell adhesion or cell migration and is regulated by phosphatidylinositol 3-kinase. *J. Cell Biol.* **143,** 1385–1398.

21. Lai, J-F., Kao, S-C., Jiang, S-T., Tang, M-J., Chan, P-C., and Chen, H-C. (2000) Involvement of focal adhesion kinase in hepatocyte growth factor-induced scatter of Madin-Darby canine kidney cells. *J. Biol. Chem.* **275,** 7474–7480.

22. Liang, C-C. and Chen, H-C. (2001) Sustained activation of extracellular signal-regulated kinase stimulated by hepatocyte growth factor leads to integrin (α2 expression that is involved in cell scattering. *J. Biol. Chem.* **276,** 21,146–21,152.

23. Hartmann, G., Weidner, K. M., Schwarz, H., and Birchmeier, W. (1994) The motility signal of scatter factor/hepatocyte growth factor mediated through the receptor tyrosine kinase met requires intracellular action of Ras. *J. Biol. Chem.* **269,** 21,936–21,939.
24. Montesano, R., Schaller, G., and Orci, L. (1991) Induction of epithelial tubular morphogenesis in vitro by fibroblast-derived soluble factors. *Cell* **66,** 697–711.
25. Montesano, R., Matsumoto, K., Nakamura, T., and Orci, L. (1991) Identification of a fibroblast-derived epithelial morphogen as hepatocyte growth factor. *Cell* **67,** 901–908.

8

Cell Migration Analyses
Within Fibroblast-Derived 3-D Matrices

Edna Cukierman

Summary

Research in cell biology often is based in tissue culturing cells on artificial substrates, such as plastic or glass. These artificial conditions are prone to distorting findings by persuading cells to adjust to artificial flat rigid surfaces. In contrast, the natural substrate for most cells in living organisms is the extracellular matrix (ECM), which is three-dimensional, complex, and dynamic in its molecular composition, and variable in pliability. Here we show how to measure the rates and directionality of fibroblasts motility within physiologically relevant in vivo-like cell-derived three-dimensional matrices.

Key Words: 2-D mix; 3-D fibronectin; 3-D matrix; cell directionality; cell motility; cell-derived 3-D matrix; collagen gels; digital imaging analyses; fibroblast; microenvironment; time-lapse microscopy.

1. Introduction

Fibroblasts are naturally nonpolar cells. Culturing them on classic two-dimensional (2-D) substrates induces artificial polarity, which results in variation between their lower and upper surfaces. Not surprisingly, fibroblast morphology and migration differ when suspended in three-dimensional (3-D) substrates like collagen gels *(1–4)* or within in vivo-like tissue- and cell-derived 3-D matrices *(5)*. The cell-derived 3-D matrices are part of our approach herein proposed. The logic behind using the proposed approach is that fibroblastic cells do not normally encounter multimillimeter-thick basement membranes or the pure microenvironment that collagen gels and artificially deposited 3-D fibronectin provide. Instead, they exist in a complex environment comprised of various molecules exhibiting a substantial degree of fibrillar mesh-like organization and complexity known as the extracellular matrix (ECM; **refs. 6,7**). Fi-

From: *Methods in Molecular Biology, vol. 294: Cell Migration: Developmental Methods and Protocols*
Edited by: J-L. Guan © Humana Press Inc., Totowa, NJ

broblasts normally secrete and organize the ECM, which provides structural support for their adhesion, migration, and tissue organization as well as regulating cellular functions such as growth and survival *(8–12)*. Cell–matrix interactions are vital for vertebrate development *(9,10,13–19)*. Disorders in these interactions have been associated with fibrosis, developmental malformations, cancer, and other diseases *(20,21)*.

This chapter describes how to measure fibroblast motility rates and directionality movements while within in vivo-like cell-derived 3-D matrices. The basic method is to acquire images of the active fibroblasts at constant intervals throughout a given period of time (**Subheading 3.1.**) and then to analyze their movement rates (**Subheading 3.2.1.**) and directional tendencies (**Subheading 3.2.2.**). We propose some low-cost options for how to perform this high-tech assay and indicate how to analyze statistically the obtained data (**Subheading 3.3.**). To successfully perform these measurements, a brief explanation of how to obtain fibroblast-derived 3-D matrix coated culture plates is described (**Subheading 3.4.**). Finally, we show how to assess the quality of the 3-D matrices (**Subheading 3.5.**) and propose the use of some control substrates for comparison purposes (**Subheading 3.6.**).

2. Materials

All solutions and equipment being exposed to living cells must be sterile, and aseptic techniques should be used accordingly. All culture incubations should be performed in a humidified incubator at 37°C and 10% CO_2 unless specified. All chemicals utilized are from Sigma unless stated.

2.1. Time-Lapse Motility Assay

1. Heat-denatured bovine serum albumin (BSA): 2% (w/v) BSA in H_2O filtered through a 0.2-μm filter; can be stored at 4°C for 3 mo. Before use, aliquot the needed amount into a 50-mL polypropylene tube and heat denature the BSA for 7 min in boiling water. Do not use if too turbid or milky; it should appear translucent for successful blocking purposes. Cool to about 37°C and use immediately.
2. PBS: 8 g of NaCl, 0.2 g of KCl, 0.24 g of KH_2PO_4, and 1.44 g of Na_2HPO_4 dissolved in H_2O to a final volume of 1 L. Store at room temperature and verify the lack of phosphate precipitates prior to use.
3. 15-cm Culture-plate containing 80% confluent fibroblast cells to be assessed.
4. Culture media: high glucose Dulbecco's modified Eagle's medium supplemented with 10% fetal bovine serum (FBS) (*see* **Note 1**) or 10% calf serum (depending on the cell line used), 100 U/mL penicillin, and 100 μg/mL streptomycin. Store media at 4°C for up to 1 mo.
5. Six 35-mm plates precoated with fibroblast-derived 3-D matrices (*see* **Subheading 3.4.**) or six 35-mm plates precoated with assorted control substrate (*see* **Subheading 3.6., Note 2**).

6. Hemacytometer or cell counter.
7. Stage micrometer.
8. Trypsin–ethylene diamine tetraacetic acid (EDTA) solution (available from Hyclone): 2.5 g of trypsin, 0.2 g of EDTA, 8 g of NaCl, 0.4 g of KCl, 1 g of glucose, 0.35 g of $NaHCO_3$, and 0.01 g of phenol red dissolved in H_2O to a final volume of 1 L. Sterilize solution by filtration through a 0.2-μm filter and store up to 3 mo at $-20°C$.

2.2. Digital Analyses

Steps 1–3 are for motility rate measurements and **step 4** is for directionality analysis.

1. Calibration image (*see* **step 15** in **Subheading 3.1.**).
2. Images to be assessed.
3. Analysis software or grided see-through paper and string.
4. Analysis software, grided and see-through paper.

2.3. Statistical Analyses

1. Microsoft Excel software, GraPhpad InStat software, or any other program for standard statistical analysis and graphs.
2. Data obtained in **Subheading 3.2.**

2.4. Coating Tissue Culture Plates With Fibroblast-Derived 3-D Matrices

1. Culture media: High glucose Dulbecco's modified Eagle's medium supplemented with 10% FBS (*see* **Note 1**), 100 U/mL penicillin, and 100 μg/mL streptomycin. Store media at 4°C for up to 1 mo.
2. Three 15-cm plates with NIH 3T3 cells at 80% confluent, which have been passaged at least 22 times in culture media (*see* **Note 3**).
3. Trypsin–EDTA solution (available from Hyclone): 2.5 g of trypsin, 0.2 g of EDTA, 8 g of NaCl, 0.4 g of KCl, 1 g of glucose, 0.35 g of $NaHCO_3$, and 0.01 g of phenol red dissolved in H_2O to a final volume of 1 L. Sterilize solution by filtration through a 0.2-μm filter and store up to 3 mo at $-20°C$.
4. Hemacytometer or cell counter.
5. 35-mm Tissue culture plates.
6. PBS: 8 g of NaCl, 0.2 g of KCl, 0.24 g of KH_2PO_4, and 1.44 g of Na_2HPO_4 dissolved in H_2O to a final volume of 1 L. Store at room temperature and verify the lack of phosphate precipitates prior to use.
7. 0.2% Gelatin precoating solution: prepare a 0.2% (w/v) gelatin solution in PBS. Autoclave the solution, let cool to room temperature, and then filter through a 0.2-μm sterilization device. The solution can be stored at 4°C for a period of 6 mo.
8. Glutaraldehyde 1% (Sigma).
9. Ethanolamine 1 *M* (Sigma).

10. L-ascorbic acid sodium salt; make a stock solution of 50 mg/mL and sterilize by filtration with a 0.2-µm filter. This stock solution should be freshly made immediately prior to use and cannot be stored.
11. Extraction buffer: 25 m*M* Tris-HCl, pH 7.4, 150 m*M* NaCl, 0.5% (v/v) Triton X-100, and 20 m*M* NH$_4$OH. Can be stored up to 1 mo at 4°C.
12. DOC Buffer: 20 m*M* Tris-HCl, pH 8.5, 1% (w/v) sodium-deoxycholate; 2 m*M* NEM (*N*-ethylmaleimide) 2 m*M* iodoacetamide, and 2 m*M* phenylmethyl-sulfonylflouride (predissolved 100 m*M* stock in 100% ethanol). DOC buffer should be prepared fresh immediately before use; reagents are not stable in aqueous solution.
13. Tris buffer: 25 m*M* Tris-HCl, pH 7.4, 150 m*M* NaCl.
14. PBS pen/strep: PBS containing 100 U/mL penicillin, and 100 µg/mL streptomycin. Can be stored at 4°C up to 6 mo.
15. Parafilm.

2.5. Quality Control Assessments

1. 35-mm plate precoated with cell-derived 3-D matrix (**Subheading 3.4.**).
2. Culture media: high glucose Dulbecco's modified Eagle's medium supplemented with 10% FBS (*see* **Note 1**), 100 U/mL penicillin, and 100 µg/mL streptomycin. Store media at 4°C for up to 1 mo.
3. One 80% confluent 15-cm plate of fibroblasts. Human foreskin fibroblasts between passages 3 and 18 or NIH 3T3 cells from ATCC can be used on this assessment (*see* **Note 3**).
4. Trypsin–EDTA solution (available from Hyclone): 2.5 g of trypsin, 0.2 g of EDTA, 8 g of NaCl, 0.4 g of KCl, 1 g of glucose, 0.35 g of NaHCO$_3$, and 0.01 g of phenol red dissolved in H$_2$O to a final volume of 1 L. Sterilize solution by filtration through a 0.2-µm filter and store up to 3 mo at –20°C.
5. Heat-denatured BSA: 2% (w/v) BSA in water filtered through a 0.2-µm filter; can be stored at 4°C for 3 mo. Before use, aliquot the needed amount into a 50-mL polypropylene tube and heat denature the BSA for 7 min in boiling water. Do not use if too turbid or milky; it should appear translucent for successful blocking purposes. Cool to about 37°C and use immediately.
6. PBS: 8 g of NaCl, 0.2 g of KCl, 0.24 g of KH$_2$PO$_4$, and 1.44 g of Na$_2$HPO$_4$ dissolved in H$_2$O to a final volume of 1 L. Store at room temperature and verify the lack of phosphate precipitates prior to use.

2.6. Pre-Coating Plates With Proposed Control Substrates

All subsections for the proposed use of different coated substrates use PBS (*see* **Subheading 2.5.**). Six 35-mm tissue culture plates are used for 2-D substrates (**steps 1–6**) and six time-lapse glass bottom 35-mm plates from MatTeck Corporation are used for 3-D substrates (**steps 7–11**).

1. For 2-D fibronectin: Fibronectin in PBS at 10 µg/mL; on the morning of the experiment, dilute fibronectin with PBS to a final concentration of 10 µg/mL

and keep at 4°C until required (*see* **Note 4**). Alternatively, 0.2% gelatin precoating solution: Prepare a 0.2% (w/v) gelatin solution in PBS. Autoclave the solution, let cool to room temperature, and then filter through a 0.2-µm sterilization device. Store solution at 4°C for a period of 6 mo (*see* **Note 5**).

2. For 2-D Collagen I: Collagen I, Vitrogen from Collagen Corporation stored at 4°C. On the morning of the experiment, dilute with PBS to a final concentration of 10 µg/mL and keep at 4°C until required.

Steps 3–6 are for making the 2-D mix:

3. Solubilization buffer: 5 M guanidine containing, 10 mM dithiothreitol. Can be stored indefinitely at 4°C.
4. Six 35-mm plates precovered with fibroblast-derived 3-D matrices (*see* **Subheading 3.4.**).
5. Plastic cell scraper (rubber policeman).
6. Twelve 1.5-mL microcentrifuge tubes.
7. Fibronectin stock; purified fibronectin in PBS at a concentration equal to or higher than 2 mg/mL.
8. 0.1 N NaOH.
9. 0.1 N HCl.
10. 10X PBS: 8 g of NaCl, 2 g of KCl, 0.24 g of KH_2PO_4, and 1.44 g of Na_2HPO_4 dissolved in H_2O to a final volume of 100 mL. Store at 4°C and verify the lack of phosphate precipitates prior to use.
11. pH paper.

3. Methods

3.1. Time-Lapse Motility Assay

1. Block six 35-mm cell-derived 3-D matrix coated plates (**Subheading 3.4.**; **ref.** *23*) or control substrates (**Subheading 3.6.**) by incubating with 2 mL of heat-denatured BSA for 1 h at 37°C (*see* **Notes 2** and **6**).
2. Carefully rinse matrices twice with 2 mL of PBS (*see* **Note 7**).
3. Remove media from the 15-cm plate containing the fibroblast and rinse the cell layer briefly with 37°C preheated trypsin–EDTA to remove trypsin inhibitors.
4. Quickly substitute with enough fresh trypsin–EDTA to cover the cell layer and immediately aspirate again.
5. Observe under a low-magnification tissue culture microscope at room temperature until the cells have detached from the culture dish (1–3 min).
6. Collect the cells into 10 mL of media by repetitive and gentle pipetting media over cells avoiding the presence of cell aggregates.
7. Count cells using hemacytometer, calculate cell number per milliliter, and dilute with media to three different cell-mixture stocks of 2.5×10^4, 1×10^4, and 4×10^3 cells/mL.
8. Add 2 mL of the above cell-mixtures per plate; do so in duplicates (*see* **Note 8**).
9. Place plates into the incubator for overnight culture.
10. The following morning a quick evaluation of the plates should be done in order to select a pair of plates to be recorded (*see* **Note 9** and **Subheading 3.5.**).

11. Place one of the selected plates onto the environmentally controlled microscope stage (*see* **Note 10**) allowing 30 min for cells to equilibrate to their new surroundings.
12. During the 30-min equilibration period, find an optimal region within the sample using a 10X objective by phase-contrast transmitted light under the microscope, predicting the observation of about nine individual cells (*see* **Note 9**).
13. At the end of the equilibration period, initiate the recording process of acquiring a phase-contrast image every 10 min for a period of 6 h, rendering 37 images, including the starting image at time zero (*see* **Note 11**).
14. Repeat **steps 11–13,** using the second 35-mm plate of the pair selected that morning.
15. Acquire an image at the same magnification used for the acquisition of the live cells utilizing a stage micrometer; this image will serve as a reference for calibration purposes (*see* **Note 12**).
16. Repeat **steps 1–14** for each experimental condition and substrate to be tested.

3.2. Digital Analyses (See *Note 13*)

1. Stack all resultant images from the 6-h recording (one plate) in chronological order.
2. Observe the cell dynamics by flipping through the images (*see* **Note 14**). When deciding which cells to analyze, you must attempt to avoid cells that collide with other cells or engage in cell division because these events interfere with the cell motility rates and directionality movements (*see* **Note 15**).
3. Follow each cell by selecting the point at the middle of their cell body for each time point and trace the path of these points during the analyzed period (*see* **Note 16** and **Fig. 1**). The resultant trail will be representative of the path or trajectory the cell moved during the period being analyzed (bottom panel of **Fig. 1**).
4. Repeat **step 3** for each cell (*see* **Note 17**).

3.2.1. Motility Rates

1. Use the stage micrometer image to calibrate the files (or prints) by calculating the number of microns per pixel (if measuring utilizing software) or to one grid unit or centimeter (for printed hard copies).
2. Measure each "path distance" in microns (*see* **Note 18**). Convert the measured paths to the correspondent micron-units multiplying the "measured path" (grid units) by the "calibration number" (μm/grid unit).
3. Calculate the motility rate per hour by dividing the "path distance" (in microns) by the elapsed time, for example, 6 h, and record this velocity outcome for each individual cell; it will later on serve as data for statistical analyses (**Subheading 3.3.**).

3.2.2. Directionality

1. Trace a straight-line starting at the point of origin at time zero (**Fig. 1**, point A) and ending at the coordinate where the cell mid-body was localized at the

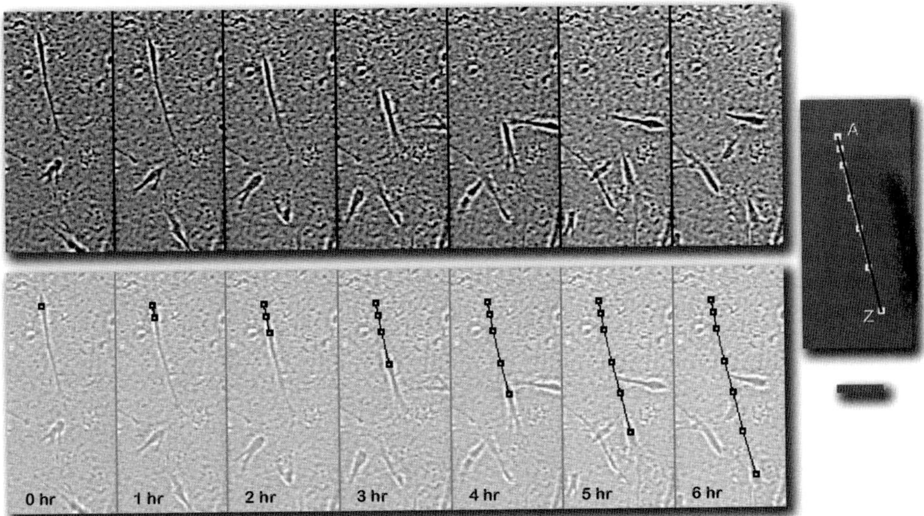

Fig. 1. Tracking the motility of a human fibroblast within NIH 3T3 cell-derived 3-D matrix for a period of 6 h. The top panel represents a chronological montage of seven digital images acquired at 1 h intervals following a specific cell. The bottom panel represents the same images on which the path trajectory of the individual cell has been systematically traced as suggested in **Subheading 3.2.** The right panel represents the final path track and relative distance. Note that the trajectory-path length and the measured relative distance (A to Z) are 349.7 μm and 349.5 μm, respectively, rendering a directionality rate of 1.0006, which is very close to 1, indicating a considerable directional motility, whereas the calculated motility rate for the given example is 58.3 μm/h. Bar represents 100 μm.

    conclusion of the 6 h period (**Fig. 1**, point Z). Measure the length of this line for each cell.

2. Repeat **step 1** for each cell path.
3. Divide the "path distance" (acquired on **step 2** from **Subheading 3.2.1.**) by the straight-line net "relative distance traveled" (from point A to point Z) in grid units or microns (*see* **Note 19**). The closer this number is to 1 the more directional the trajectory of the path of the cell in question (*see* example **Fig. 1**).

3.3. Statistical Analyses

1. Make sure to start with a count greater than 12 (individual path-rates); there should be approx 54 motility rate numbers (**Subheading 3.2.1.**) or directionality ratios (**Subheading 3.2.2.**) from each experimental condition (*see* **Note 20**).
2. Calculate the mean velocity or mean directionality by dividing the sum of the velocity rates or directionality ratios (Σx) by the total sample number (n).

3. The standard deviation can be calculated using Microsoft's Excel software, any statistic analysis software, or by following the formula (*see* **Note 21**):

$$\text{STDEV} = \sqrt{\frac{n\sum x^2 - \left(\sum x\right)^2}{n(n-1)}}$$

4. Repeat **steps 1–3** for each experimental condition.
5. Submit all array data (all the individual rate results) from **Subheading 3.2.** (not outcome mean average) for statistical significance analysis. Use the student double tailed and unpaired Welch corrected (Gaussian distributed) '*t*-test' for motility rate assessment or Kruskal–Wallis (prenormalized because of the ratio calculation lacking Gaussian distribution) for the directionality *t*-test (*see* **Note 22**).
6. The resultant number will be indicative of the *p* value:
   a. a *p* value greater than 0.5 should be stated not significant;
   b. a *p* value between 0.5 and 0.001 should be stated statistically significant;
   c. a *p* value between 0.001 and 0.0001 should be stated statistically very significant; and
   d. a *p* value equal to 0.0001 or smaller should be stated statistically 'extremely significant.'
7. Present results as table or bar graph (*see* **Note 23**).

3.4. Fibroblast-Derived 3-D Matrix Coated 35-mm Plates

It has been shown that fibroblast- and tissue-derived 3-D matrices successfully mimic in vivo mesenchymal matrices *(5,19,24)* while still providing the advantages of cell biology research in the modern laboratory. The protocols for the production of fibroblast-derived 3-D matrices have recently been published *(23)* and therefore the herein protocol is brief, although it contains key details and updated modifications.

1. Add 2 mL of sterile 0.2% gelatin solution to each 35-mm plate and incubate for 1 h at 37°C.
2. Wash plates with 2 mL of PBS.
3. Add 2 mL of glutaraldehyde to each well and incubate plate for 30 min at room temperature.
4. Wash three times, 5 min each, with 2 mL of PBS.
5. Add 2 mL of ethanolamine and incubate plate for 30 min at room temperature.
6. Wash plates three times, for 5 min each with 2 mL of PBS.
7. Wash plates once for 5 min with 2 mL of media.
8. Remove media from NIH 3T3 cells and rinse the cell layer briefly with 37°C preheated trypsin–EDTA to remove trypsin inhibitors (*see* **Note 3**).
9. Quickly substitute with enough fresh trypsin–EDTA to cover the cell layer, and immediately aspirate again.
10. Observe under a low magnification tissue culture microscope at room temperature until the cells have detached from the culture dish (1–3 min).

11. Collect the cells into 10 mL of media per plate by repetitive pipetting avoiding the presence of cell-aggregates.
12. Determine the cell concentration (cell/mL).
13. Dilute cells with media to a final concentration of 2.5×10^5 cells per milliliter.
14. Add 2 mL of diluted cell suspension (5.0×10^5 cells) to each 35-mm plate.
15. Incubate plates for 24 h.
16. Change media on the mornings of d 2 and 4 and in the evening of d 5 to that containing 50 µg/mL ascorbic acid.
17. On the morning of d 8, wash cultures carefully with 2 mL of PBS.
18. Extract cultures with 0.7 mL of extraction buffer preheated to 37°C and observe under a low magnification tissue culture microscope until complete cell lysis is achieved (5–10 min).
19. Without discarding the extraction buffer, dilute extremely carefully avoiding liquid turbulence by adding 2 mL of Tris buffer.
20. Discard approx 2.5 mL from the top of the cell-derived 3-D matrices with great care to avoid damaging the matrices while leaving some liquid on top of the matrices.
21. Carefully and slowly, add 2.5 mL of Tris buffer per plate.
22. Repeat **steps 20** and **21** once.
23. Incubate for 45 min at room temperature with 2 mL of DOC buffer (*see* **Note 24**).
24. Wash three times (5 min each) with 2 mL of Tris buffer.
25. Store matrices with 2 mL of PBS Pen/strep sealed with parafilm at 4°C for up to 3 mo.

3.5. Quality Control Assessment

To ensure that cells intercalating within the fibroblast-derived 3-D matrices will sense only the 3-D microenvironment as opposed to the rigidity of the 2-D plate, two very easy procedures should be followed. The first one deals with the fact that the matrices could be contaminated, altered, turned, or somehow damaged while in storage. Therefore, a thorough visual examination should be performed before plating cells by observing matrices under the tissue culture microscope thus ensuring not only that matrices are intact, but that they are firmly attached to the culture plate. The second control is ensuring that the fibroblast to be assessed acquired spindle shape morphology while in contact with the matrices as opposed to the triangular spread shape observed when fibroblasts are cultured overnight on tissue culture plates (*see* fibroblast morphology in **Fig. 1** on top panel).

3.6. Control Substrates

Performing motility-rate measurement experiments when compared with control substrates ensures that conclusions being made are because of the substrate and not to the changing conditions with the experimental settings (like the addition of pharmacological agents etc.). If 2-D, as well as 3-D, controls are needed the following options should be considered followed by protocols in **Subheading 3.1.**

3.6.1. 2-D Fibronectin Control

1. Precoat control plates with 0.8 mL of 10 µg/mL fibronectin incubated for 1 h at 37°C (*see* **Notes 4** and **5**).
2. Wash twice with 2 mL of PBS.
3. Proceed to **Subheading 3.1.**

3.6.2. 2-D Collagen I Control

1. Precoat control 35-mm plates with 0.8 mL of 10 µg/mL collagen I incubated for 1 h at 37°C.
2. Wash twice with 2 mL of PBS.
3. Proceed to **Subheading 3.1.**

3.6.3. 2-D Mix Control

Solubilized fibroblast-derived 3-D matrix to produce a protein mixture containing all the proteins derived from the 3-D matrix.

1. Aspirate PBS from 3-D matrix-covered 35-mm plates (from **Subheading 3.4.**).
2. Tip the plates at approximately a 30° angle relative to bench top for 1 min to accumulate the excess PBS on one side of the plate.
3. Aspirate the excess PBS carefully to avoid detaching the matrix layer.
4. Place the plates on ice and add 300 µL of solubilization buffer.
5. Incubate on ice for 5 min.
6. Scrape the dish with a rubber policeman toward one side of the dish and pipet the mixture into a 1.5 mL of microcentrifuge tube.
7. Add an additional 200 µL of solubilization reagent to each tube.
8. Rotate at 4°C for 1 h.
9. Spin at 12,000*g* in a microcentrifuge for 15 min at 4°C.
10. Transfer the supernatant into a fresh microcentrifuge tube.
11. Calculate the protein concentration.
12. Store 2-D mix at 4°C or precoat control 35-mm plates with 0.8 mL of 2-D mix at a final concentration of about 30 µg/mL (stock is normally about 3 mg/mL) incubated for 1 h at 37°C.
13. Wash twice with 2 mL of PBS.
14. Proceed to **Subheading 3.1.**

3.6.4. 3-D Collagen Gel Control (See **Note 25**)

1. Mix 8 mL of collagen I with 1 mL of 10X PBS and 1 mL of NaOH.
2. Check the pH and adjust to pH 7.4 ± 0.2 (*see* **Note 26**).
3. To form a gel, add a 200-µL drop of the collagen solution onto the glass of the time-lapse plates and carefully place the plates into an incubator at 37°C lacking CO_2 for about 1 h or until polymerization is achieved.

4. Wash twice carefully with PBS.
5. Proceed to **Subheading 3.1.**

3.6.5. 3-D Fibronectin Control (See **Note 25**)

1. Add an approx 200-µL drop of fibronectin stock (concentration grater or equal to 2 mg/mL) to the glass well at the bottom of the 35-mm time-lapse plate.
2. Allow the fibronectin drop to dry by placing the plate uncovered into a tissue culture laminar flow hood for about 1 h or until completely desiccated.
3. Repeat **steps 1** and **2** twice more onto the same spot of dried fibronectin.
4. Wash carefully with PBS.
5. Confirm the presence of the 3-D fibronectin coated glass by observing under a low magnification tissue culture microscope.
6. Proceed to **Subheading 3.1.**

4. Notes

1. The serum quality in the United States has periodically been changing for some years now. If low in quality, fetal bovine serum can greatly affect the production of fibrillar adhesions therefore affecting the resultant matrix, which depends on this fibrillogenesis process *(5,19,24)*. In the past, we had experienced some difficulties in generating high-quality 3-D matrices by utilizing diverse sources of serum. Therefore, we propose to use only serum that has been pre-tested for its quality.
2. The protocol in **Subheading 3.1.** can be followed as proposed or alternatively can be followed using other substrates as comparison controls (*see* **Subheading 3.6.**).
3. NIH 3T3 cells must be routinely cultured in high-glucose Dulbecco's modified Eagle's medium supplemented with 10% calf serum, 100 U/mL penicillin, and 100 µg/mL streptomycin. Never allow cultured NIH 3T3 cells to become completely confluent while maintaining stock cultures. When cells reach 80% confluence (about once per week), subculture at 1:20 dilution. However, if the matrix deposition at confluence is performed in the presence of calf serum, the resultant matrices are less stable and more likely to detach from the surface than matrices obtained after culture in fetal bovine serum. Therefore, NIH 3T3 cells should be grown in media containing fetal bovine serum prior to matrix deposition. A preadaptation of the cells in fetal bovine serum-containing medium after at least 22 passages is empirically recommended.
4. Fibronectin aliquots should be stored frozen at –80°C. Place aliquot in ice and allow it to thaw slowly. Always keep fibronectin on ice. Store the working solution at 4°C and avoid freezing thawing. If exact amounts of fibronectin are needed (not for following these procedures), then double-checking the fibronectin concentration after thawing and prior to use is recommended.
5. Fibronectin is commercially available but fairly expensive; therefore, it can be extracted from plasma utilizing gelatin affinity columns *(22)*. Alternatively, 0.2% gelatin precoat can be used; the fibronectin from serum (and other serum contents)

will bind to the pre-coated gelatin. In case gelatin is used, then gelatin-coated plates (following the same procedure as in **Subheading 3.6.1.**) should be incubated with 30% bovine serum for 20 min and rinsed with PBS prior to BSA block.

6. If setting up time-lapse recording of fibroblast onto control substrates, then the assorted control coated plates should be used in **Subheading 3.1.**

7. Always handle matrices with great care; all processes handling matrices should be conducted gently to prevent turbulence that may cause the matrix-layer to detach from the surface. Never attempt to aspirate the entire liquid volume; this will prevent removing or damaging the matrix layer.

8. The cell concentration plated needs to be such that it will result in scattered cells after overnight culture avoiding contact among cells as much as possible.

9. The recommended number of cells to attempt to follow should be approx 9 (10X objective is suggested for most microscopes); cells should appear scattered within the field of view therefore increasing the odds of avoiding collisions among each other during the tracking time.

10. Living cells will be followed in real time for a period of 6 h; therefore, a constant environmental control to maintain stable conditions is needed. There are many commercially available boxes or incubators on the market that will maintain constant 37°C, 100% humidity, and 10% CO_2 levels. In the absence of such equipment, using a fan-assisted heater at a pre-set distance can be used while monitoring the stage temperature to remain constant at 37°C. The humidity should not be of concern if the 35-mm plate containing 2 mL of media is pre-sealed with parafilm. In order to ensure a neutral pH, media should appear red at all times (as opposed to high pH purple in the absence of CO_2). Maintaining the pH can be accomplished by adding 25 m*M* HEPES buffer to the culturing media. Finally, make sure the light source is set at low light level because too much illumination may result in cell damage.

11. Lack of technologically advanced equipment (which normally comes with instructions) will require some creativity from the individual performing the task. Placing a transparent film over a TV monitor (if video system is used) and manually following (every 20 min for a period of only 4 h) three or four cells at a time can assist in obtaining data. Alternatively, one could acquire phase contrast images with a normal or digital camera and store them sequentially to be printed onto see-through paper (grided or not).

12. A stage micrometer containing marks representing about 10, 50, and 100 microns should be used. To ensure the most accurate calibration, it is of great importance that the stage micrometer is placed straight and that great attention is given to focusing correctly before acquiring the image.

13. This section could automatically be completed by a distinct variety of programs that are commercially available in the market programmed to digitally analyze images. Nevertheless, we opted to explain the process to ensure that even in the absence of automation, the measurements could be performed manually (*see* **Fig. 1**). We have proposed one technique; however, the performer is encouraged to improvise depending on the photographic and analysis equipment available.

14. Flipping through the images can be achieved either by software as a movie or stack file where each frame corresponds to one time point (10 min suggested). Alternatively, if hard copies are used, then printing files onto see-through paper is suggested to allow ease in flipping through pages for observational purposes.

15. If you must include cells that come in contact with each other or engage in cell division then stop the measurement of these cells at the frame (file or page) before the mentioned event.

16. If the tracking is conducted manually, it is recommended to select the files representative for each hour or half hour as opposed to every 10 min (rendering only seven images; *see* **Fig. 1**). When hesitation arises regarding a specific cell path or recognizing the location of the cell being tracked, flipping through all files within the interval in question is recommended. If a computer program will perform the tracking automatically instead, then using all files following the tutorial instructions will be more accurate.

17. Individual tracks can be followed simultaneously by assigning specific numbers, patterns, or colors to each cell.

18. If using hard copies, a string can be used to follow the points and its length can be measured for the total path length and then converted to microns. For a more accurate result, the hard copies can be enlarged, in which case the control calibration image should be proportionally enlarged as well (*see* **Fig. 1**). Otherwise, the software should render a path distance when instructed.

19. The resulting values have no units because they are ratios. Nevertheless, to ensure a correct result it is of great importance to use numbers with equal units for the arithmetic division.

20. It is recommended that the total count (velocity rates or directionality values) be greater than 12 for statistical normality tests; nevertheless, for biological statistical significance, we suggest repeating three times rendering a minimum sample size (n) of 54 for each experimental condition.

21. The standard deviation should be calculated using the "nonbiased" or "$n - 1$" method as shown by the formula.

22. If using, for example, Microsoft's Excel for these calculations, then "array 1" and "2" will be all the accumulated data (minimum 54; velocity rates or directionality ratios) for each condition to be compared. The "tail" space should contain the number "2," which states two tailed data meaning that there is no prediction for faster or more directional conditions. Finally the "type" space is "3," indicating nondependent variance among the sample conditions. There are numerous statistical programs on the market like GraphPad Instat, which contain easy-to-follow instructions and will provide statistical analyses including short explanations and graphical representation.

23. To avoid misleading information, it is recommended that both the statistical significance statement and the numerical P value be posted together with the mean results and standard deviations.

24. **Steps 23** and **24** (in **Subheading 3.4.**) are optional. Performing these steps will ensure clearing the matrices from deoxyribonucleic acid debris and other soluble

molecule residues; however, it may alter the matrices. In order to ensure reproducibility the inclusion or exclusion of these steps should be consistent throughout all experiments and stated when reporting the experimental procedures.

25. 3-D fibronectin and collagen gels will assist in testing whether the effects observed in the motility rates and directionality are to the result of the three dimensionality of the substrate as opposed to the molecular composition of the substrate like in the case of 2-D mix. 3-D fibronectin is achieved by desiccation of relatively high quantities of fibronectin onto a small area.

26. Use NaOH or HCl by adding drop wise and monitoring the solution's pH with pH paper after mixing and adding a drop of the solution to the pH paper. Do not dip the paper into the mixture, which should be aseptic for culturing cells. Make sure to keep the mixture at 4°C during the entire procedure or until otherwise stated.

Acknowledgments

Thanks to M. D. Amatangelo, K. B. Buchheit, M. R. Chintalapudi, and especially K. Clark for critical comments and proofreading of this manuscript.

References

1. Elsdale, T. and Bard, J. (1972) Collagen substrata for studies on cell behavior. *J. Cell Biol.* **54,** 626–637.
2. Elsdale, T. and Bard, J. (1972) Cellular interactions in mass cultures of human diploid fibroblasts. *Nature* **236,** 152–155.
3. Elsdale, T. and Bard, J. (1974) Cellular interactions in morphogenesis of epithelial mesenchymal systems. *J. Cell Biol.* **63,** 343–349.
4. Friedl, P. and Brocker, E. B. (2002) Tcr triggering on the move: Diversity of t-cell interactions with antigen-presenting cells. *Immunol. Rev.* **186,** 83–89.
5. Cukierman, E., Pankov, R., Stevens, D. R., and Yamada, K. M. (2001) Taking cell-matrix adhesions to the third dimension. *Science* **294,** 1708–1712.
6. Streuli, C. (1999) Extracellular matrix remodelling and cellular differentiation. *Curr. Opin. Cell Biol.* **11,** 634–640.
7. Badylak, S. (2002) The extracellular matrix as a scaffold for tissue reconstruction. *Semin. Cell Dev. Biol.* **13,** 377–383.
8. Buck, C. A. and Horwitz, A. F. (1987) Cell surface receptors for extracellular matrix molecules. *Annu. Rev. Cell Biol.* **3,** 179–205.
9. Hay, E. D. (1991) *Cell Biology of Extracellular Matrix*, 2nd Ed., Plenum Press, New York, NY.
10. Hynes, R. O. (1999) Cell adhesion: Old and new questions. *Trends Cell Biol.* **9,** M33–M77.
11. Geiger, B., Bershadsky, A., Pankov, R., and Yamada, K. M. (2001) Transmembrane crosstalk between the extracellular matrix and the cytoskeleton. *Nat. Rev. Mol. Cell Biol.* **2,** 793–805.
12. Walpita, D. and Hay, E. (2002) Studying actin-dependent processes in tissue culture. *Nat. Rev. Mol. Cell Biol.* **3,** 137–141.

13. Bissell, M. J., Radisky, D. C., Rizki, A., Weaver, V. M., and Petersen, O. W. (2002) The organizing principle: Microenvironmental influences in the normal and malignant breast. *Differentiation* **70,** 537–546.
14. Miranti, C. K. and Brugge, J. S. (2002) Sensing the environment: A historical perspective on integrin signal transduction. *Nat. Cell Biol.* **4,** E83–E90.
15. Bosman, F. T. and Stamenkovic, I. (2003) Functional structure and composition of the extracellular matrix. *J. Pathol.* **200,** 423–428.
16. Clegg, D. O., Wingerd, K. L., Hikita, S. T., and Tolhurst, E. C. (2003) Integrins in the development, function and dysfunction of the nervous system. *Front. Biosci.* **8,** D723–D750.
17. Schmeichel, K. L. and Bissell, M. J. (2003) Modeling tissue-specific signaling and organ function in three dimensions. *J. Cell Sci.* **116,** 2377–2388.
18. Wakatsuki, T. and Elson, E. L. (2003) Reciprocal interactions between cells and extracellular matrix during remodeling of tissue constructs. *Biophys. Chem.* **100,** 593–605.
19. Yamada, K. M., Pankov, R., and Cukierman, E. (2003) Dimensions and dynamics in integrin function. *Braz. J. Med. Biol. Res.* **36,** 959–966.
20. Kunz-Schughart, L. A., Heyder, P., Schroeder, J., and Knuechel, R. (2001) A heterologous 3-D coculture model of breast tumor cells and fibroblasts to study tumor-associated fibroblast differentiation. *Exp. Cell Res.* **266,** 74–86.
21. Olumi, A. F., Grossfeld, G. D., Hayward, S. W., Carroll, P. R., Tlsty, T. D., and Cunha, G. R. (1999) Carcinoma-associated fibroblasts direct tumor progression of initiated human prostatic epithelium. *Cancer Res.* **59,** 5002–5011.
22. Akiyama, S. K. (1999) Purification of fibronectin, in *Current Protocols in Cell Biology* (Bonifacino, J. S., Dasso, M., Lippincott-Schwartz, J., Harford, J. B., and Yamada, K. M., eds.) John K. Wiley & Sons, New York, NY, Vol 1, pp. 10.15.11–10.15.13.
23. Cukierman, E. (2002) Preparation of extracellular matrices produced by cultured fibroblasts, in *Current Protocols in Cell Biology* (Bonifacino, J. S, Dasso, M., Lippincott-Schwartz, J., Harford, J. B., and Yamada, K. M., eds.) John K. Wiley & Sons, New York, NY, Vol. 1, pp. 10.19.11–10.19.14.
24. Cukierman, E., Pankov, R., and Yamada, K. M. (2002) Cell interactions with three-dimensional matrices. *Curr. Opin. Cell Biol.* **14,** 633–640.

III

Analysis of Cell Migration in Specialized Cell Systems

9

Tumor Cell Invasion Assays

Leslie M. Shaw

Summary

The ability of tumor cells to invade is one of the hallmarks of the metastatic pheno-type. To elucidate the mechanisms by which tumor cells acquire an invasive phenotype, in vitro assays have been developed that mimic the in vivo process. The most commonly used in vitro invasion assay is a modified Boyden chamber assay using a basement membrane matrix preparation, Matrigel, as the matrix barrier and NIH 3T3 conditioned media as the chemoattractant. The results obtained using this assay show a strong corre-lation between the ability of tumor cells to invade in vitro and their invasive behavior in vivo, which validates this assay as a measure of invasive potential. The methods pre-sented in this chapter outline how the Matrigel in vitro invasion assay is performed.

Key Words: Invasion; transwell; Matrigel; chemoattractant; proteases; basement membrane.

1. Introduction

Invasion of tumor cells from a primary tumor site into the surrounding stromal microenvironment is a defining step in tumor progression. Once the normally restrictive basement membrane boundaries of the tumor have been breached, invasive cells can gain access to the vasculature and lymphatic sys-tem and metastasize. As a result, the ability to invade is one of the hallmarks of the metastatic phenotype *(1)*. Invasion is a multistep process that involves adhesion, directed migration, and proteolytic activity to degrade extracellular matrix (ECM) barriers *(2,3)*. To understand the mechanisms involved in tumor cell invasion and to evaluate approaches to inhibit this important pro-cess, in vitro assays have been developed that mimic the in vivo process. The most commonly used in vitro invasion assay is a modified Boyden chamber assay using a basement membrane matrix preparation as the matrix barrier

From: *Methods in Molecular Biology, vol. 294: Cell Migration: Developmental Methods and Protocols*
Edited by: J-L. Guan © Humana Press Inc., Totowa, NJ

and NIH 3T3-conditioned media as the chemoattractant. This assay was first developed in 1987, and early studies using this technique demonstrated that a strong correlation existed between the invasive behavior of tumor cells in vivo and their ability to invade in vitro using this assay system *(4)*. In the ensuing years, many investigators have used this in vitro invasion assay to demonstrate the involvement of specific adhesion receptors, intracellular signaling pathways, and proteases in tumor cell invasion *(5–11)*. It should be noted that the basement membrane matrix preparation used in these assays, Matrigel, does not recapitulate all structural aspects of an actual basement membrane and results should be interpreted with this caveat in mind. However, many of the conclusions drawn from the results of the in vitro invasion assays have been confirmed by subsequent in vivo studies, which supports the validity of this assay as a measure of invasive potential *(11,12)*. The methods presented in this chapter outline how to prepare for and perform the Matrigel invasion assay and how to utilize this assay to investigate mechanisms of tumor cell invasion.

2. Materials

1. 8 µM Pore, 6.5-mm diameter Transwells (Costar; cat. no. 3422).
2. NIH 3T3 fibroblast cells.
3. NIH 3T3 culture medium: Dulbecco's modified Eagle's medium (DMEM) containing 10% newborn calf serum and 1% penicillin–streptomycin.
4. Trypsin–ethylene-diamine tetraacetic acid.
5. Matrigel (BD Biosciences; cat. no. 356234).
6. Crystal Violet stain: 2.0% Ethanol containing 0.2% Crystal Violet.
7. Methanol.
8. Bovine serum albumin (BSA).
9. 24-Well cluster plates.
10. Q-Tip swabs.
11. Inverted microscope.
12. Eyepiece reticule (optional).

3. Methods

Transwell inserts will be used as the modified Boyden chamber for the invasion assays described in this chapter. The methods in this section outline 1) the preparation of conditioned NIH 3T3 culture media, 2) the coating of Transwell assay chambers, 3) the invasion assay.

3.1. 3T3 Conditioned Medium

To stimulate directed invasion through matrix-coated membranes, a chemoattractant is added to the bottom well of the Transwell assay chamber. The most common source of chemoattractant that is used for these assays is conditioned culture medium from NIH 3T3 fibroblasts. These cells secrete a variety of growth factors and therefore the conditioned medium from these cells is a rich source of chemoattractants for tumor cells (*see* **Note 1**).

1. Plate NIH 3T3 cells in complete medium in 100-mm tissue culture plates and grow until they are confluent.
2. Trypsinize the cells and split them 1:10 into new 100-mm dishes. The following day, replace the medium in each plate with fresh complete growth medium and allow the cells to grow for 2 d. At this time the cells will be near confluency.
3. Remove and filter the culture medium. The conditioned 3T3 medium should be stored at 4°C until needed (*see* **Note 2**).

3.2. Coating Filters With ECM

To assay the invasive capacity of tumor cells, a barrier of ECM must be established through which the cells are required to invade. The ECM barrier used in these assays is the basement membrane matrix preparation Matrigel. Matrigel is a mixture of basement membrane ECM proteins, primarily laminins, and collagen IV, which is isolated from the Englebreth–Holm–Swarm mouse sarcoma. At high concentrations, Matrigel will form a gel when warmed to 37°C and in doing so it forms a barrier that is similar in composition to a basement membrane. At lower concentrations, Matrigel does not gel but will still form a barrier when it is dried down and reconstituted in culture medium (*see* **Note 3**). The next steps will describe each of these approaches.

3.2.1. Thin Coating Matrigel Method

The thin coating method involves coating the upper surface of the porous membrane of the Transwell insert with a dilute sample of Matrigel. Although this method does not generate a gel, it does provide a thin barrier through which the cells must invade. This method is the easiest to quantitate and the most reproducible between assays because a consistent coating is obtained from assay to assay.

1. Thaw Matrigel at 4°C and keep on ice.
2. Dilute Matrigel into cold H_2O to a final concentration of 0.125 µg/µL using precooled pipet tips. Prepare a large enough volume to coat the total number of wells that will be used (*see* **Note 4**).
3. Add 40 µL of diluted Matrigel to the upper chamber of each Transwell (*see* **Fig. 1**). Tap the sides of the plate gently to ensure the liquid covers the bottom of each well completely.
4. Allow the wells to sit without a cover until the Matrigel has completely dried onto the porous membranes. This usually takes 4–5 h but the wells can also be left overnight.

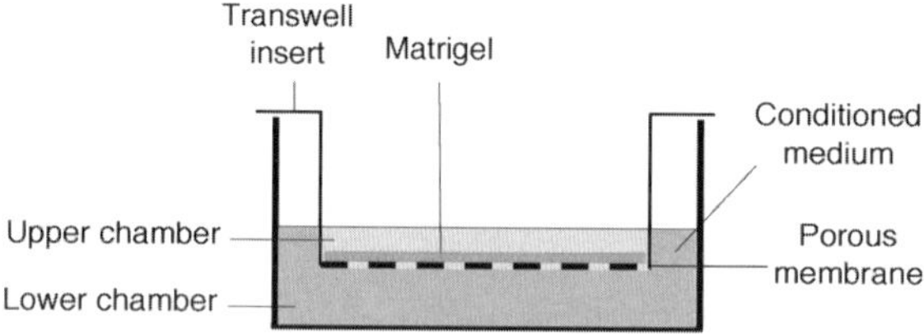

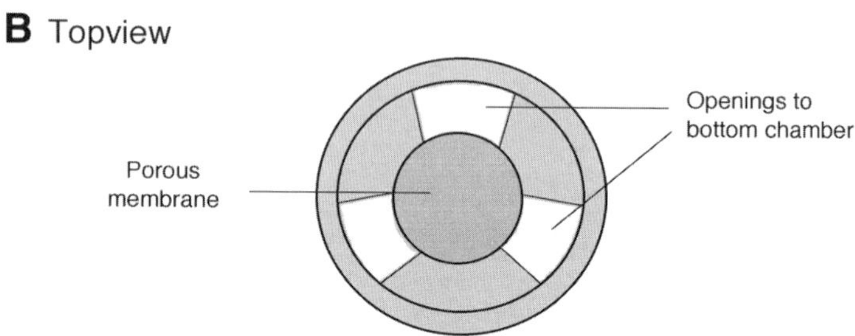

Fig. 1. Schematic of the Transwell assay system. (**A**) Side view of the Transwell assay chamber showing the Transwell insert containing the Matrigel-coated porous membrane and the lower well containing conditioned medium. (**B**) Top view of the Transwell insert showing the porous membrane and the openings into the bottom well of the assay chamber. The conditioned medium can be added to the bottom well by placing the pipet tip through the openings in the Transwell insert.

5. Reconstitute the Matrigel by adding 40 μL of DMEM to the upper chamber of each Transwell. Tap the sides of the plate gently to completely cover the bottom of the wells with the medium.
6. Incubate the wells at 37°C for 1 h prior to the invasion assay.

3.2.2. Gel Matrigel Method

The gel method involves coating the porous membranes with a concentrated sample of Matrigel, which forms a gel at 37°C. The density and thickness of the gel can be varied by altering the concentration of Matrigel and total volume of Matrigel that is used to coat the wells, respectively (*see* **Note 5**).

1. Thaw Matrigel at 4°C and keep on ice.
2. Place cluster plates containing the Transwell inserts on ice to chill.
3. Aliquot 15–45 μL of Matrigel into the upper chamber of each well using pre-cooled pipet tips. Tap the sides of the plates to completely cover the bottom of the wells.
4. Incubate the plates at 37°C for at least 30 min before starting the invasion assay.

3.3. Invasion Assay

The invasion assay is performed by adding cells to the upper, coated surface of the Transwell inserts and NIH 3T3-conditioned medium to the lower well of the assay chamber. Cells that are invasive will invade through the matrix barrier, pass through the pores of the membrane, and migrate onto the lower surface of the filter. The basic assay is described subsequently. To investigate specific mechanisms of invasion, such as the involvement of individual surface receptors, intracellular signaling pathways, or proteases in the invasive process, the cells can be treated with inhibitory compounds prior to being added to the Transwell chambers.

1. Trypsinize the experimental cells and add them to a 15 mL conical tube containing 4 mL of DMEM containing 1.0% BSA.
2. Collect the cells by centrifugation (200g). Wash the cells one time by resuspending the cell pellet in DMEM containing 0.1% BSA. Take an aliquot of cells to count.
3. Collect the cells by centrifugation (200g) and resuspend the cell pellets at a final concentration of 1×10^6 cells/mL in DMEM containing 0.1% BSA (*see* **Note 6**).
4. Remove the coated Transwells from the 37°C incubator. Add 100 µL of the cell suspensions to the upper chamber of each coated Transwell insert. Be careful not to disrupt the Matrigel when adding the cell suspension by pipetting the cells gently down the inner wall of the upper Transwell chamber.
5. Add 600 µL of NIH 3T3-conditioned medium to the bottom well of the Transwell chamber by placing the pipet tip through one of the openings of the Transwell insert (*see* **Fig. 1**).
6. Incubate the plates containing the Transwells at 37°C for 5 h (*see* **Note 6**).
7. Prepare cotton swabs to remove the cells from the upper portion of the Transwell filters by pulling on the top of the swab to make a "flat surface." By doing so, the entire surface of the well will be swabbed and all of the cells that have not invaded will be removed. Prepare both ends of one swab for each experimental well.
8. Add 500 µL of methanol to the wells of a new 24-well cluster plate.
9. Without removing the liquid from the upper chamber, place one end of the cotton swab into the well and swab the filter using a rotating motion while applying moderate pressure. The cotton swab will absorb the liquid at this time. Repeat this process with the other end of the swab. Caution should be taken in applying too much pressure because the swab can detach the membrane from the Transwell insert if it is pushed too hard. The optimal pressure that will remove the cells that have not invaded but that will not damage the membrane must be learned through experience.
10. Place the swabbed Transwell inserts into the wells containing methanol to fix the cells that have invaded onto the lower surface of the porous membrane. Fix for 15 min at room temperature.
11. Add 500 µL of Crystal Violet stain to the wells of a new 24-well cluster plate. After fixing, transfer the Transwell inserts into these wells. As an alternative, the

methanol can be removed and the Crystal Violet stain added to the same wells. Incubate for 15 min at room temperature.

12. Wash the stained Transwell inserts in H_2O to remove the excess stain. To do so, immerse the inserts in a beaker of H_2O and rinse several times. Invert the Transwell inserts and allow to air dry.

13. Quantify invasion by counting the cells that have invaded onto the lower surface of the porous membrane. Visualize the cells with an inverted microscope using brightfield optics. The use of an eyepiece reticule that delineates an area to count is preferred, but not necessary. If a reticule is not available, the cells in the full field should be counted. The pores of the filters are small and colorless; do not count these! The cells will be stained pink/purple and will appear fuzzy (**Fig. 2**). If the cells are in sharp focus and a distinct cell border is apparent, the cells are on the top of the filter and were not removed completely by the cotton swabs. If this occurs, the filters can be cleaned again by using a swab that has been pre-moistened with water (*see* **Note 7**).

4. Notes

1. NIH 3T3-conditioned medium is a rich source of chemoattractants and it is optimal for use when the specific sensitivity of an experimental cell line is not known or a general stimulus is desired. In addition, matrix proteins in the conditioned medium, including fibronectin (FN) and vitronectin (VN), coat the underside of the membranes and provide a matrix substrate upon which the cells can adhere after they invade from the upper surface of the filter. However, individual chemoattractants can be tested using this assay by substituting a specific factor for the 3T3-conditioned medium in the bottom well of the Transwell chamber. Dilute the chemoattractants in DMEM containing 0.1% BSA to prevent absorption to the bottom well.

2. The concentration of chemoattractants in the 3T3-conditioned medium will vary each time the medium is collected. Serial dilutions of the medium should be tested each time a new batch is harvested to determine the optimal concentration of the conditioned medium for the invasion assays. Generally, a 1:1 or 1:2 dilution is optimal for promoting chemoinvasion. However, it should be noted that the specific dilution to use can be cell type dependent and should be determined empirically. Dilute the conditioned medium in plain DMEM medium. Large volumes of the 3T3-conditioned medium can be harvested at one time and aliquoted and frozen at −20°C. By doing this, the conditions for the invasion assays will be normalized across multiple experiments and the results should be more reproducible.

3. Invasion assays can also be performed using a collagen gel as the barrier through which the cells must pass. Although Matrigel assays are an in vitro correlate to in vivo invasion through the basement membrane, collagen assays are an in vitro correlate to invasion within the collagen-rich interstitial matrix. The choice of which assay to perform will depend upon the specific facet of invasion that is of interest. Rat tail collagen will form a gel at 37°C and the assays are performed in a similar manner as with a Matrigel gel *(13)*.

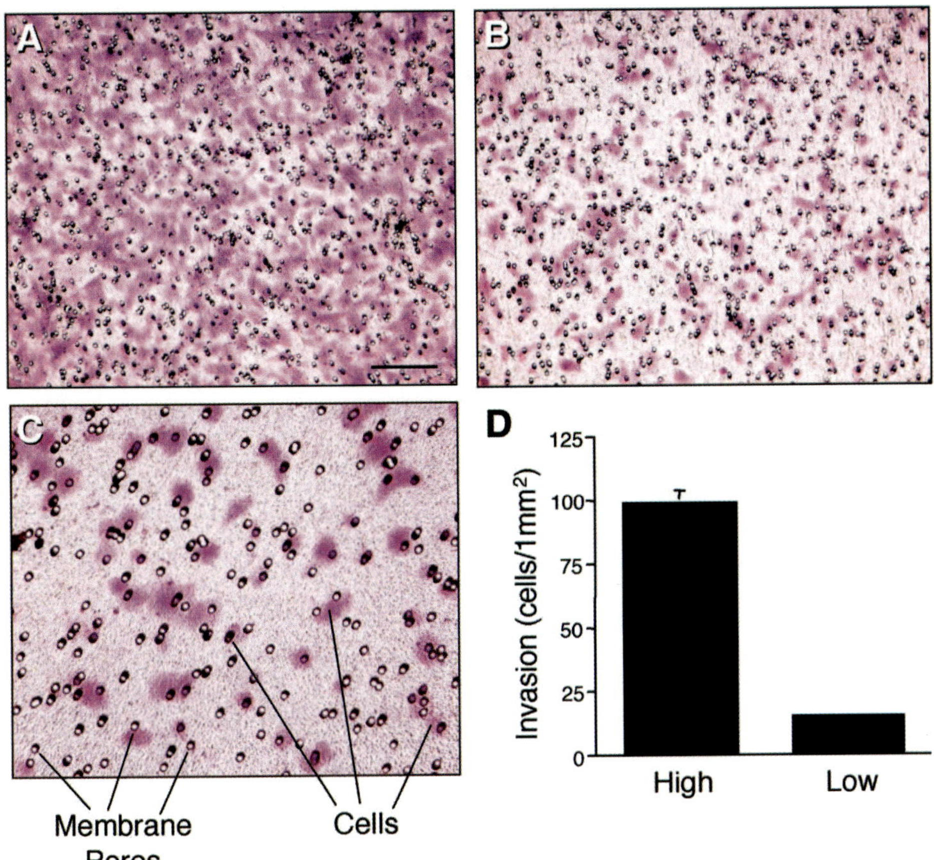

Fig. 2. Quantitation of invasion. Cells that have invaded through the Matrigel and onto the lower surface of the porous membrane are stained purple. (**A**) The cells in this invasion assay are too dense to count because it is difficult to distinguish individual cells. Either the number of cells should be decreased or the length of the assay should be shortened so that this experiment can be accurately quantitated. (**B**) Individual cells can be identified in this invasion assay and as a result, the assay can be accurately quantitated. (**C**) Higher magnification image showing the cells and the pores of the membrane. (**D**) Representative graph of an invasion assay comparing tumor cells that have high and low invasive potentials. (**A,B**) Bar, 200 μ*M*; (**C**) bar, 100 μ*M*.

4. To increase reproducibility between experiments, multiple wells for several experiments can be coated at the same time and stored at 4°C after the Matrigel has dried completely.
5. Coating with a Matrigel gel is a more rigorous test of invasion than the thin coating method because a thicker barrier is formed through which the cells must penetrate. However, it can be difficult to accurately quantitate these assays because

the cells tend to invade through channels that are made by a few early cells. As a result, the cells are often clumped together on the underside of the filter and it is difficult to identify individual cells to count. The thickness of the Matrigel gel is determined by the volume of Matrigel that is used to coat the wells. The density of the gels can also be altered by diluting the Matrigel stock. Dilutions of Matrigel of up to 1:3 will retain the ability to form a gel at 37°C. The commercial stock preparations of Matrigel vary in concentration, between 10 and 15 mg/mL, and this variation needs to be taken into consideration when optimizing the conditions for each experimental assay. Matrigel should be diluted into cold DMEM using pre-chilled pipet tips and mixed thoroughly by gentle pipetting.

6. The optimal cell number and length of assay should be determined empirically for each cell line or experimental condition. For most tumor cell lines 1×10^5 cells is the maximum number of cells that can adhere to the surface area of the 6.5-mm^2 Transwell membrane. Cells that have a high invasive capacity will invade well within a 5-h assay. However, these conditions should be considered only a starting point and it will likely be necessary to alter both the number of cells and the length of the assay to establish the optimal assay conditions for each new cell line or treatment to be tested.

7. To accurately count the cells that have invaded, it is necessary to distinguish individual cells from one another. Cells that are too dense will appear as large aggregates and the exact cell number will be difficult to determine (**Fig. 2A**). As mentioned previously, the conditions must be adjusted (i.e., cell number, length of assay and Matrigel concentration) to obtain cell densities that can be quantitated (**Fig. 2B,D**).

References

1. Hanahan, D. and Weinberg, R. A. (2000) The hallmarks of cancer. *Cell* **100,** 57–70.
2. Liotta, L. A. and Stetler-Stevenson, W. G. (1990) Metalloproteinases and cancer invasion. *Semin. Cancer Biol.* **1,** 99–106.
3. Liotta, L. A. (1988) Gene products which play a role in cancer invasion and metastasis. *Breast Cancer Res. Treat.* **11,** 113–124.
4. Albini, A., Iwamoto, Y., Kleinman, H. K., Martin, G. R., Aaronson, S. A., Kozlowski, J. M., et al. (1987) A rapid in vitro assay for quantitating the invasive potential of tumor cells. *Cancer Res.* **47,** 3239–3245.
5. Albini, A., Melchiori, A., Santi, L., Liotta, L. A., Brown, P. D., and Stetler-Stevenson, W. G. (1991) Tumor cell invasion inhibited by TIMP-2. *J. Natl. Cancer Inst.* **83,** 775–779.
6. Shaw, L. M., Chao, C., Wewer, U. M., and Mercurio, A. M. (1996) Function of the integrin alpha 6 beta 1 in metastatic breast carcinoma cells assessed by expression of a dominant-negative receptor. *Cancer Res.* **56,** 959–963.
7. Shaw, L. M., Rabinovitz, I., Wang, H. H., Toker, A., and Mercurio, A. M. (1997) Activation of phosphoinositide 3-OH kinase by the alpha6beta4 integrin promotes carcinoma invasion. *Cell* **91,** 949–960.

8. Shaw, L. M. (2001) Identification of insulin receptor substrate 1 (IRS-1) and IRS-2 as signaling intermediates in the alpha6beta4 integrin-dependent activation of phosphoinositide 3-OH kinase and promotion of invasion. *Mol. Cell Biol.* **21,** 5082–5093.

9. Jauliac, S., Lopez-Rodriguez, C., Shaw, L. M., Brown, L. F., Rao, A., and Toker, A. (2002) The role of NFAT transcription factors in integrin-mediated carcinoma invasion. *Nat. Cell Biol.* **4,** 540–544.

10. Bachelder, R. E., Wendt, M. A., and Mercurio, A. M. (2002) Vascular endothelial growth factor promotes breast carcinoma invasion in an autocrine manner by regulating the chemokine receptor CXCR4. *Cancer Res.* **62,** 7203–7206.

11. Mamoune, A., Luo, J. H., Lauffenburger, D. A., and Wells, A. (2003) Calpain-2 as a target for limiting prostate cancer invasion. *Cancer Res.* **63,** 4632–4640.

12. Wewer, U. M., Shaw, L. M., Albrechtsen, R., and Mercurio, A. M. (1997) The integrin alpha 6 beta 1 promotes the survival of metastatic human breast carcinoma cells in mice. *Am. J. Pathol.* **151,** 1191–1198.

13. Eble, J. A., Niland, S., Dennes, A., Schmidt-Hederich, A., Bruckner, P., and Brunner, G. (2002) Rhodocetin antagonizes stromal tumor invasion in vitro and other alpha2beta1 integrin-mediated cell functions. *Matrix Biol.* **21,** 547–558.

10

Analysis of Endothelial Cell Migration Under Flow

Song Li

Summary

The migration of endothelial cells (ECs) play an important role in embryonic vasculogenesis, angiogenesis, and post-angioplasty reendothelialization. ECs are constantly subjected to fluid shear stress (the tangential component of hemodynamic forces) from blood flow, but the effects of shear stress on EC migration have not been well characterized until recently. We have used an in vitro flow system to apply shear stress to EC cultures and used imaging and biochemical techniques to analyze EC migration under well-defined flow conditions. Time-lapse microscopy and cell tracing analysis generated quantitative information on EC migration under flow, and this assay was used to dissect the signaling events involved in shear stress-induced EC migration. By expressing green fluorescent protein-tagged molecules in EC, cell focal adhesions and cytoskeleton were visualized and quantified in living cells under flow, which provides temporal and spatial resolution for mechanotransduction at the molecular level. The studies on EC migration under flow will advance our understandings on vascular repair and remodeling in vivo, and provide a rational basis for the development of strategy to promote endothelialization and vascularization in tissue-engineered constructs.

Key Words: Endothelial cell; migration; flow; shear stress; time-lapse microscopy; green fluorescent protein (GFP); molecular dynamics; focal adhesion; cytoskeleton.

1. Introduction

The migration of endothelial cells (ECs) plays an important role in embryonic vasculogenesis, angiogenesis, and postangioplasty re-endothelialization. EC migration can be modulated by environmental factors through different mechanisms, such as chemotaxis, haptotaxis, and mechanotaxis (mechanical force-induced directional migration). The focal adhesions (FAs), cytoskeleton, and signaling molecules inside cells need to respond to diverse extracellular signals and translate them into coordinated intracellular responses to mediate

From: *Methods in Molecular Biology, vol. 294: Cell Migration: Developmental Methods and Protocols*
Edited by: J-L. Guan © Humana Press Inc., Totowa, NJ

cell migration *(1,2)*. ECs are constantly subjected to fluid shear stress (the tangential component of hemodynamic forces) from blood flow, but the effects of shear stress on EC migration have not been well characterized until recently. By using in vitro flow system, we and others have shown that the physiological level of shear stress (approx 12 dyn/cm^2) enhances EC wound healing *(3,4)* and promotes lamellipodial protrusion and EC migration in the flow direction *(4–7)*. The directional migration of EC under flow is regulated by integrins *(8–10)*, the cytoskeleton *(5,6)*, and signaling molecules, including Rac, Rho, focal adhesion kinase (FAK), Shc, phosphoinositide-3 kinase, and extracellular-regulated kinase *(5–7,9,11)*.

Multiple imaging techniques and biochemical assays have been used to analyze EC migration under flow. Time-lapse microscopy and cell tracing analysis generated quantitative information (speed, distance, direction, and persistence) on EC migration, and this assay was used to analyze the signaling events involved in shear stress-induced EC migration. By expressing green fluorescent protein (GFP)-tagged molecules in ECs, cell focal adhesions and cytoskeleton (actin, microtubule, and intermediate filaments) were visualized in living cells *(5,6,12,13)*. The molecular dynamics was monitored by confocal microscopy and quantified by image analysis. The temporal and spatial molecular events at subcellular level provided important insights regarding the directional migration of EC under flow. In this chapter, we describe the procedures for time-lapse microscopy of EC migration (wound closure model and individual cell model) and molecular dynamics (using GFP-FAK as an example) under flow.

2. Materials

1. Flow chamber base and gasket (made from silicone sheets from Specialty Manufacturing Products; Saginaw, MI).
2. Metal plates, screws, and screwdriver.
3. Glass slides and cover glasses (Lab-Tek 155361, VWR International).
4. Chamber slides and chamber cover glasses (Lab-Tek 177372, VWR International).
5. Histology slide rack (for autoclaving slides).
6. Culture-medium reservoirs for flow system.
7. Clamps and clamp hold for the reservoirs.
8. Tubing for flow system (Pharmed; cat. no. NSF-51) or Transparent Tygon tubing from Masterflex; cat. no. 0642916.
9. Peristaltic pump (Masterflex L/S) and pumphead (Easy-load; cat. no.7518-60) (Cole-Parmer Instrument; Vernon Hills, IL).
10. Three-way stopcocks (Baxter; cat. no. 2C6241), connectors and adaptors for flow system.
11. Syringes (20 mL).

12. Weighing balance.
13. Temperature hood.
14. Heater, fan, and temperature controller.
15. Compressed air with 5% CO_2.
16. Flowmeter for airflow control (Cole-Parmer Instrument Co.).
17. Cell culture dishes and tubes.
18. Pipets (1–25 mL).
19. Cell lifters (scrapers).
20. 70% Ethanol.
21. Bovine aortic endothelial cells (BAECs). BAECs were isolated and characterized (*see* **Note 1**) and were maintained in Dulbecco's modified Eagle's medium (DMEM) supplemented with 10% fetal bovine serum (FBS) and penicillin (100 units/mL)–streptomycin (100 µg/mL) (complete medium). All cell cultures were kept in a humidified 5% CO_2–95% air incubator at 37°C. BAECs between passage 4 and 10 were used in experiments.
22. DMEM, high glucose, with L-glutamine and sodium pyruvate (stored at 4°C).
23. FBS (stored at –20°C in 50-mL aliquots).
24. Phosphate-buffered saline (PBS).
25. Penicillin–streptomycin solution (stored at –20°C in 5 mL of 100X aliquots).
26. 10X Trypsin–ethylene diamine tetraacetic acid (EDTA; 0.5% trypsin and 5.3 mM EDTA; stored at –20°C in 5-mL aliquots).
27. 0.1 mg/mL Fibronectin (stored at –20°C in aliquots).
28. 2% Gelatin.
29. 1% Heat-denatured bovine serum albumin in PBS.
30. GenePorter reagent (GeneSystems; San Diego, CA).
31. Inverted microscope with image acquisition system for time series.
32. Microscope stage: motorized X-Y stage and Z motor for auto-focusing.
33. Image analysis software with particle tracing algorithm (e.g., C-Imaging System software, Compix; Cranberry Township, PA; and Dynamic Image Analysis System (DIAS), Solltech; Oakdale, IA).
34. Confocal microscopy system with Argon laser.

3. Methods

In this section, quantitative analysis of EC migration under flow is described in **Subheading 3.1.**, and the study on FA dynamics during EC migration under flow is described in **Subheading 3.2.**

3.1. Analysis of EC Migration Under Flow

We have used two different models (wound closure and individual cell migration model) to study EC migration under flow. Wound closure model provides information on the remodeling of EC monolayer and individual cell migration at the wound edge. In this model, EC migration is mediated by both cell–matrix and cell–cell interactions. To isolate different factors and study

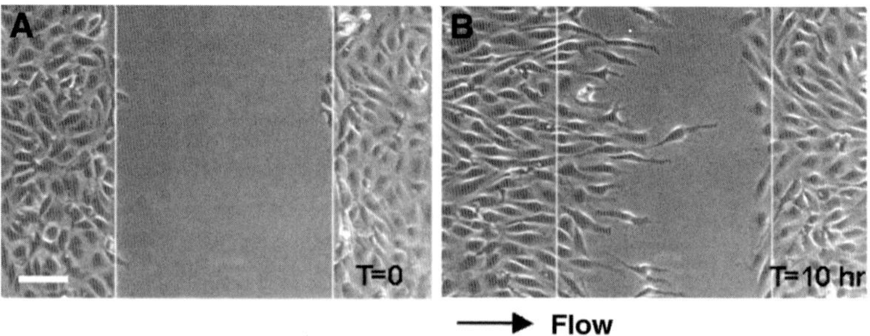

Fig. 1. Analysis of EC wound closure. (**A**) EC monolayer on a glass slide was mechanically wounded with a cell lifter, and the slide was assembled into a parallel flow chamber for time lapse-microscopy ($t = 0$). (**B**) The wounded EC monolayer was subjected to shear stress at 12 dyn/cm^2 for 10 h. Note the directional migration of EC in the flow direction. Bar, 100 µm.

cell–matrix interaction, individual cell migration model can be used. Cell seeding for both models is described in **Subheading 3.1.1.** The setup of flow system, time-lapse microscopy and image analysis are described in **Subheadings 3.1.2.–3.1.4.**, respectively. EC migration under static condition (as control) is described in **Subheading 3.1.5.**

3.1.1. Seeding Cells for Experiments

In this section, the cell seeding for wound closure model and individual cell migration model is described in **Subheadings 3.1.1.1.** and **3.1.1.2.**, respectively.

3.1.1.1. WOUND CLOSURE MODEL

1. Glass slides (22 mm × 75 mm) were rinsed with deionized water, loaded on slide racks and autoclaved.
2. The slides were coated with 2% gelatin for 30 min at room temperature (RT), and washed with PBS.
3. Confluent ECs were washed with PBS (without Ca^{2+}), treated with 0.5X trypsin–EDTA in PBS for 2 min at 37°C, detached and suspended in complete medium, spun down at 200g for 3 min, and resuspended in complete medium.
4. The cells were seeded on slides at 80% confluency and cultured for 2 d to form confluent monolayer.
5. Wounds in EC monolayer were created by removing strips of the cells (400- to 800-µm wide) with the tip of a cell scraper or a 200-µm micropipet tip. The wound could be made either perpendicular to the flow direction (to observe the directional migration of ECs under flow) or parallel to the flow direction. An example of wound closure under flow is shown in **Fig. 1**.
6. The culture medium was changed, and the slide was ready for flow experiment.

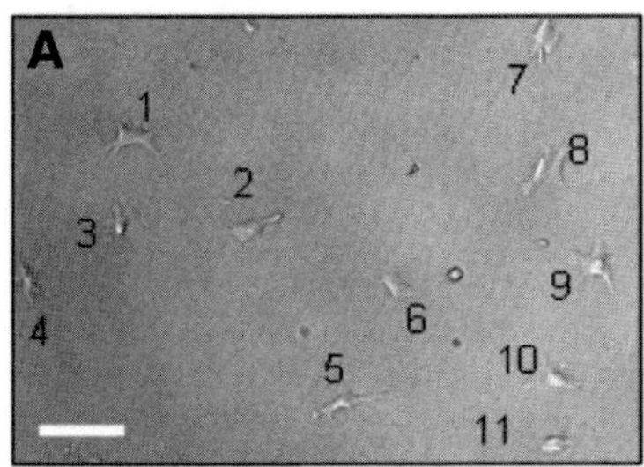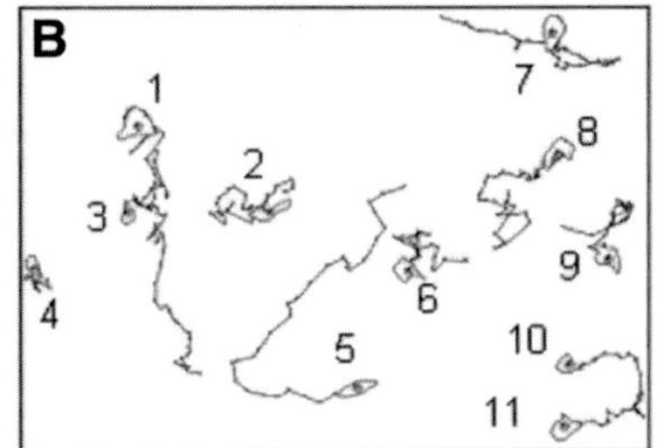

Fig. 2. Quantitative analysis of individual EC migration. Subconfluent BAECs were seeded on fibronectin-coated surface and the migration was monitored at 10-min intervals. (**A**) Phase-contrast microscopy of ECs on the surface at the beginning of the experiment. (**B**) The migration paths of ECs in (**A**) were reconstituted from the digitized movie recorded by time-lapse microscopy. Each dot represented the position of the centroid of the cell at that moment, and the time interval between two successive dots on a path was 10 min. The speed and direction of cell migration were calculated from the cell paths. Cells in contact with each other were excluded from analysis. Bar, 100 μm.

3.1.1.2. INDIVIDUAL EC MIGRATION MODEL

1. Glass slides were cleaned and autoclaved as in **Subheading 3.1.1.1.**
2. The slides were coated with fibronectin (or matrix proteins of interest, e.g., collagen I and laminin) at 0.1 mg/mL for 2 h at RT. The slides were then washed with PBS.
3. To study the specific interactions between matrix proteins and cell adhesion receptors and minimize the nonspecific interactions between cells and the slide surface, the coated slides were blocked with 1% heat-denatured BSA in PBS for 30 min at RT.
4. Confluent ECs were trypsinized as described in **Subheading 3.1.1.1.** and resuspended in DMEM with 0.5% FBS. The low concentration of FBS was to keep ECs in a quiescent state and to minimize the effects of growth factors and matrix proteins in FBS on cell growth and matrix deposition/synthesis.
5. The cells were seeded on slides at 5–10% confluency, incubated for 3 h in the incubator to allow for adhesion and spreading, and used for individual cell migration studies. An example of individual cell migration and cell tracing analysis is shown in **Fig. 2**.

3.1.2. Setting Up the Flow System for Microscopy

A temperature hood was built around the microscope to maintain the temperature at 37°C (**Fig. 3**). The flow system was set up inside the temperature hood for microscopy. The medium in circulation was ventilated with air/5% CO_2 to keep the pH level of the medium at 7.4–7.5.

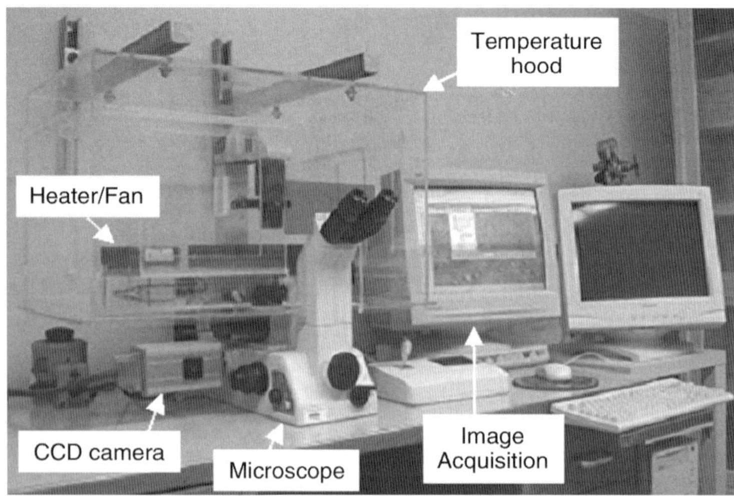

Fig. 3. A temperature hood built around the microscope for time-lapse micros-copy. The temperature hood was built with clear acrylic plates (3/8 in. in thickness). A fan heater (cat. no. 20055K21; McMaster Carr Supply Company; Los Angeles, CA), an equipment cooling fan (cat. no. 1976K41, McMaster Carr Supply Company), and a digital temperature controller (cat. no. 61161–300, VWR International) were assembled to form a feedback system to keep the temperature at 37°C inside the hood during the experiments. The space in the hood allowed the setup of a flow system. A water bath was placed in the temperature hood to humidify the air and decrease me-dium evaporation. To avoid the transmission of the vibration from the heating system and flow system to the microscope stage, the hood was attached to the wall without touching the microscope and the table.

1. The heater, fan, and temperature controller were turned on to warm up the tem-perature hood.
2. The circulation system was set up in the temperature hood as shown in **Fig. 4** (*see* **Note 2**). All tubing, connectors, and reservoirs were autoclaved before the experiment. The two custom-made glass reservoirs were fixed at different heights using clamps and clamp holder. Transparent Tygon tubing could also be used to connect the circulation system except the part loaded in the pumphead.
3. The polycarbonate flow chamber base and gasket were sterilized by soaking in 70% ethanol for 1 min and air dry, and the flow chamber was assembled as shown in **Fig. 5** (*see* **Note 3**) using a sterilized slide without cells.
4. A 20-mL syringe was attached to the side port of the three-way stopcock at the upstream of the flow chamber, and medium was slowly infused (to avoid air bubbles) into the upstream and downstream tubing and reservoirs.

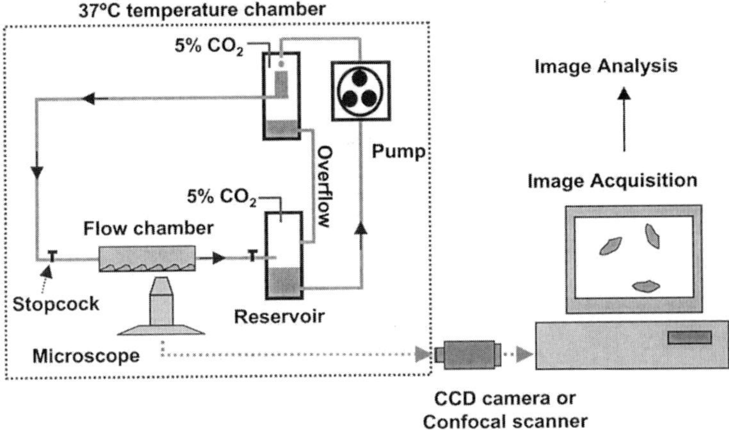

Fig. 4. The setup of the flow system for time-lapse microscopy. A glass slide cultured with ECs was assembled into a flow chamber connected to a flow system. The laminar flow across the flow chamber was driven by hydrostatic pressure between the two reservoirs, and the hydrostatic pressure was maintained by a peristaltic pump. The medium in circulation was ventilated by air with 5% CO_2. The flow system was set up in a 37°C temperature hood built around the microscope. For phase-contrast time-lapse microscopy, the stage was motorized in X, Y, and Z direction, which allowed the scanning of the cells at different locations. The images were acquired by a Hamamatsu Orca100 digital CCD camera and stored in the computer for image analysis. For high-resolution fluorescence confocal microscopy, a confocal scanner was used to collect fluorescence images at different time points.

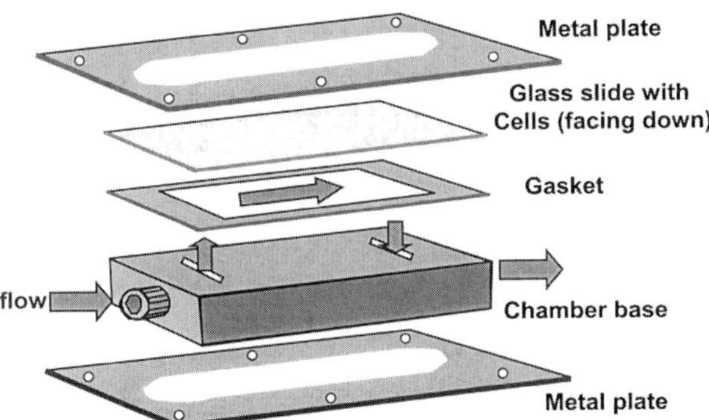

Fig. 5. The setup of a parallel flow chamber. A polycarbonate chamber base was connected to the tubing of the flow system (*see* **Fig. 4** for details). A silicone gasket was placed onto the chamber base with the two slits exposed. A glass slide with cells facing down was placed on the gasket to form a chamber. The flow chamber was sandwiched by two metal plates and tightened with six screws.

5. The peristaltic pump was turned on at a speed that was high enough to maintain the hydrostatic pressure between the two reservoirs and the flow through the system. The system was checked to make sure there were no air bubbles and leaking in the flow system.
6. The 5% CO_2 ventilation was connected to the medium reservoirs (*see* **Note 4**). A flowmeter was used to control the airflow.
7. The flow rate was calculated for specific shear stress: $Q = \tau W h^2/6\,\mu$, where Q is the flow rate (e.g., 3.7 mL/min) across the flow chamber, τ is the shear stress on the slide surface (e.g., 12 dyn/cm^2), W is the width of the flow chamber (the inner width of the gasket, e.g., 15 mm), h is the height of the flow chamber (the thickness of the gasket, e.g., 0.125 mm), and μ is the viscosity of the medium (e.g., 0.00755 dyn·s/cm^2 at 37°C).
8. The flow rate across the flow chamber was measured. The stopcock, at the same height as the entry of the lower reservoir, was turned to stop the flow to the reservoir and allow the medium to flow from the side port into a tube. The medium was collected for 30 s, and the weight of the medium was measured using a weighing balance. The volume of the medium was obtained by dividing the weight with the medium density (approx 1 g/mL), and the flow rate was calculated as volume/30 s.
9. The relative height of the two medium reservoirs was adjusted according to the flow rate measurement, and this step was repeated until the desired flow rate was obtained.
10. After the flow rate was calibrated, the flow across the flow chamber was stopped by adjusting the three-way stopcocks, the flow chamber was disassembled, and the empty slide was replaced with a slide with cells. The surface of the chamber base should be wetted with the medium to avoid air bubbles during assembly. After assembly, the flow chamber was turned upside-down, with the cells facing up and the slide next to the objective.

3.1.3. Time-Lapse Phase Contrast Microscopy

1. The flow chamber was secured on the microscope stage (e.g., using adhesive tape), and the flow was started.
2. A 10X phase objective was used to collect images. The exposure time for image acquisition was adjusted.
3. By using image acquisition software, a list of areas of interest (e.g., 10 areas for each sample) was generated by scanning the sample manually and recording the X, Y, and Z position of each area.
4. The time interval between two cycles of image acquisition (e.g., 10 min) and the number of cycles (e.g., 37 cycles for 6 h) were defined, and the time-series of image acquisition was started. Because prolonged exposure of cells to the transmission light might have an adverse effect, we selected an area outside the chamber to be the first scanning area that was exposed to the transmission light during the waiting period between two cycles of scanning. Alternatively, an automatic shutter for the transmission light could be installed to block the light path during waiting time.

5. At the end of the experiment, the flow chamber was disassembled, and the cells on the slide could be used for further biochemical analysis if needed (*see* **Note 5**). The time-series images were either output as an Avi movie or individual image files.

3.1.4. Image Analysis

The cell migration in wound closure and individual cell model was quantified by different methods, as described in **Subheading 3.1.4.1.** and **3.1.4.2.**, respectively.

3.1.4.1. QUANTITATIVE ANALYSIS OF WOUND CLOSURE

Wound closure was quantified by measuring the cell migration at the wound edge as described below. Alternatively, wound closure could be quantified by measuring the area of the wound covered by the cells (*see* **Note 6**).

1. The time-series movie was replayed to make sure there was no sample/stage movement. For any movement, the displacement of three reference points in the area were measured and this information was used for correction in **step 3**.
2. The image of the first frame ($t = 0$) was thresholded using NIH Image software, and the centroid position (X, Y) of each cell at the wound edge was measured.
3. The centroid positions of the same cells (in **step 2**) were measured after a certain time period (e.g., 6 h). The sample/stage movement was corrected if there was any.
4. The migration distance of each cell was calculated, and the average migration distance of the cells at the wound edge (AMD) was obtained.
5. **Steps 1–4** were repeated to calculate AMD for both sides of the wound at different areas. The average and standard deviation of AMD were calculated, and AMD were compared with those in other samples by using analysis of variance /multiple comparison testing (more than two samples) or *t*-test (between two samples).

3.1.4.2. QUANTITATIVE ANALYSIS OF INDIVIDUAL EC MIGRATION

Individual cell migration was quantified by using software with particle tracing function. From the path of each cell, the migration distance, speed, direction, and persistence (migration distance/change of direction/time interval) were calculated.

1. The time-series movie of an area was replayed, and the sample/stage movement was corrected in DIAS by manually defining the reference points in the frames.
2. The movie was thresholded to identify individual cells, and each cell was traced throughout the frames automatically. The path of each cell was generated from the positions of the centroids at different time points.
3. The cell paths were edited according to the movie. Cells that were too close to each other might confuse the cell-tracing algorithm, and their paths were either

manually corrected or excluded from data analysis. Dividing cells and cells in contact with each other during migration were excluded. The cells that went out of focus plane were also excluded.

4. The data were output from the software in spreadsheet format. The pixel number was converted into µm by scale calibration. The migration distance, speed, direction, and persistence were calculated for each time interval and each cell.

5. **Steps 1–4** were repeated for different areas. The migration distance, speed, direction, and persistence at particular time points or within a certain period could be compared with those in other samples using analysis of variance/multiple comparison testing or t-test. The migration direction of cells under flow was described by Cos θ (θ was the angle between the cell migration direction and the flow direction).

3.1.5. EC Migration Under Static Condition

EC migration under static direction was used as the control for EC migration under flow. The experiment procedure was similar to that for flow experiments except that cells were grown on chamber slides.

1. Cells were seeded on chamber slides either for wound closure or individual cell migration experiments, as described in **Subheading 3.1.1.**

2. The chamber slide was placed and secured (using adhesive tape) in a 100-mm culture dish (with a hole drilled in the cover), and the dish was secured on the microscope stage. Then CO_2 ventilation was connected to the dish through the hole in the cover.

3. Time-lapse microscopy was performed as described in **Subheading 3.1.3.**

4. Quantitative analysis of cell migration was performed as described in **Subheading 3.1.4.**

3.2. Analysis of the Dynamics of Focal Adhesions Under Flow

1. Expression vector encoding GFP-FAK was transfected into ECs by using GenePorter reagent following manufacturer's protocol (*see* **Note 7**). Briefly, BAECs were seeded in a well (in a six-well plate) at 50% confluency on the day before transfection. For each well (approx 10 cm^2) of BAECs, 4 µg of deoxyribonucleic acid in 0.5 mL of DMEM was mixed with 20 µL of GenePorter reagent in 0.5 mL of DMEM and incubated for 45 min. Then the mixture was added to BAECs and incubated for 5 h. Transfected ECs were used for experiments after 2 d. Usually GFP proteins were visible in BAECs within 5 d after transfection. An EC expressing GFP-FAK is shown in **Fig. 6**.

2. Transfected ECs were seeded on glass slides (24 × 60 mm) that had been coated with matrix proteins as described in **Subheading 3.1.1.** (either wound closure or individual cell migration model).

3. As described in **Subheading 3.1.2.**, the flow system was set up in the temperature hood around the inverted microscope (connected to a confocal microscopy system, e.g., Leica TCL SL).

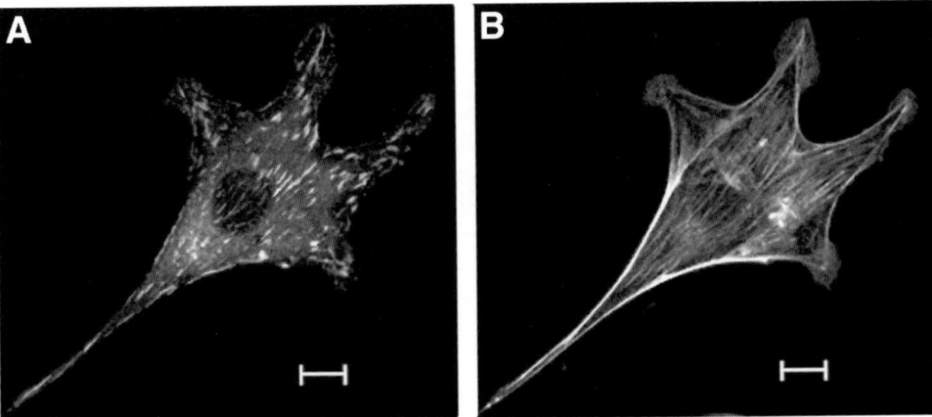

Fig. 6. A cell-expressing GFP-FAK. ECs were transfected with GFP-FAK. Two days after transfection, the cells were fixed with 4% paraformaldehyde, permeabilized with 0.5% Triton X-100 in PBS, and stained on F-actin with rhodamine-conjugated phalloidin. Confocal microscopy was used to visualize the GFP-FAK (**A**) and actin filaments (**B**). GFP-FAK was localized at focal adhesions for the anchorage of actin filaments, or at the leading edge of the lamellipodia, with slight background in cytosol. Bar, 10 μm.

4. The cover glass with cells was assembled into the flow chamber, and the chamber was flipped upside-down and secured on the stage with adhesive tape.

5. A 40X oil immersion objective was used to visualize GFP molecules. The transfected cells were quickly identified by epi-fluorescence microscopy (prolonged exposure of cells to excitation light might cause cell retraction), and a cell with medium expression level of GFP-FAK was selected for the experiment.

6. The light path was switched to confocal microscopy system, and the laser intensity, gain, focus plane and digital zoom were adjusted to optimize the image of GFP-FAK. A section approx 0.3- to 0.5-μm thick was scanned at each time point. The laser intensity and scanning time should be set as low as possible to minimize the adverse effect on the cell.

7. The time-series was started (e.g., with 5-min interval), and the cell was monitored for 30 min under static condition (*see* **Note 8**).

8. The flow across the chamber was started (e.g., 12 dyn/cm^2) with the time-series continued (*see* **Note 9**). The change of cell lamellipodia and FAs could be observed within seconds to minutes, and a shorter time interval (e.g., 1 min) could be used for time-series if necessary (*see* **Note 9**).

9. After the experiment, the images were output as gray-scale TIF files. The movie of FA dynamics could be viewed as time-series with confocal system software or NIH Image (Scion Image) software.

10. By using NIH Image (or Scion Image) software, the image at a certain time point was thresholded, and the FAs were identified as particles. The outline of the particles was edited to correct the artifacts (e.g., because of background) if needed.
11. The centroid (X, Y), area and number of FAs were measured.
12. By playing the movie, the new FAs that appeared in each frame were identified, and the centroids (X, Y), area and number of new FAs were grouped.
13. The time course of total area and number of FAs was generated. The time course of the area and number of new FAs was also generated. The data on FAs under flow was compared with data on FA in the same cells before the application of the flow, or compared with the data on FA in other cells under static conditions.

4. Notes

1. BAEC was isolated from bovine aorta. The aorta was washed in PBS, cut into pieces (approx 3 × 3 cm), and placed in culture dishes. The inner surface of the vessel was incubated with 0.5% collagenase in DMEM for 30 min and the detached cells were collected, spun down, and re-suspended in complete medium. The cells were seeded in 12-well plates and cultured until confluency with medium changed every 2–3 d. The cells that formed a nice monolayer with cobblestone cell shapes were selected and further characterized. The cells incorporated acetylated low-density lipoproteins conjugated with DiI (Molecular Probes; Eugene, OR), and were stained positive for vascular endothelial–cadherin with a vascular endothelial–cadherin antibody (Santa Cruz Biotechnologies; Santa Cruz, CA), suggesting that the cells were ECs. Bovine, porcine, and human ECs from aorta or other vascular beds are available commercially from companies such as Cambrex Corp. (East Rutherford, NJ) and Cascade Biologics (Portland, OR). To dissect the signaling pathways involved in shear stress-induced EC migration, ECs could be transfected with mutants of signaling molecules or pre-treated with specific inhibitors *(5,7)*, and then used for flow experiments.
2. Hydrostatic pressure-driven laminar flow was steady and minimized the vibration of the flow chamber during microscopy, especially for experiments on molecular dynamics that require high magnification (e.g., using 40X objective). For experiments using 10X objective, an alternative setup to hydrostatic system was to use an air chamber to damp the flow oscillation at the downstream of the peristaltic pump. For short-term experiments, syringe perfusion pump instead of peristaltic pump could be used to generate steady laminar flow. Besides steady laminar flow, pulsatile laminar flow could also be generated as needed *(4)*. For example, a syringe pump with oscillatory infusion/withdrawal function could be added to the upstream of the flow chamber. If the experiment was designed for biochemical analysis, multiple flow chambers could be run in parallel, and the flow system could be set up in a cell culture incubator or a temperature hood.
3. Similar flow chambers were available from CytoShear (San Diego, CA) and Bioptechs (Butler, PA). This chamber could be modified to create eddy flow by adding a gasket (with wider boundary at upstream side) between the existing gasket and the slide *(4)*.

4. An alternative to CO_2 ventilation is to use CO_2-independent DMEM (Gibco-BRL). The CO_2-independent DMEM is composed of mono- and di-basic sodium phosphate and a small amount of sodium bicarbonate and can maintain the pH of the medium for at least 24 h without CO_2.

5. After flow experiments, the cells could be used for biochemical assays such as immunostaining of FAs and cytoskeleton and immunoblotting analysis of protein activity. For immunostaining, the cells were washed with PBS, fixed in 4% paraformaldehyde in PBS for 15 min, and permeabilized with 0.5% Triton X-100 in PBS for 10 min, followed by staining with specific antibodies *(6)*. For the analysis of protein amount and activity, the cells were washed with ice-cold PBS and lysed in a buffer containing 50 m*M* Tris-HCl, pH 7.4; 1% Triton X-100; 0.1% sodium dodecyl sulfate; 1% deoxycholate; 150 m*M* NaCl, 10 µg/mL leupeptin, 2 m*M* Na_3VO_4, and 1 m*M* phenylmethyl sulfonyl fluoride, and the lysates were subjected to sodium dodecyl sulfate-polyacrylamide gel electrophoresis and immunoblotting analysis *(14)*.

6. The wound closure could also be quantified by measuring the area of the wound covered by migrating cells. By using NIH Image software, the image at a certain time point t1 was thresholded to identify cells, and the total area of the cells at one side of the wound was measured. The same measurement was performed for the image of this area at the beginning of the experiment t0. The difference of the total area at time t1 and t0 was the area of the wound covered by migrating cells during t1–t0 period.

7. Transfection reagents such as lipofectAMINE or lipofectAMINE PLUS (Gibco-BRL) were also used to transfect deoxyribonucleic acid plasmids into BAECs with similar efficiency. Other GFP proteins, such as GFP-actin and GFP-tubulin, were transfected into BAECs with the same protocol.

8. If the cell would be observed under static condition for more than 30 min, a syringe pump could be used to perfuse the flow chamber (through the side port of the stopcock at upstream) at a very low flow rate (e.g., 0.1 dyn/cm^2) to provide nutrients and O_2/CO_2. If the experiment was to observe FA dynamics under static condition, transfected ECs were seeded in chamber coverglass and used for time-lapse confocal microscopy.

9. When using 40X oil objective to visualize the subcellular structure, the samples might go out of focus due to several factors: stage or flow chamber movement/vibration, flow system-caused vibration and fan-caused vibration. To minimize movement/vibration, the microscope should be placed on an anti-vibration air table. The peristaltic pump and heater/fan should be placed in the temperature hood that does not have contact with the microscope (e.g., attaching the temperature hood to the wall). The flow chamber needs to be secured on custom-made stage adapter. If the cell is out of focus, manual correction with confocal scanning is needed.

Acknowledgment

The author thanks the researchers in his laboratory (especially Steven Hsu and James Moon) at UC Berkeley and Dr. Shu Chien's laboratory at UC San

Diego for their excellent technical assistance during cell migration studies. This work was supported by grants (to S.L.) from Whitaker Foundation and American Heart Association.

References

1. Lauffenburger, D. A. and Horwitz, A. F. (1996) Cell migration: a physically integrated molecular process. *Cell* **84,** 359–369.
2. Sheetz, M. P., Felsenfeld, D. P., and Galbraith, C. G. (1998) Cell migration: regulation of force on extracellular-matrix-integrin complexes. *Trends Cell Biol.* **8,** 51–54.
3. Albuquerque, M. L., Waters, C. M., Savla, U., Schnaper, H. W., and Flozak, A. S. (2000) Shear stress enhances human endothelial cell wound closure in vitro. *Am. J. Physiol. Heart Circ. Physiol.* **279,** H293–H302.
4. Hsu, P. P., Li, S., Li, Y. S., Usami, S., Ratcliffe, A., Wang, X., et al. (2001) Effects of flow patterns on endothelial cell migration into a zone of mechanical denudation. *Biochem. Biophys. Res. Commun.* **285,** 751–759.
5. Hu, Y. L., Li, S., Miao, H., Tsou, T. C., del Pozo, M. A., and Chien, S. (2002) Roles of microtubule dynamics and small GTPase rac in endothelial cell migration and lamellipodium formation under flow. *J. Vasc. Res.* **39,** 465–476.
6. Li, S., Butler, P., Wang, Y., Hu, Y., Han, D. C., Usami, S., et al. (2002) The role of the dynamics of focal adhesion kinase in the mechanotaxis of endothelial cells. *Proc. Natl. Acad. Sci. USA* **99,** 3546–3551.
7. Wojciak-Stothard, B. and Ridley, A. J. (2003) Shear stress-induced endothelial cell polarization is mediated by Rho and Rac but not Cdc42 or PI 3-kinases. *J. Cell Biol.* **161,** 429–439.
8. Chen, K. D., Li, Y. S., Kim, M., Li, S., Yuan, S., Chien, S., et al. (1999) Mechanotransduction in response to shear stress–roles of receptor tyrosine kinases, integrins, and Shc. *J. Biol. Chem.* **274,** 18,393–18,400.
9. Tzima, E., Del Pozo, M. A., Kiosses, W. B., Mohamed, S. A., Li, S., Chien, S., and Schwartz, M. A. (2002) Activation of Rac1 by shear stress in endothelial cells mediates both cytoskeletal reorganization and effects on gene expression. *EMBO J.* **21,** 6791–6800.
10. Jalali, S., del Pozo, M. A., Chen, K., Miao, H., Li, Y., Schwartz, M. A., Shyy, J. Y., et al. (2001) Integrin-mediated mechanotransduction requires its dynamic interaction with specific extracellular matrix (ECM) ligands. *Proc. Natl. Acad. Sci. USA* **98,** 1042–1046.
11. Urbich, C., Dernbach, E., Reissner, A., Vasa, M., Zeiher, A. M., and Dimmeler, S. (2002) Shear stress-induced endothelial cell migration involves integrin signaling via the fibronectin receptor subunits alpha(5) and beta(1). *Arterioscler. Thromb. Vasc. Biol.* **22,** 69–75.
12. Albuquerque, M. L. and Flozak, A. S. (2001) Patterns of living beta-actin movement in wounded human coronary artery endothelial cells exposed to shear stress. *Exp. Cell Res.* **270,** 223–234.

13. Helmke, B. P., Goldman, R. D., and Davies, P. F. (2000) Rapid displacement of vimentin intermediate filaments in living endothelial cells exposed to flow. *Circ. Res.* **86,** 745–752.
14. Li, S., Kim, M., Hu, Y. L., Jalali, S., Schlaepfer, D. D., Hunter, T., et al. (1997) Fluid shear stress activation of focal adhesion kinase—Linking to mitogen-activated protein kinases. *J. Biol. Chem.* **272,** 30,455–30,462.

11

Angiogenesis Assays in the Chick CAM

Chris Storgard, David Mikolon, and Dwayne G. Stupack

Summary

The growth of new blood vessels from pre-existing vascular elements, or angiogenesis, involves coordinated signals to the adhesion, migration, and survival machinery within the target endothelial cell. Agents that interfere with any of these processes may therefore influence angiogenesis. Here, we describe the angiogenesis assay in the chick chorioallantoic membrane (CAM). The CAM is a useful tool to studying angiogenesis because 1) it is amenable to both intravascular and topical administration of study agents, 2) it is a relatively rapid assay, and 3) it can be adapted very easily to study angiogenesis-dependent processes, such as tumor growth. Importantly, the CAM provides a physiological setting that permits investigation of pro- and anti-angiogenic agent interactions in vivo.

Key Words: Chorioallantoic membrane; angiogenesis; inhibitor; blood vessel; chick.

1. Introduction

Angiogenesis is a complex biological process involving the generation of new blood vessels from the pre-existing vasculature. The process involves the cooperation of several different cell types, and in particular involves proliferation, survival, and tissue invasion by activated endothelial cells *(1)*. Angiogenesis plays a role in the pathogenesis of cancer, rheumatoid arthritis, and retinal disease but also may be a clinically important tool for the treatment of ischemia *(2)*. Although isolated elements of angiogenesis may be modeled in vitro, such as in endothelial cell migration, proliferation, or "tube-forming" assays, angiogenesis remains by definition an in vivo process. Accordingly, most investigations of angiogenesis include at least one in vivo study *(3)*.

In this respect, the chick chorioallantoic membrane (CAM) assay possesses several advantages. It is robust, inexpensive, and may be adapted to varied applications, including assessments of tumor, inflammation, or cytokine-

From: *Methods in Molecular Biology, vol. 294: Cell Migration: Developmental Methods and Protocols*
Edited by: J-L. Guan © Humana Press Inc., Totowa, NJ

induced angiogenesis *(4–7)*. The accessibility of the chick CAM allows for topical treatment as well as standard intravenous injections to be used to introduce test compounds ranging from antibodies *(4,7)* to drugs/small molecules *(7–9)*, to viral vectors encoding exogenous genes *(10)*. Here, we describe the use of the CAM as an in vivo assay to study neovascularization. The preparation and placement of cytokine-saturated discs to induce angiogenesis on the CAM as well as the administration of topical and intravenous compounds that impact angiogenesis will be discussed.

2. Materials

1. Egg incubator.
2. Embryonated eggs.
3. Rotary tool with drill bits and cutting wheel.
4. Safety glasses and a dust mask.
5. Filter paper (similar to Whatman type I).
6. Standard 6-mm hole punch.
7. Hydroxycortisone acetate (2.5 mg/mL) dissolved in 95% ethanol.
8. Bent-tip microdissecting forceps.
9. Pointed forceps.
10. Dissecting scissors.
11. Nonpyrogenic saline solution.
12. Sterile gauze.
13. Egg-candling lamp.
14. Dynalyte (Dynalume, McCormick Scientific) or similar flexible hooded lamp.
15. Basic fibroblast growth factor (bFGF; or an alternative proangiogenic growth factor or compound).
16. Test compounds or antibodies suitable for intravenous administration.
17. Mineral oil.
18. 0.5-mL Syringes with 29- to 31-gage 0.5 in. needles.
19. Dissecting microscope (7–10X).

3. Methods

This section describes the set up and analysis of angiogenesis in the CAM using both topical and intravenous administration of angiogenesis-altering compounds. Depending upon whether a given investigation requires intravenous administration, **Subheading 3.2.** and portions of **Subheading 3.5.** may not be required (*see* **Note 1**).

3.1. Preparation

1. Fill the lower pan of the egg incubator with ddH$_2$O to provide a constant humid environment and allow it to reach constant temperature (approx 38°C) by the time the eggs are received.

2. Obtain 10-d-old embryonated eggs for use within the next day.
3. Early on the day of use, dissolve cortisone acetate in 95% ethanol. It may be necessary to vortex and incubate in a 50°C water bath to permit solvation. Four milliliters are typically prepared, which is sufficient for about 72 filter discs.
4. Prepare filter discs using a standard hole punch and then place them in a 70% ethanol solution for 20 min. Allow filter discs to air dry in a laminar flow hood.
5. Place the filter discs in the hydroxycortisone acetate solution, permitting saturation, and then allow the discs to dry in an open Petri dish in a laminar flow hood (**Fig. 1A**). The cortisone coating the discs acts as an anti-inflammatory agent, preventing nonspecific angiogenesis associated with a general inflammatory response *(11)*.

3.2. Candling the Eggs for Intravenous Injection

1. Candle the eggs in a dark room. Eggs are candled by directing the light down the axis of symmetry from either end (apex or air sac) and rotating the egg to locate blood vessels suitable for injection (**Fig. 1B**). For this purpose, a straighter section of vessel with few "branches" originating from the air sac works well (*see* **Note 2**).
2. Using a pencil, trace the appropriate section of vessel, then draw a box framing the region of interest that is approx 10–12 mm × 15–20 mm. This serves as a guide, allowing the shell to then be cut with a rotary tool (**Fig. 1C**).
3. Take care to cut only the shell without penetrating the underlying membrane to prevent hemorrhage and subsequent loss of embryo viability. All four sides of the window should be cut; however, the shell should not be removed at this point. The shell will remain tightly attached via its association with the underlying membrane and will be removed immediately prior to injection. The CAM may now be "dropped."

3.3. Dropping the CAM

The transfer of the air sac from the base of the egg to the upper lateral surface is termed "dropping the CAM."

1. Position the egg so that the vessel selected for injection during candling (*see* **Subheading 3.2.**) is located in the horizontal midline of the egg (*see* **Note 3**). Swab the upper surface and the end of the air sac with 70% ethanol.
2. Drill a small hole into the air sac found at the round end of the egg with the rotary tool (**Fig. 2A,B**) and then drill a second hole on the top surface of the egg (as it lays sideways; **Fig. 2C**). This second hole should be made gently with a rounded drill bit (also called a cluster bit) to prevent damage to the underlying CAM. Depending on the type and source of the eggs as well as local environmental conditions such as humidity, the air sac may transfer spontaneously, either immediately or when the egg is set aside to drill further eggs. This occurs as the air sac evacuates and the entire CAM "drops," to fill this space, in turn creating a new air pocket on the upper surface of the CAM (as air enters through the top hole).

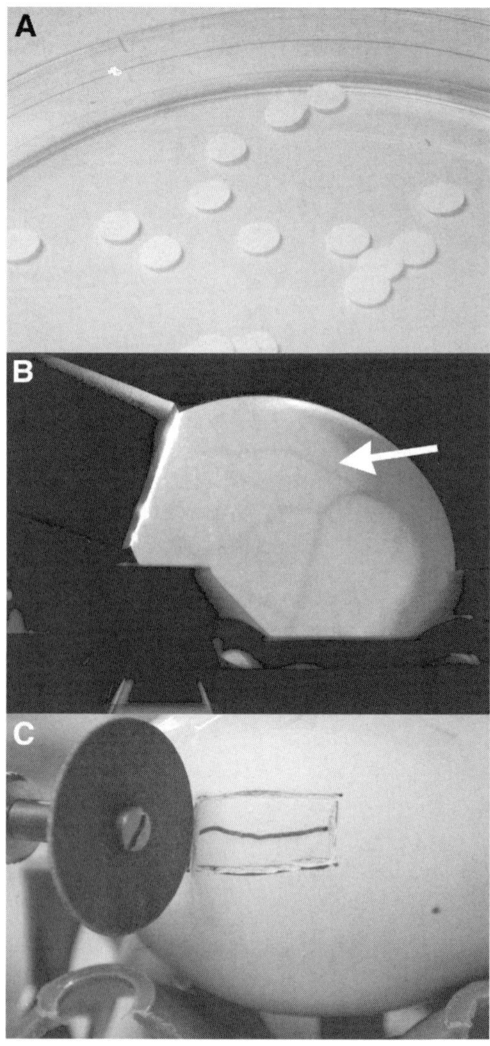

Fig. 1. Initial steps in the CAM assay. Filter disks are punched out, and placed in 70% ethanol, then dried sterilely in a laminar flow hood. (**A**) Dried disks are immersed in hydroxycortisone dissolved in 95% ethanol with sterile forceps and again allowed to air dry in the laminar flow hood. (**B**) Eggs are candled in a dark room to locate vessels suitable for injection. Select an unbranched, lateral vessel traveling along the horizontal midline that is slightly larger than the caliber of the gage of the needle to be used. The arrow shows a potential candidate vessel. The vessel is marked with a pencil and a margin drawn. (**C**) A rotary tool with a cutting bit is used to carefully cut along the pencil line. Care is taken not to penetrate the underlying CAM, which will cause bleeding. After all four sides have been cut, the shell will remain in place and adherent to the underlying CAM until the time of injection.

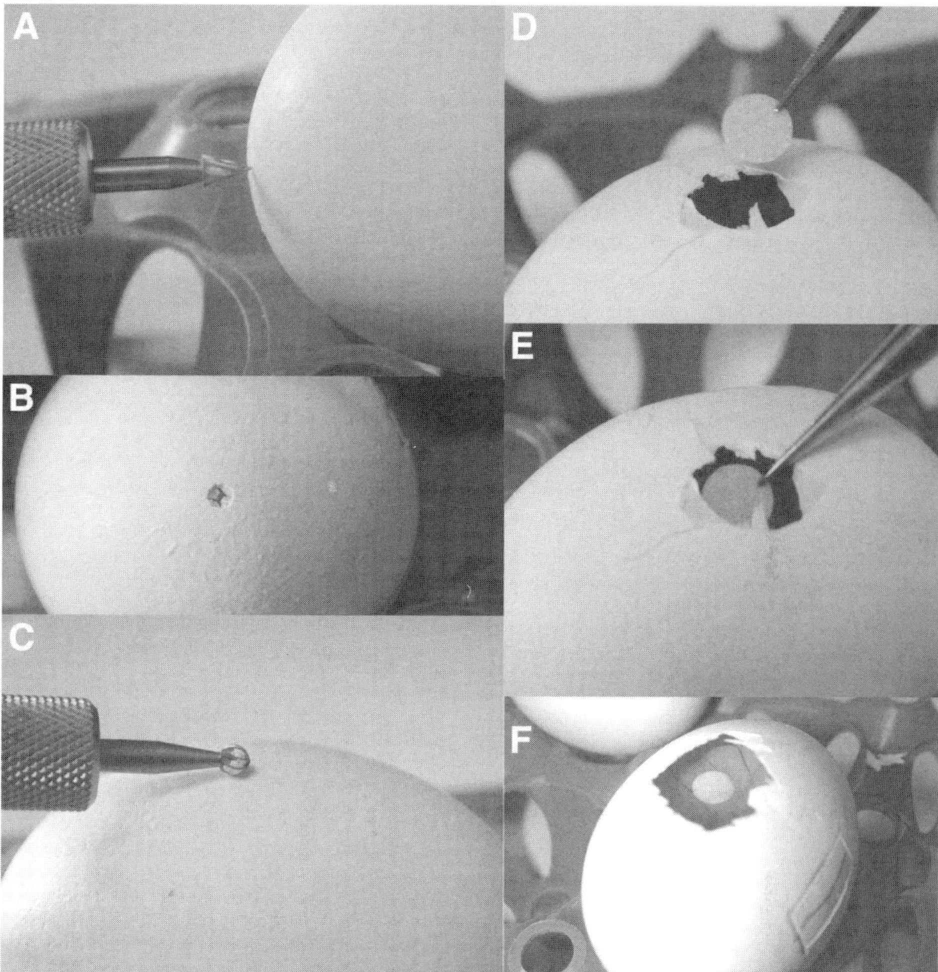

Fig. 2. Preparing the CAM for the placement of the filter disk. Using a sharp drill bit, the initial hole is made in the rounded end of the egg (**A**), in the center of the air sac (**B**). A second hole is then made using a rounded bit (also called a cluster bit) on the side of the egg selected to be the upper surface (**C**). The CAM is then dropped and a small window (10- to 15-mm diameter) is created on the upper side of the egg. Saline or saline containing growth factor (15 µL) is then pipetted onto the filter disk (**D**) and the filter placed without touching the sides of the shell (**E**). Place the disk onto an avascular area containing a pre-existing blood vessel, and seal the upper chamber with transparent tape. (**F**) A disk is shown placed on a dropped CAM in this figure, although the window has been expanded much larger than standard size to allow the filter disk to be seen clearly. The precut window for intravenous injection is clearly visible along the horizontal midline.

3. It is common for the CAM to fail to drop spontaneously. In this case, initiate this process by one of the following methods. Press an evacuated Pasteur pipet bulb against the first (air sac) hole drilled and, allowing it to inflate apply an aspiration pressure. This will initiate the collapse of the air sac, and subsequently the CAM will drop. Alternatively, the bulb may be used to force air into the second (upper) hole on the top of the egg (similarly causing the CAM to drop and evacuation of the air sac). Several additional methods to drop the CAM may also be used, as is convenient (*see* **Note 4**).

4. If uncertain as to whether a CAM has dropped, this can be readily determined by briefly candling the eggs; the new air pocket will be prominent and visible below the second hole.

3.4. Placement of the Angiogenic Filter Disc

1. Gently remove the shell surrounding the upper hole above the dropped CAM, using forceps, to create a small window (approx 10–15 mm diameter) that allows access to the upper chamber and the surface of the CAM (*see* **Fig. 2D**). This window can also be made with a rotary tool; however this tends to introduce eggshell dust onto the surface of the CAM.

2. Place the eggs in the incubator for 30–60 min prior to the addition of the filter discs. This time period allows evaporation of excess moisture from the surface of the CAM, which might otherwise cause displacement of the applied filter discs or permit dilution of cytokines impregnated in the filter discs. However, take care that the CAM does not become too dry during this period, as this may interfere with the induction of angiogenesis. In general, when handling and transferring the eggs, care must be taken to avoid movement of the egg contents, and the window should remain on the upper surface of the egg.

3. Pipet 15 µL of sterile saline containing 100–300 ng of bFGF (or vascular endothelial cell growth factor, VEGF, or an alternative test cytokine), or saline vehicle alone (control) onto each filter disc to saturate it, and place the disc immediately on the CAM in a region that is relatively avascular, yet contains a pre-existing blood vessel (**Fig. 2D–F**).

4. Seal the window closed with transparent tape.

3.5. Topical or Intravenous Administration of Test Compounds

If a test compound is being compared with bFGF or VEGF to determine its capacity to promote neovascularization, the eggs are simply incubated for three days and the assay quantitated (*see* **Subheading 3.6.**). However, many studies involve examining how specific interventions (drugs or antibodies) influence angiogenesis. The administration of these compounds is described here (**Subheading 3.5.**). As an alternative to using a filter disk impregnated with growth factors to induce angiogenesis, a variety of growth factor-secreting tumors may also be grown directly on the CAM (*see* **Note 5**).

3.6. Topical Application

Typically, topical or intravenous administration of test compounds is performed 24 to 48 h after placement of the filter disks. The simpler of these approaches is topical application, and this technique can be used to introduce drugs *(7–9)*, virus *(10)*, or even antibodies *(4,7)* to the CAM to influence angiogenesis. The upper window provides an easy and direct means of introducing test compounds into the immediate microenvironment of the CAM.

1. Remove the tape sealing the upper window. Pipet the compound to be tested slowly onto the filter disk in a total volume of 15 µL, allowing it to soak into the disk.
2. Reseal the tape on the window and record the event on the outer eggshell. Control disks should be treated with an equal volume of the appropriate diluent. Keep the windows open for the minimal time possible during these manipulations. Note that since the CAM is readily accessible, it is possible to repeat topical administration frequently—each day or every 8–12 h, as required.

3.7. Intravenous Injection

Intravenous injection is used to introduce a test agent directly into the chick's circulatory system. This approach is technically more challenging, however, the lack of excretion in the chick embryo system can provide an advantage over other in vivo systems. Compounds that are poorly metabolized, such as monoclonal antibodies, will typically remain in circulation at effective concentrations for the duration of the assay.

1. To perform the intravenous injection, remove the pre-cut shell covering the vessel window (performed in **Subheading 3.2.**). To prevent ripping the underlying CAM during removal of the shell, apply a few drops of mineral oil to the top corners of the precut window and allow the oil to wick under the shell (**Fig. 3A**). Use forceps to gently pry the shell away from the CAM, starting with one of the top corners (**Fig. 3B**). If the shell does not easily separate from the CAM, additional mineral oil should be added behind the shell to help loosen the CAM–shell junction.
2. After the shell is removed, add a final drop of mineral oil to the exposed surface of the CAM. This will render the CAM transparent and will improve visualization of the vessel.
3. Top-light the egg through the upper window to improve resolution of the blood vessel for injection. This is important, since there is very little tactile resistance when injecting a CAM. Injection relies almost completely upon visual cues.
4. Use a 29- or 31-gage needle on a 0.5-mL syringe and inject with the bevel of the needle rotated towards the injector (i.e., "up" relative to the CAM). It may be helpful to bend the needle 20 to 40° before injection. The test compound, diluted in nonpyrogenic saline, should be injected in a volume of about 50 µL. During injection, care must be taken to minimize rotation and movement of the egg.

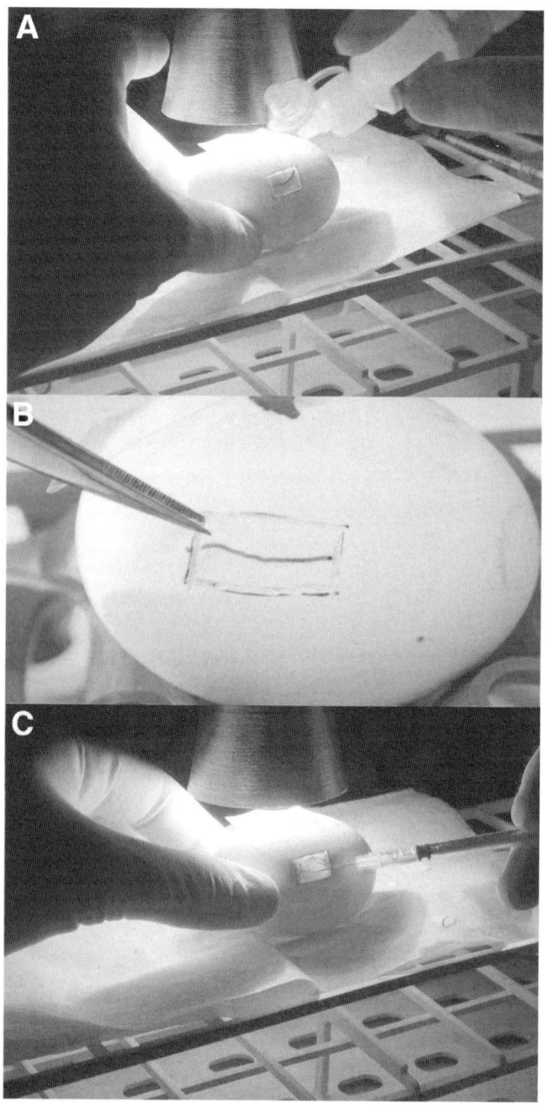

Fig. 3. Intravenous injection of the egg. (**A**) The egg is top lit through the upper window to insure that the vessel initially candled remains accessible. A drop of mineral oil is applied to the upper corners of the precut eggshell (*see* **Fig. 1C**) to loosen the adhesion between the shell and the CAM. Allow a few seconds for the oil to penetrate. Next, place the tip of one side of the forceps behind one corner of the shell obstructing the window (**B**), and gentle pressure is applied to release the shell from the CAM. If resistance is encountered, a few additional drops of mineral oil may be added behind the shell as it is pulled toward you. The shell will often come off in a single piece, but may be removed in several pieces if the underlying CAM is not damaged. *(Continued)*

It is often helpful to position the egg on a platform (egg carton, Styrofoam tray, plastic holder) to free both hands to stabilize the syringe during injection (**Fig. 3C**). Advance the needle slowly and do not go through the opposite vessel wall as this will result in uncontrolled bleeding and loss of the embryo. Note that it can be difficult to differentiate between artery and vein during the candling and thus the injection may occur against the direction of blood flow. (The CAM veins typically attach to the membrane on the "top" surface of an incubator-matured egg, whereas the major CAM arteries run laterally along either side.) It is often possible to determine the direction of blood flow before injection. However, successful injection can be accomplished in either case. The injection is initially performed very slowly, and the vessel observed for "clearing." Injection "with flow" results in a transient clearing of the blood vessel, while injection "against flow" results in clearing that is very brief, or unobservable. If injection has occurred against blood flow, take care and inject extremely slowly to prevent local vessel damage or hemorrhage and loss of embryo viability.

5. After the injection, staunch the bleeding with a small piece of sterile gauze and seal the injection window with tape. Return the egg to the incubator, maintaining the original window on top and the injection window in the horizontal plane. Note that if appropriate controls are used, it is possible to use both topical and intravenous administration in the same egg to examine the combined influence of two interventions.

3.8. Filter Disc Harvest and Quantitation of the CAM Assay

1. On the third day, examine the eggs to confirm embryo movement and viability. Eggs that are nonviable tend to have lost vasculature normally visible on the surface of the CAM, and should not be included in further analysis.
2. Transfer the eggs to ice for 30 min.
3. Expand the upper window with forceps and/or scissors to increase access to the CAM. Remove the filter by cutting the CAM around the perimeter of the filter disk, leaving a margin of about 5 mm. The unused egg and embryo material is considered to be biological waste. Dispose according to institutional guidelines.
4. Rinse the filter plus CAM tissue briefly in PBS, then place on a 35-mm tissue culture dish (or similar) inverted to reveal the tissue below the filter. Take care not to disrupt the adhesion between the filter and the underlying CAM, as angiogenesis will be quantified solely in the area of the CAM covered by the filter. CAMs should be kept moist and on ice in tissue culture dishes until counted.
5. The most common approach to quantification of angiogenesis in the chick CAM is branch counting (*see* **Note 6**). In this approach, the number of vessel branches is counted on a dissecting microscope at a magnification of 7–10X. Begin the

Fig. 3. *(Continued)* (**C**) The egg is injected intravenously at a shallow angle with a 29- or 31-gage needle attached to a 0.5-mL syringe. In this photo, one hand has been moved to allow observation of the window and injection. Both hands are commonly used on the syringe to stabilize it and to allow fine control.

branch counting by selecting a major vessel and tracing it to the first junction. As the originally selected vessel branches, it is considered to "create" two new vessels. Count each of these. This typically involves branches of similar size, however, within angiogenic tissues a small neovessel will often branch from a much larger vessel. Both the small vessel and the large vessel are nevertheless counted as "new" vessels beyond the brand. This convention will increase consistency among counts by different observers.

6. Repeat this process, following the initial vessel as it continues to branch, until it becomes impossible to observe further branches (*see* **Note 7**) or the vessel exits the CAM tissue above the filter disk. Backtrack on the current vessel to the nearest uncounted branch and similarly follow it, counting branches, until again no further branching is seen. The approach is repeated in each area of the CAM until all branches have been counted (examples are shown in **Fig. 4**).

7. It may be convenient to take digital images of the CAM. However, for purposes of counting branch points the limited depth of focus of a single image reduces the number of vessels resolved and counted, resulting in reduced accuracy and sensitivity. Optimally, count the CAMs when fresh, and capture images for archival purposes.

3.9. Observed Variation

Eggs and embryos vary with respect to development, basal vascularity of the CAM, and responsiveness to growth factor stimulation depending upon their source. This may result from environmental as well as genetic factors but also can be influenced by the presence of seasonal or endemic pathogens. Thus, both unstimulated and stimulated CAMs may exhibit some variation in the number of branch points counted. This can be minimized by maintaining a consistent local egg vendor, and by performing a growth factor titration to determine optimal concentrations for neovessel formation prior to embarking on an experimental series.

4. Notes

1. In assays where no intravenous injection is to be performed, there is not an absolute need to candle the eggs; therefore, **Subheading 3.2.** may often be omitted. This will depend on the local egg supply and relative ease with which an avascular region can be located on the surface of the CAM. Moreover, the use of the rotary tool may be unnecessary in these cases because the two holes described can be created with simple needles or dedicated "egg picks." However, the relative availability and accessible cost of cordless rotary tools permits most labs to use these as the "tool of choice" for egg work.

2. We tend to choose the straightest vessels that run roughly along the longitudinal axis of the egg because these permit easiest the injection. When candling the eggs, make certain that the vessels observed are anchored within the CAM tissues. Vessels within the yolk are not viable for injections but may appear so in the candling attempts if they happen to lay against the side of the egg. These

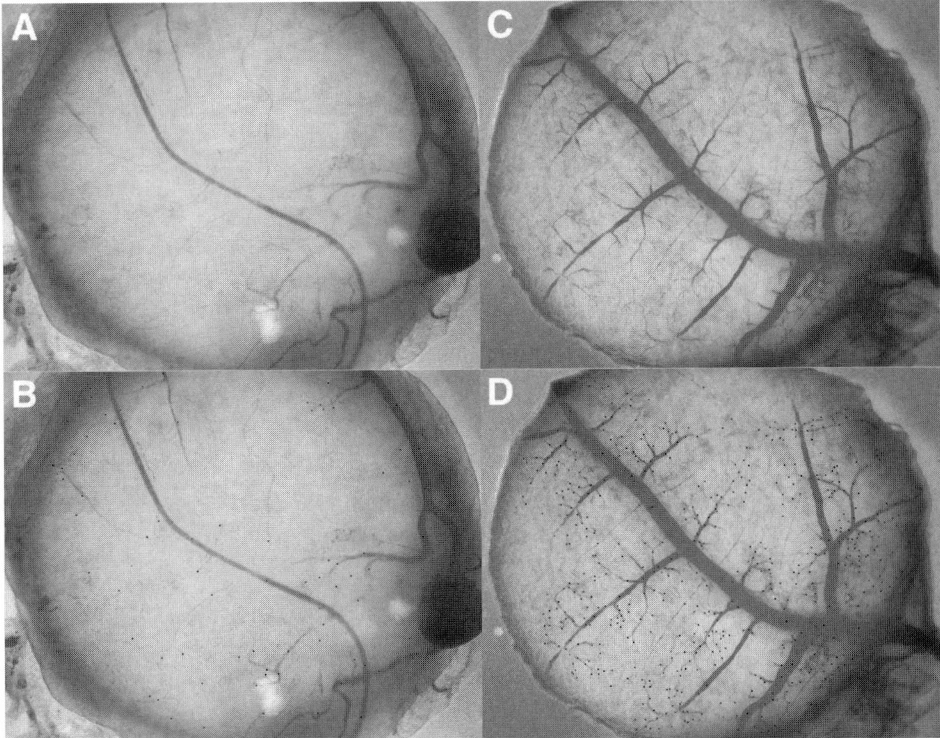

Fig. 4. Quantitation of angiogenesis in the chick CAM. Saline- (**A**) or bFGF- (**C**) impregnated filter disks are placed on blood vessels in otherwise avascular sections of CAM for 72 h to induce angiogenesis. The disks and underlying CAM tissue are then harvested, washed in PBS, and the margins trimmed for photography. Neovascularization was then scored by counting vessel branches present in the CAM tissue below the filter (**B** and **D**). Over 80 branches were counted in the saline control in this case, and more than 500 in the bFGF-treated CAM. If the semiquantitative angiogenic index system was used, "A" would be scored as a 0 to 1, whereas "C" would be scored as a 2–3, depending on the observer and the relative index scale used.

vessels can be made to move by rotating the egg back and forth. Finally, do not select a vessel that is too large in caliber, as it is often very difficult to stop the bleeding from these vessels after the injection. This can sometimes be the case during direct injection of the major CAM vein in older chicks. If eggs are received earlier than d 10–11 and incubated in a rocking incubator within the laboratory, the CAM veins will typically attach to the CAM very near the top surface of the egg. Tracing these vessels during candling will allow a determination of blood flow prior to injection, and is typically indicated by marking an arrow below the window. The optimal vessels to select are those that are slightly larger than the gage of the needle to be used. Vessels of the same size as the needle will usually also accommodate injection, but take extra care during injection not to damage these small vessels (i.e., by pushing the needle through the entire vessel).

3. The window is positioned along the horizontal midline, such that dropping the CAM will not result in evacuation of the region behind the window (i.e., fluid will remain behind the window). If the window is positioned too high, dropping the CAM may influence the position of the vessel, and will at the very least complicate the removal of the eggshell window and the subsequent injection. Alternatively, a window that is too low will present difficulties in accessibility during injection, and may also bleed for longer periods after injection.

4. The ease with which a CAM drops is also dependent on egg maturity. If eggs are received on d 10 but not used for several days, it will become progressively harder to drop the CAM. If investigators are examining spontaneous blood vessel formation in the CAM, younger eggs may be used, and CAMs will tend to drop more easily. There are additional methods available to assist in dropping the CAM. Instead of using a Pasteur pipet to evacuate the air sac as described, brief and gentle vacuum may be applied. Alternatively, mechanical stimulation with forceps can also be used. Bent-tip forceps are inserted into the upper hole such that the point of the forceps is under the shell, but the bent end of the forceps run parallel to the surface of the CAM. The forceps are used as a probe to gently release adhesion between CAM and shell. Simultaneous, gentle downward pressure can be applied to help initiate the drop. Another common alternative for dropping the CAM is to use a 31-gage syringe to penetrate the air sac and CAM itself through the first hole drilled and slowly remove albumin directly. Since these eggs are not allowed to progress to maturity, the loss of 1–2 mL of albumin will not generally influence angiogenesis or embryo development during the assay period.

5. Angiogenesis is commonly induced with cytokine-laden filter disks, as described, or cytokine-impregnated collagen or Matrigel plugs on a physical supporting matrix *(12)*. However, the CAM supports the growth of a variety of tumor types from several species in an angiogenesis-dependent manner *(7,13)*. Tumors can be seeded from tissue culture cells or tumor fragments. In the first case, a cell slurry of $5–10 \times 10^6$ cells are suspended in 20–50 µL (total volume) and spotted onto the CAM. Tumors are allowed to develop for 5–7 d in this case, and intravenous injection or topical administration of test compounds should be delayed for 48–72 h after the tumor is seeded. Alternatively, 50-mg fragments from tumors grown on the CAM 1 wk can be re-seeded onto new CAMs for continued growth or analysis. In these experiments, tumors are generally not scored for angiogenesis by branch counting, but rather overall tumor mass is determined as an "indirect measure" of angiogenesis. The relative vascularity of the tumors in question is determined by subsequent immunohistochemical analysis of tumor sections using endothelial specific markers, such as von Willebrand Factor.

6. Variations in branch counting are sometimes used, for example, counting only the blood vessels that penetrate upwards out of the CAM *(6)*. As a more rapid

alternative to branch counting, some investigators may prefer to simply score each CAM as having a relative angiogenic index as an integer ranging from 0 to 4. This is less time intensive than counting each branch point, and also less sensitive. However, increased experience in branch counting, and the use of multiple blinded observers can partially offset the loss of sensitivity associated with this approach. A typical scale might be CAMs with estimates of "less than 75 branches" as "0," from 75 to 150 as "1," 150 to 300 as "2," 300 to 600 as "3," and more than 600 as "4." However, this will depend on the local supply of eggs. Highly vascular eggs might use 100 as the bottom cutoff, and double up from 100 rather than 75 to form the other four categories. Using this method, saline stimulated CAMs give an average angiogenic index of approx 0.4–1.0, whereas bFGF-stimulated CAMs would have an index of 2.0–3.0, depending on the background level of vascularization in the batch of eggs used and the relative responsiveness to cytokine.

7. If difficulty is encountered resolving small vessels, investigators may wish to fix the blood vessels *in situ* by adding 0.5 mL of 25% glutaraldehyde to the surface of the CAM during the 30-min incubation on ice. This will limit leakage of the red blood cells during the excision and manipulation of the filter disk and adjoining CAM, and can permit greater resolution of small vessels.

References

1. Keshet, E. and Ben-Sasson, S. A. (1999) Anticancer drug targets: approaching angiogenesis. *J. Clin. Invest.* **104,** 1497–1501.
2. Folkman, J. (1995) Angiogenesis in cancer, vascular, rheumatoid and other disease. *Nat. Med.* **1,** 27–31.
3. Auerbach, R., Auerbach, W., and Polakowski, I. (1991) Assays for angiogenesis: a review. *Pharmacol. Ther.* **51,** 1–11.
4. Brooks, P. C., Clark, R. A. F., and Cheresh D. A. (1994) Requirement for vascular integrin $\alpha v\beta 3$ for angiogenesis. *Science* **264,** 569–571.
5. Ingber, D., Fujita, T., Kishimoto, S., Sudo, K., Kanamaru, T., Brem, H., and Folkman, J. (1990) Synthetic analogues of fumagillin that inhibit angiogenesis and suppress tumor growth. *Nature* **348,** 555–557.
6. Nguyen, M., Shing, Y., and Folkman, J. (1994) Quantitation of angiogenesis and antiangiogenesis in the chick embryo chorioallantoic membrane. *Microvasc. Res.* **47,** 31–40.
7. Brooks, P. C., Mongomery, A. M. P., Rosenfeld, M., Reisfeld R. A., Hu, T., Klier, G., and Cheresh, D. A. (1994) Integrin $\alpha v\beta 3$ antagonists promote tumor regression by inducing apoptosis of angiogenic blood vessels. *Cell* **79,** 1157–1164.
8. Eliceiri, B. P., Klemke, R., Stromblad, S., and Cheresh, D. A. (1998) Integrin $\alpha v\beta 3$ requirement for sustained mitogen-activated protein kinase activity during angiogenesis. *J. Cell Biol.* **140,** 1255–1263.
9. Kusaka, M., Sudo, K., Fujita, T., Marui, S., Itoh, F., Ingber, D., et al. (1991) Potent anti-angiogenic action of AGM-1470: comparison to the fumagillin parent. *Biochem. Biophys. Res. Commun.* **174,** 1070–1076.

10. Boudreau, N., Andrews, C., Srebrow, A., Ravanpay, A., and Cheresh, D. A. (1997) Induction of the angiogenic phenotype by Hox D3. *J. Cell Biol.* **139,** 257–264.
11. Jakob, W., Jentzsch, K. D., Mauersberger, B., and Heder, G. (1978) The chick embryo choriallantoic membrane as a bioassay for angiogenesis factors: reactions induced by carrier materials. *Exp. Pathol. (Jena)* **15,** 241–249.
12. Kim, J., Yu, W., Kovalski, K., and Ossawski, L. (1998) Requirement for specific proteases in cancer cell intravasation as revealed by a novel semiquantitative PCR-based assay. *Cell* **94,** 353–362.
13. Zijlstra, A., Mellor, R., Panzarella, G., Aimes, R. T., Hooper, J. D., Marchenko, N. D., and Quigley, J. P. (2002) A quantitative analysis of rate-limiting steps in the metastatic cascade using human-specific real-time polymerase chain reaction. *Cancer Res.* **62,** 7083–7092.

12

Investigations of Neuronal Migration in the Central Nervous System

Michael E. Ward and Yi Rao

Summary

The migration of neuronal precursor cells is essential for the formation of the embryonic nervous system and for the maintenance of the adult nervous system. Modern approaches have greatly facilitated molecular and cellular studies of mechanisms underlying neuronal migration. Here we use the cells migrating from the anterior subventricular zone to the olfactory bulb as a model to discuss in some detail how neuronal migration can be studied. These methods can be adapted to other models of neuronal (or somatic cell) migration.

Key Words: Cell migration; neuronal migration; guidance; molecular cues; imaging.

1. Introduction

Interneurons in the olfactory bulb (OB) are not made in the OB but are derived from neuronal precursor cells born in the anterior subventricular zone (SVZa; **Fig. 1**, just rostral to the lateral ventricles). SVZa cells migrate to the OB via a migratory pathway called the rostral migratory stream (RMS) **refs.** *1–3*; **Fig. 1**, arrows). SVZa migration in the RMS has served as a model for neuronal migration that has been successfully used to uncover cellular *(4)* and molecular *(5)* mechanisms guiding neuronal migration. Our laboratory has used this model and expanded on it by adapting several approaches to study neuronal migration. Here we will describe three methods, which can be used for other systems as well.

The first of these methods is the SVZa migration assay in three-dimensional gels *(4,5)*. This technique makes use of SVZa cells to study neuronal migration in controlled conditions. In this assay, the caudal portion of the RMS is isolated, cut into small explants, and embedded in a collagen/Matrigel mixture.

From: *Methods in Molecular Biology, vol. 294: Cell Migration: Developmental Methods and Protocols*
Edited by: J-L. Guan © Humana Press Inc., Totowa, NJ

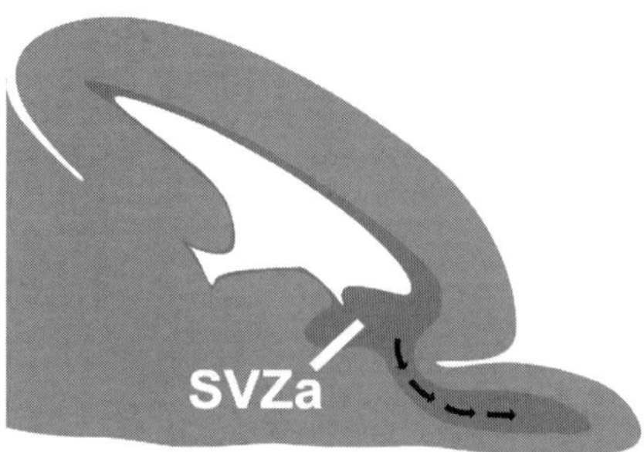

Fig. 1. Diagram showing the anterior subventricular zone (SVZa) and the rostral migratory stream (RMS, arrows). Cells are born in the SVZa and migrate into the OB via the RMS, whereupon they differentiate into olfactory interneurons.

After an overnight culture, numerous SVZa neurons can be seen migrating radially out of the explants into the gel. Cells continue to migrate as preneurons for the next 3–4 d in vitro under these conditions, whereupon they begin to differentiate into postmitotic neurons. Furthermore, these cells can be repelled by the secreted protein Slit *(6)* or attracted by attractants *(7)*, making this an excellent system for the study of directed cell migration. Although this assay has been optimized for the study of SVZa cells, the techniques described here can be used to study migration/axon guidance of other cell types (e.g., cortical or cerebellar neurons) with minimal modifications. We also describe a variation of this protocol that allows for plating of dissociated SVZa cells, which also migrate. This particular variation is often useful for live-cell microscopy and studies involving antibody staining.

We then describe a method to introduce complimentary deoxyribonucleic acid (cDNA) into SVZa cells by transfection of dissociated SVZa cells via Nucleofection (Amaxa). Although we have used retroviruses before *(8)*, our experiences with nucleofection make us feel that this is an easy, fast, and reliable way to transfect primary neurons. Transfection efficiencies between 30% and 60% are almost always obtained (when visualized by Venus EYFP expression). With minor modifications (e.g., a change in the nucleofector program), we have found that this method can be used to transfect other types of primary neurons (e.g., E15 cortical neurons). The main drawback is the price ($300–$400 per kit for 25 transfections), and necessity of a nucleofector machine. After transfection, neurons can be immediately plated onto collagen/Matrigel coated cover slips (*see* **Subheading 3.1.4.**) or reaggregated and co-cultured as explants with HEK293 aggregates in collagen/Matrigel gels.

Lastly, we provide notes on how to image neuronal migration using live-cell microscopy. These are tips and tricks that we have accumulated that may be helpful to those who are new to live cell microscopy. Although most of these guidelines are optimized for imaging of live neurons, they should apply to most mammalian cell types. We attempt to overview the necessary equipment needed to image living neurons, as well as methods to provide the required environment for keeping neurons alive and healthy on a microscope stage.

2. Materials

2.1. 3-D Gel Cultures

1. Hanks balance salt solution (HBSS) with 5 m*M* HEPES and 1X penicillin–streptomycin.
2. Dulbecco's minimal Eagle's medium (DMEM) + 10% heat inactivated fetal bovine serum (FBS), 1X penicillin–streptomycin. Leibovitz-15 media (L-15; Gibco-BRL).
3. Matrigel (BD Biosciences). This Matrigel must be thawed overnight on ice and aliquoted in a cold room into approx 0.5-mL aliquots using cold pipet tips and cold Eppendorf tubes. Store these aliquots at –20°C. Thaw individual tubes overnight on ice when necessary. In our hands, SVZa neurons do not migrate well in Matrigel stored for >2 wk at 4°C after thawing.
4. Collagen purified from rat tails (or from BD Biosciences, Collagen I, rat tail: diluted 1:1 with 0.1X DME, pH 4.0).
5. P1-P6 rat or mouse pups (P3-P5 is optimal).
6. Forceps (no. 5, Fine Science Tools).
7. Small scissors (two pairs, Fine Science Tools).
8. Small rounded metal spatula (Fine Science Tools). Spatula should be cupped, and bent at a 15-degree angle at the base of the flat portion to allow for the easy transfer of tissue.
9. Sharpened tungsten needles (approx 20-µm diameter at tip) and needle holders (Fine Science Tools). Tungsten needles can be sharpened in an approx 1 *M* NaOH bath by putting the negative electrode of a gel power supply (low voltage) into the bath and the positive electrode onto the end of the needle holder. Dip in and out to make a pointed, tapered tip, or submerge for longer periods of time to make a longer, thinner needle. The optimal molarity of the NaOH bath will have to be adjusted empirically, and depends on the particular power supply used. If the NaOH is too concentrated or if the voltage is too high, poor needles will be made. Proper needles should look very smooth under a high-powered dissection scope with a very fine, gradually tapered tip. When the solution/power supply is optimized, a smooth, faint buzzing sound should be heard as the needle is dipped into the solution. It should take approx 1 min to make a single needle (if the needle is sharpened faster than this, dilute the NaOH solution slightly with water).
10. Dissecting scope and Fiber optics light source.
11. Sterile 35-mm dishes.

12. DNaseI (Sigma) 50 mg/mL in 5 mM $CaCl_2$.
13. Trypsin (2.5%; Gibco-BRL).
14. Vibratome Series 800 McIlwain Tissue Chopper (Vibratome Company).
15. Cover slips, assistant grade (Carolina Biological). For all neurons: wash extensively with distilled water, followed by three washes with 100% εtOH. Bake at 500°F for 6 h. For SVZa neurons, coat with collagen/Matrigel (*see* **Subheading 3.1.4.**). For cortical neurons, coat with 50 µg/mL poly-L-lysine for 2 h at 37°C followed by three washes with water. Air dry before plating neurons by placing each cover slip at an angle on the walls of a sterile, dry 10-cm Petri dish in the hood.

2.2. Transfection of Dissociated SVZa Cells by Nucleofection

1. Rat neuron nucleofector solution (Amaxa) and nucleofector device (Amaxa). The use of mouse nucleofector solution with rat neurons or vice versa will work, but expect an approx 10% drop in transfection efficiency. Do not add all of the included supplement to the glass vial of nucleofector solution because of the reduced half-life after supplement addition. We usually add 200 µL of supplement to 900 µL of nucleofector solution at a time—once mixed, this solution is stable for 3 mo at 4°C.
2. 4×10^6 Freshly dissociated SVZa neurons (as per **Subheading 3.1.4.**). P0-P4 pups yield the greatest number of neurons/pup, but pups up to age P8 will work.
3. 3 µg of Qiagen purified DNA (genes driven by a cytomegalovirus promoter works well in these cells). The use of non-Qiagen-purified DNA will result in more cell death and lower transfection efficiencies. Endotoxin-free DNA is usually not necessary.

2.3. Imaging

1. Imaging chamber: this can be as simple as a 35-mm dish with cells plated on the bottom that is heated from underneath or as complex as a perfusion chamber inside of a temperature and CO_2-controlled microincubator built into the microscope itself.
2. Media: phenol-red free L-15 + 10% fetal calf serum (FCS) in ambient air conditions.
3. Temperature control: heated stage and objective heater (if using a high N.A. oil/water immersion lens).
4. Microscope: for phase-contrast transmitted light microscopy, almost any upright or (preferably) inverted scope can be used without any special modifications. All that is required is a heated stage, a CCD camera, and software that allows for time-lapse images to be taken. For wide-field fluorescence microscopy, a more elaborate setup is required. The most important features to consider include: 1) An automated shutter (Uniblitz or Sutter) to limit exposure of cells to light to only the image acquisition periods, 2) A sensitive, cooled CCD camera (e.g., Photometrics Coolsnap HQ or Micromax 512BFT), 3) high-quality objectives, and 4) software (Metamorph, or Open Lab).

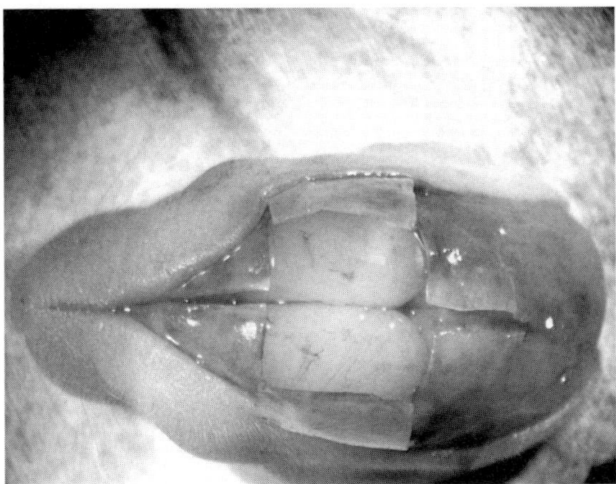

Fig. 2. After euthanizing a rat pup, the skin overlying the skull and the skull itself is cut down the midline. Two perpendicular cuts are then made, and the skull between these cuts is peeled back using forceps.

5. Fluorophore: EYFP, Venus, ECFP, RFP.
6. Specific equipment required for multicolor imaging: either a filter wheel-based system (e.g., Lambda 10-2, Sutter) or a device that allows for simultaneous multicolor imaging (e.g., Dual-View, Optical Insights).

3. Methods

3.1. SVZa Cultures in 3-D Gels

3.1.1. Brain Removal

1. Euthanize a single rat/mouse pup. Make an incision in the skin overlying the skull from the nose to the base of the skull with a pair of scissors. Peel open skin and, using a separate pair of scissors, make the same cut through the skull. Make two additional horizontal cuts on each side of the skull, one rostral and one caudal. This will allow you to peel open the skull, exposing the underlying brain (**Fig. 2**). Be very careful not to damage the delicate OBs: they should be visible at this point. Use the forceps to remove any adhesions between the dorsal part of the OB and the overlying skull (**Fig. 3**). Peel back the skull overlying the OBs. You may also have to use the forceps tips to remove any midline bone that is situated between the OBs before laterally peeling back the skull that directly overlies each OB.
2. Using a sweeping motion, with the forceps roughly parallel to the peeled-back skull that were used to overlay each OB, sever the ventral olfactory nerves (**Fig. 4**). Starting caudally, gently remove the entire brain, using the flat part of the spatula parallel to the base of the skull under the brain (**Fig. 5**). Place into HBSS on ice.

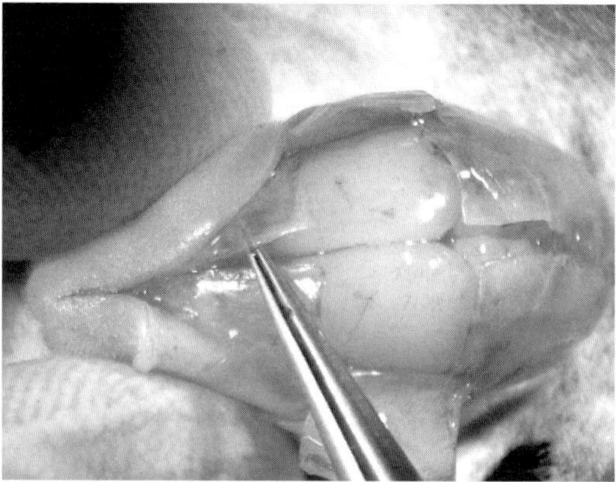

Fig. 3. Dorsal attachments between the OBs and the skull are removed with a superficial caudal to rostral sweeping motion of the forceps.

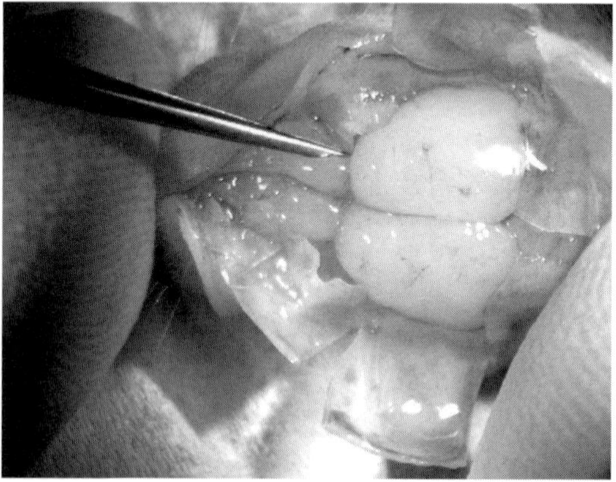

Fig. 4. Ventral attachments between the OB and skull are removed with a lateral to medial sweeping motion of the forceps.

3. Remove the meninges from the OBs by using two pairs of forceps. This is a rather difficult step because the bulbs themselves are more delicate than the meninges, and can easily be torn. However, failure to remove the meninges will make the subsequent steps difficult, if not impossible. This is most easily performed at the lowest magnification of the scope.

3.1.2. Isolating the RMS

1. The RMS, which contains migrating SVZa neurons, is most easily isolated by making a series of coronal slices of the OB with a tungsten needle. First, remove

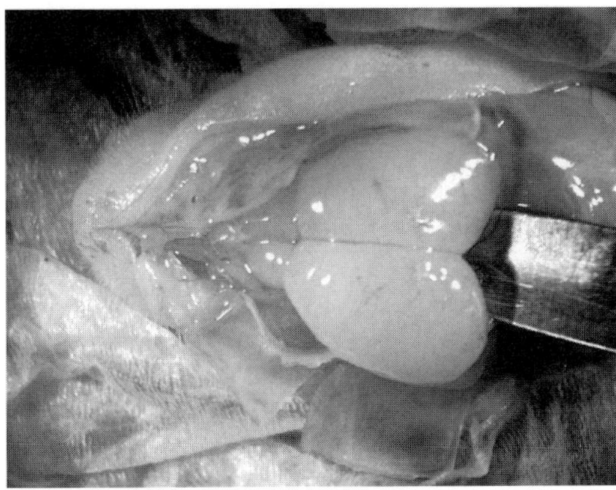

Fig. 5. The brain is removed using a spatula, taking care not to damage the delicate OBs.

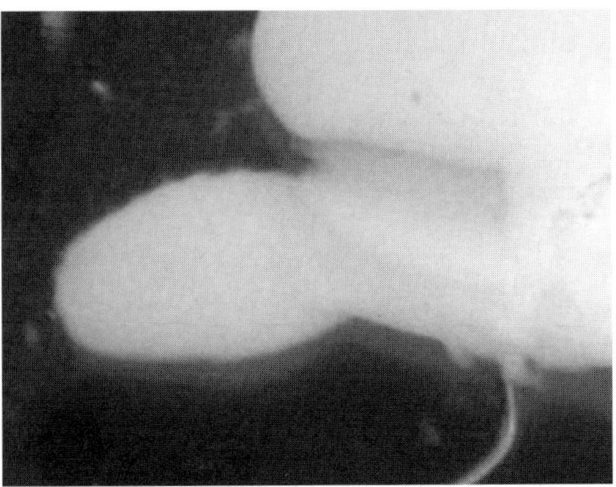

Fig. 6. After removing the meninges from the OB, it is separated from the rest of the brain as shown using a tungsten needle.

the most rostral one third of the OB and throw this away: it doesn't contain much RMS. Then, make three to four coronal slices starting rostrally. The last slice will contain the rostral tip of the lateral ventricle. Using a gentle back-and-forth sawing motion as you cut will help preserve the morphology of the tissue. Transfer the slices into another 35-mm dish containing HBSS. Alternatively, a tissue chopper set at 500 μm can be used for this step, which results in cleaner cuts and a higher yield of tissue. If using a tissue chopper, first cut off the entire olfactory bulb at its most caudal point (**Fig. 6**) and place this tissue directly on the round plastic block of the tissue chopper. Multiple bulbs can be chopped at once by

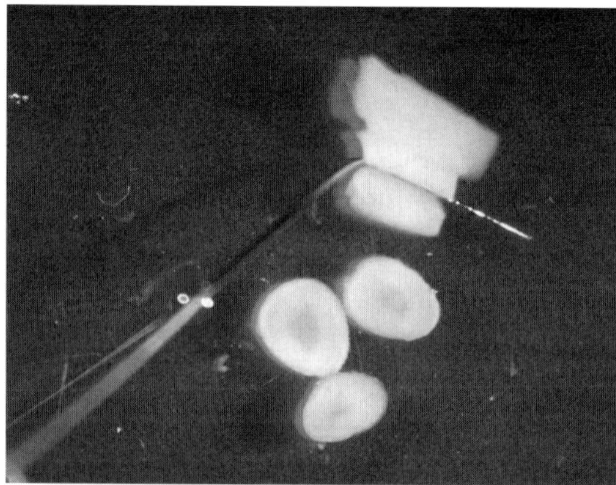

Fig. 7. Coronal slices are then made using a tissue chopper, and a tungsten needle is used to separate these slices after they are submerged in HBSS after the chopping procedure.

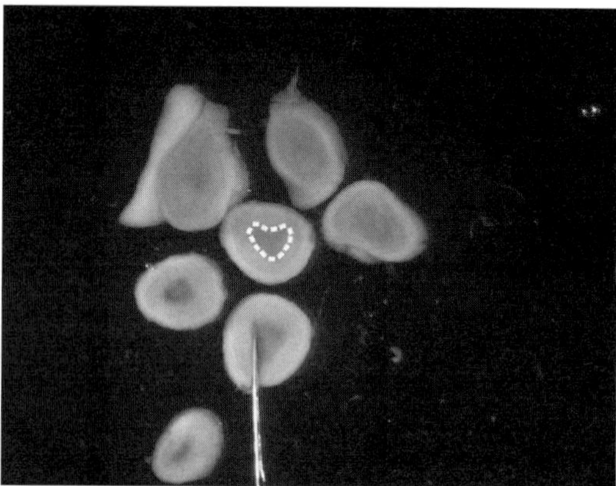

Fig. 8. The RMS is dissected from the rest of the olfactory bulb slice (darker tissue, dotted lines) using a tungsten needle.

lining up the bulbs on the block so that they are all perpendicular to the blade. Remove any excess media before chopping. After chopping, add a drop of HBSS to the cut bulbs on the block. If necessary, use the needle to gently free any slices that have adhered to the block. Add more media to wash the slices into a 35-mm dish containing HBSS. Use the tungsten needle to separate the olfactory bulb slices (**Fig. 7**).

2. Visualize the RMS. This next step is tricky and takes some practice. Arrange the light sources so that the RMS can be distinguished from the surrounding OB tissue. In the early postnatal rat/mouse, the RMS will appear more translucent/

Fig. 9. Explants of the desired size are cut using a sharp tungsten needle. A 200-µL plastic micropipet tip is shown for reference.

dark/shiny than the surrounding tissue and oftentimes resembles a triangle with rounded corners in the middle of each coronal slice (**Fig. 8**, tissue inside of dotted line). In P3 pups, the RMS should be one third to one half the width of each coronal slice at its widest point. Also, because the lateral ventricle actually protrudes into the OB in early postnatal pups, you can sometimes see a small hole in the center of the RMS in each slice, which can be used as a point of reference.

3. If you are planning on embedding the explants in collagen/Matrigel, proceed with the next section (**Subheading 3.1.3.**). If you are planning on dissociating these SVZa cells, proceed to **Subheading 3.1.4.** If you are planning on transfecting these SVZa cells, proceed to **Subheading 3.2.**

3.1.3. Embedding the Explants in Collagen/Matrigel

1. Once you can see the RMS, make cuts with a tungsten needle on the inside of its border (**Fig. 8**): in this manner, you should have six to eight chunks of RMS per pup. Remove all of the extraneous OB tissue with a plastic pipet to prevent confusion later.
2. Cut each RMS chunk into many small (approx 200- to 300-µm wide) explants with a very sharp tungsten needle (**Fig. 9**, yellow pipet tip shown as reference). This is most easily done at the highest power of a dissecting scope. You will probably have >100 explants by the time you are finished. Some protocols benefit from very small explants (more individual cells can be seen), whereas some benefit from larger explants (more cells/explant migrate out, but the spacing is denser).

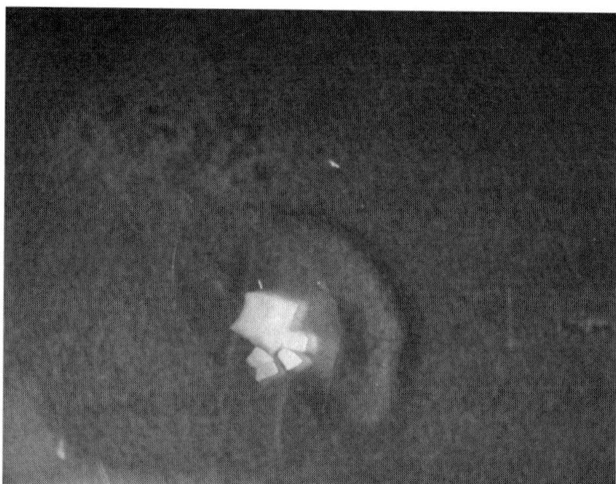

Fig. 10. Explants and a single HEK cell aggregate (if performing a co-culture assay) are pipetted onto a 35-mm plastic dish in a minimal volume of media.

3. Prepare the collagen/Matrigel mixture. We make our own collagen from adult rat tails using standard protocols and dilute it with 0.1X DME, pH 4.0. We have also found that collagen purchased from BD Biosciences (Collagen I, rat tail, 3–4 mg/ mL) works well for explant outgrowth and tends to exhibit less batch-to-batch variation in quality than home-made collagen. The amount you need to dilute it varies slightly from batch to batch and must be determined empirically but is usually one part collagen to one part 0.1X DME for BD Biosciences collagen. Just before embedding explants, prepare collagen in the following manner in the cold room using chilled pipet tips and tubes: mix by pipetting 80 µL of diluted collagen (dilution determined empirically) with 10 µL of 10X DME (without Na Bicarb). Add 1 M NaOH until the color changes to an orange/light pink color (usually around 2.5 µL). Keep on ice when not in use—this collagen will now be good for 30–45 min. Immediately before embedding the explants, mix diluted collagen: matrigel:DMEM 10% FCS in a 3:2:1 ratio thoroughly by pipetting and keep on ice. This ratio can also be varied. Using more Matrigel results in more chain migration of SVZa neurons. Using more collagen results in more individual cell migration. Using too little Matrigel, however, results in fewer cells migrating out into the gel, and can sometimes cause problems with the gelling of the mixture.

4. Transfer explants to the middle of a 35-mm dish in a minimal volume of L-15 (2– 4 µL; **Fig. 10**). Depending on your experiment, you can also transfer an aggregate of HEK293 cells secreting your favorite protein (e.g., Slit) to this dish at this time, also in a minimal volume of media (**Fig. 10**). This is a crucial step: too little media results in explant drying/dying. Too much media results in diluted gel that easily peels off the plate later. When we embed explants in the absence of HEK293 aggregates, we usually use 10–20 explants/dish. When co-culturing

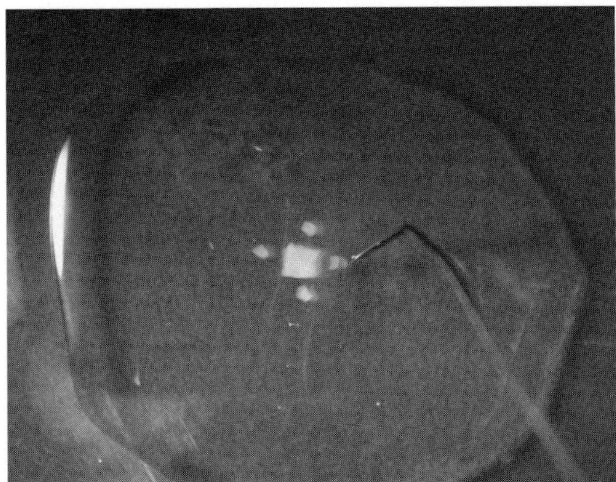

Fig. 11. 15 μL of collagen/Matrigel is added to the explants/aggregate, mixed carefully by pipetting, and the explants are positioned around the HEK aggregate using a dulled tungsten needle. The gel is allowed to harden for 15 min in a tissue culture incubator, followed by the addition of 2 mL of DMEM + 10% FCS.

explants with HEK293 cell aggregates, use one HEK293 cell aggregate and four SVZa explants.

5. Add 10–20 μL of the gel mixture to the top of the explants in the middle of the dish. Gently suck up the explants three to four times, being careful not to make air bubbles. This step ensures that each explant is surrounded by gel, not diluted medium, and increases reproducibility. Using a small pipet tip, smear the gel into a rectangular pad that takes up approx one half of the dish. This is important for two reasons: 1) it reduces the chance of the gel lifting off the dish later and 2) it makes imaging the migrating neurons easier later because more of them migrate out of the explants horizontally. Use a thicker tungsten needle to push the explants around (**Fig. 11**). We try to evenly space them in rows. Try to work quickly during this step because the gel will start to harden as it warms up. The light source can be turned down during this step to reduce its heat output. In addition, we have found that placing ice below the removable dissecting scope stage (with a layer of saran wrap underneath) allows one to manipulate explants more easily and for a longer time before the gel hardens.

6. Optional: if you are co-culturing explants with an HEK293 cell aggregate, use the following procedure. Add 2 mL of DMEM to a 35-mm dish. Use the top of this dish to make "hanging drops" of HEK293 cells: trypsinize an entire confluent plate of HEK293 cells. Spin down at 1000 rpm for 5 min. Remove medium. Resuspend in 200 μL of DMEM. Pipet up and down gently to mix. Add 16-μL drops of these dense HEK293 cells to the inside-top of the dish spaced evenly apart. Carefully invert the top of the dish and place it back onto the dish contain-

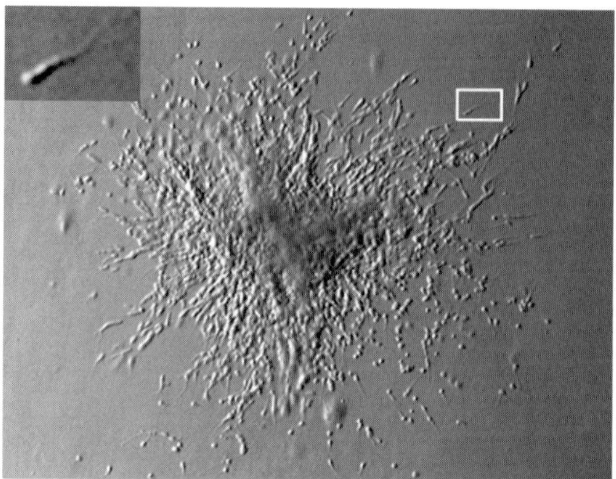

Fig. 12. 12–24 h after embedding the explants, numerous cells should be seen migrating out of the explants into the collagen/Matrigel. Inset shows a single migrating SVZa cell.

ing the DMEM. Incubate at 37°C in a CO_2 incubator for 2 to 6 h. You will see round aggregates of cells in each drop. A stream of DMEM from a plastic pipet can be used to wash these aggregates into the bottom of the dish containing DMEM. Cut into square pieces, which should be roughly the width of the opening of a yellow pipet tip. When placing explants next to aggregates in the gel, position four explants around a single aggregate/dish. The space between the explant and aggregate should be roughly the width of an explant. Approximately 5 min after placing the explants around the aggregate, check to make sure that the explants haven't moved—if they have, gently push the explants back in place surrounding the aggregate with a dulled tungsten needle.

7. Place the 35-mm dishes containing the explants in gel in a 15-cm dish containing a wetted Kim wipe (this keeps the gel from drying out in the next step). Incubate for 12–15 min at 37°C in a CO_2 incubator (although in theory non-CO_2 incubators accelerate collagen gelling). Add 2 µL of prewarmed DMEM/10% heat-inactivated FCS (slowly) to each dish. Incubate overnight. You should see many SVZa neurons migrating out of each explant the next morning (**Fig. 12**; individual neurons shown in inset). They will continue to migrate out of the explants for 3–4 d. If you see the gel pads lifting off, there is probably something wrong with your collagen, or the Matrigel has been stored at 4°C too long. Sometimes, longer gel times solve this problem, but it is likely a problem with the reagents, which will reduce cell migration/axon extension even if gelling occurs.

3.1.4. SVZa Neuron Dissociation and Plating

1. Once you can see the RMS, make cuts with a tungsten needle on the inside of its border. Remove all of the extraneous olfactory bulb tissue with a plastic pipet to prevent confusion later.

2. Dissociate the cells in the following media: 4.5 mL of HBSS, 0.5 mL of 2.5% trypsin, 15 µL of DNaseI. Incubate at 37°C for 15 min. After the incubation, add 5 mL of HBSS + 10% FCS to inactivate the trypsin. Using a large-bore plastic pipet, dissociate the tissue pieces into single cells by pipetting approx 10 times. Fire-polished glass pipettes are not necessary and lead to diminished cell viability. Spin down and wash with 5 mL of HBSS + 10% FCS (FCS helps keep the cells from sticking to each other). Repeat once. Resuspend in DMEM + 10% FCS (heat inactivated) at 5×10^5 cells/mL. Place tube containing cells at 37°C in the CO_2 incubator while cover slips are being coated.

3. Prepare the collagen/Matrigel mixture. We make our own collagen from adult rat tails using standard protocols and dilute it with 0.1X DME, pH 4.0. We have also found that collagen purchased from BD Biosciences (Collagen I, rat tail, 3–4 mg/mL) works well for explant outgrowth and tends to exhibit less batch-to-batch variation in quality than home-made collagen. The amount you need to dilute it varies slightly from batch to batch and must be determined empirically but is usually one part collagen to one part 0.1X DME for BD Biosciences collagen. Just before embedding explants, prepare collagen in the following manner in the cold room using chilled pipet tips and tubes: Mix by pipetting 80 µL of diluted collagen (dilution determined empirically) with 10 µL of 10X DME (without Na Bicarb). Add 1 *M* NaOH until the color changes to an orange/light pink color (usually around 2.5 µL). Keep on ice when not in use—this collagen will now be good for 30–45 min. Immediately before embedding the explants, mix diluted collagen:Matrigel:DMEM 10% FCS in a 3:2:1 ratio thoroughly by pipetting and keep on ice. This ratio can also be varied.

4. Place one cover slip into a single 35-mm dish. Spread 5–6 µL of the gel mixture evenly onto an 18-mm round cover slip with pipet tip.

5. Place the 35-mm dishes containing the cover slips in a 15-cm dish containing a wetted Kim wipe and seal the 35-mm dish with Parafilm (this keeps the gel from drying out in the next step). Incubate for 12–15 min at 37°C in a CO_2 incubator. Do not allow the gel to dry! Alternatively, five cover slips can be coated at a time to expedite the coating procedure: place five baked, untreated 18-mm cover slips in a 60-mm dish with even spacing between cover slips. Spread 5 µL of the collagen/Matrigel mixture using a 10 µL pipet tip onto each cover slip. Try to coat all five cover slips as quickly as possible; this can be facilitated by using the flat portion of the pipet to spread the collagen/Matrigel mixture instead of the tip itself. Cover the 60-mm dish containing the coated cover slips. Moisten a small Kim wipe and drape over the 60-mm dish so that the moistened Kim wipe covers the junction between the top half and the bottom half of the plate. Seal with Parafilm and incubate for 12–15 min at 37°C.

6. Drop 150 µL of the SVZa neurons onto each cover slip. Incubate overnight. We typically use 150 µL of 50×10^4 cells per 18-mm cover slip. By the next morning, you will see that most of the cells have attached and migrated under the surface of the gel (they will have processes). The cells will migrate in this thin three-dimensional environment.

3.2. Transfection of Dissociated SVZa Neurons Via Nucleofection

1. Isolate and dissociate SVZa neurons as described in **Subheading 3.1.4.** After dissociation, wash twice with HBSS + 10% FCS as per these instructions. It is best to use $4–5 \times 10^6$ total neurons per transfection. A minimum of 2×10^6 neurons per transfection will work but with slightly increased cell death. You can expect approx 2×10^6 SVZa neurons from a single P2 rat pup (if using cortical neurons, expect approx 7×10^6 cortical neurons per E15 mouse embryo).

2. Prepare the collagen/Matrigel-coated cover slips, as described in **Subheading 3.1.4.**, during the spin/wash steps if you are plating dissociated cells.

3. After the second wash, resuspend in 5 mL of HBSS + 10% FCS. Put cells on ice until you are at a centrifuge located near the nucleofector device that you can use to pellet your cells. Pellet cells as before, and carefully remove most of the media using a 5-mL pipet. Remove the rest of the media using a P200/yellow pipet tip, being careful not to aspirate any cells in the pellet.

4. Prewarm DMEM + 10% FCS (heat-inactivated).

5. Immediately add the 100 µL of nucleofector solution per 4×10^6 cells (i.e., 100 µL per transfection) and gently resuspend the cells using a P200/yellow pipet tip by pipetting four to five times. Do not press the tip against the bottom of the tube because this will damage the cells, and avoid making air bubbles. We have found that the volumes for each nucleofection can be reduced from 100 µL per cuvet to 50 µL per cuvet, with only a small decrease in transfection efficiency and increase in cell death. If reaction volumes are reduced by half, decrease the total cell number and DNA concentration by half as well. Each electroporation cuvet can be re-used once by steriligation with εtOH and UV treatment in a tissue culture hood.

6. Transfer 100 µL of this cell suspension to an Eppendorf tube containing the 3 µg of DNA. Mix by pipetting four to five times. Transfer to new nucleofector cuvet.

7. Within 5 min, nucleofect these cells using program G-13 (SVZa neurons) or program O-03 (cortical neurons). Cell-specific Nucleofector programs have been designed by Amaxa, and new programs are constantly added by the company. If transfection efficiency is low or cell death is high, to determine if a better/newer program is available, contact Amaxa technical support.

8. Immediately suck up cells using the provided plastic pipet and transfer to the appropriate amount of prewarmed DMEM. The amount of DMEM is different depending on whether you are plating dissociated neurons or reaggregating SVZa neurons (*see* **steps 9** and **10**, respectively).

9. For plating dissociated neurons: assume 50% loss of neurons during this entire process. Resuspend cells so that they are approx 1×10^6 cells/mL. Note that this is a higher cell number than in the nontransfection protocol because some of these cells will die after nucleofection. Plate 150 µL of this solution onto the premade collagen/Matrigel coated cover slips. Incubate overnight at 37°C in 5% CO_2 and then add an additional 2 mL of DMEM + 10% FCS per 35-mm dish. If using cortical neurons, for single cells plate 200 µL of 5×10^4 cells/mL onto 50 µg/mL poly-L-lysine-coated 18-mm cover slips.

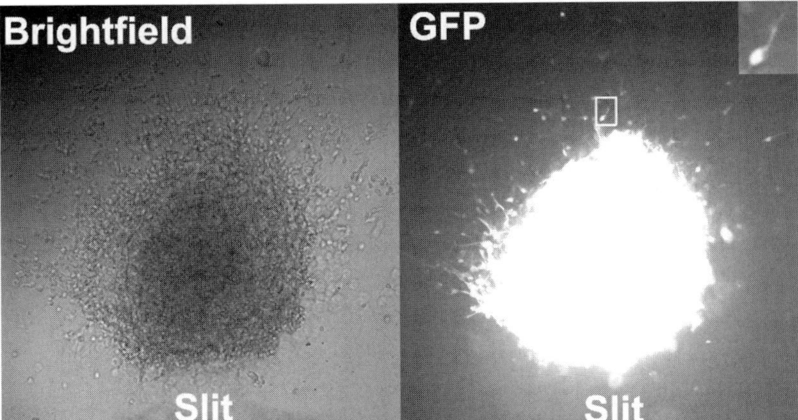

Fig. 13. SVZa cells that have been dissociated, nucleofected with Venus EYFP, reaggregated overnight, embedded in collagen/Matrigel, and co-cultured with Slit-secreting HEK cell aggregates for a further 24 h. Numerous transfected cells (inset) can be seen migrating out of the explant along with untransfected cells away from the Slit source.

10. For reaggregating dissociated neurons and co-culturing in collagen/Matrigel: after nucleofection, resuspend in 5 mL of prewarmed DMEM + 10% FCS (heat inactivated). Spin at 1000g for 5 min. Gently remove excess media as before, leaving approx 30 µL at the bottom of the tube. Resuspend cells in this residual media by pipetting four to five times. Pipet this solution as a bead onto the inside of the top cover of a 35-mm dish (avoid bubbles). Invert this top cover gently and place over a dish containing 2 mL of DMEM + 10% FCS. Incubate this hanging-drop for 2–4 h at 37°C in 5% CO_2. After incubation, gently add 100 µL of prewarmed DMEM + 10% FCS (heat inactivated) to the drop so that the reaggregated cells are beneath the surface of the solution. Suck up this aggregate using a 1000-mL pipet tip and add to a dish containing 2 mL of prewarmed DMEM + 10% FCS (heat inactivated). Incubate these aggregated cells overnight at 37°C in 5% CO_2 to allow for expression of exogenous DNA before co-culture. Cut into explants and co-culture with HEK293 aggregates in three-dimensional collagen/Matrigel as described in **Subheading 3.1.** We usually cotransfect with Venus EYFP at a ratio of 3:1 construct-of-interest, Venus, and then count the green cells on the proximal and distal sides of the explant after co-culture to examine effects on Slit-mediated repulsion (**Fig. 13**, 24 h after co-culture with Slit-secreting HEK293 cell aggregate).

3.3. Imaging Neurons Using Live-Cell Microscopy

Here we describe procedures that can be used to visualize neuronal migration in three-dimensional culture, as well as the effects of chemoattractants and chemo repellents on neuronal migration, through the use of time-lapse micro-

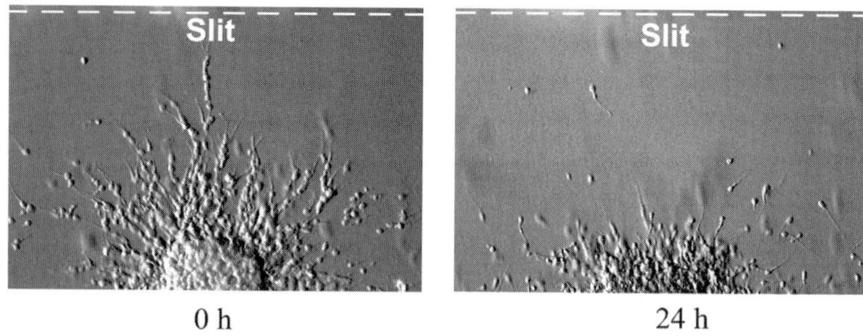

Fig. 14. Time-delayed co-culture of SVZa neurons with Slit-secreting HEK cell aggregates. SVZa neurons were allowed to migrate out of explants in collagen/Matrigel for 24 h, media was aspirated, and then a Slit-secreting HEK cell aggregate was placed next to the SVZa explant (dotted line) and covered with more collagen/Matrigel. The explant was then cultured for another 24 h in DMEM + 10% FCS. Cells that were initially close to the Slit source at the time of aggregate placement were repelled after 24 h of co-culture with Slit. This repulsion can be visualized with live-cell time-lapse microscopy.

scopy. Additional tips related to live-cell microscopy that new users may find helpful are included (*see* **Notes 1–10**).

1. Prepare and embed explants as described in **Subheading 3.1.**, except culture explants without the addition of any HEK293 cell aggregates. If you are planning on transfecting these neurons via nucleofection, proceed as described in **Subheading 3.2.** We usually culture 10–15 explants in each gel pad in evenly-spaced rows in a minimal gel volume. We have found that thinner gels result in more neurons migrating in a single plane of focus during subsequent live-cell microscopy. We usually use plastic 35-mm dishes for transmitted light microscopy, but SVZa explants can be cultured in collagen/Matrigel pads on glass cover slips for high N.A. imaging.

2. After embedding the explants in gel, culture overnight in DMEM + 10% FCS (heat inactivated). The following morning, prepare hanging drops of control or Slit-secreting HEK293 aggregates as described in **Subheading 3.1.** Wait at least 2 h for aggregates to form.

3. Prewarm the microscope imaging chamber and L-15 (phenol-red free) + 10% FCS (heat inactivated). Under optimal conditions, individual cells have usually migrated a distance equal to the diameter of the explant 15–24 h after embedding the explants. At this time, cut the HEK293 aggregates into squares approximately the size of the internal diameter of a yellow pipet tip. Choose the explant that you want to image during the time-lapse experiment, noting its position and orientation in the 35-mm dish. We try to choose explants in which many individual cells on a single side of an explant are in the same plane of focus. Prepare collagen/Matrigel mixture, using the same formula that was used to embed the explants on d 1.

4. Remove all media from the dish containing the explant. Using a yellow pipet tip on a vacuum aspirator, make a dry circle on the dish around the gel pad. Add the HEK293 aggregate to the top of the gel pad in a minimal volume of media. Add 20 µL of collagen/Matrigel mixture to the top of the existing gel pad, and use the pipet tip to smear the gel into the dry circle on the dish-this step will reduce the tendency of the new gel to lift off of the old gel. Using a blunt tungsten needle, quickly position the HEK293 aggregate next to the selected SVZa explant in the proper orientation. Allow the gel to set in a moist chamber at 37°C for 15 min.

5. Slowly add 2 mL of prewarmed L-15 + 10% FCS (heat inactivated) to this dish. Add 2 mL of light mineral oil to the top of this media, making sure that the mineral oil completely covers the L-15. Place dish in imaging chamber and begin taking time-lapse images. SVZa neurons usually turn in response to a newly formed Slit gradient 4–14 h after aggregate placement (**Fig. 14**, SVZa cells before and 24 h after Slit aggregate placement. Note that cells have been repelled by the Slit source).

4. Notes

1. Sterilize all tools for dissecting embryos by spraying with EtOH and air drying.
2. For imaging, we recommend one to try to make the Imaging chamber as simple as possible, given what your cells can tolerate.
3. We have found that most neurons (e.g. cortical, SVZa, cerebellar EGL cells) grow well in phenol red free L-15 + 10% FCS in ambient air conditions. This media is more complete than simple HBSS, allowing for longer imaging sessions (sometimes multiple days). For low-density cell cultures, 2 mL of L-15 per 3-mm dish covered with a thin layer of light mineral oil works well. For slice cultures, perfusion may be necessary due to high oxygen demands, but we have seen good cell migration in slices cultured in non-perfused L-15/mineral oil.
4. For long working distance, dry objectives, heated stages can be used for temperature control. For short working distance, high N.A. oil/water immersion lenses, it is best to use a metal cover slip holder (Warner Instruments makes one). This can either be placed on top of a heated stage, or heated with built in resistive elements. When using immersion lenses, an objective heater (Bioptechs Inc.) must be used in addition to the heated stage, since the objective will act as a heat sink and lower the temperature of the cell being imaged by up to approx 5°.

   For imaging primary neurons (either dissociated or as explants embedded in gels), we have found that the best chambers for transmitted light microscopy can be made as follows: plate cells on 35-mm dishes or cover slips that are then dropped into 35-mm dishes. Create a reusable mineral oil bath by shaving off the plastic ring of the bottom of a plastic 60-mm dish to make a flat surface. Add 1–2 mL of light mineral oil to this 60-mm dish. Place the 35-mm dish containing your cells/2 mL L-15 covered with 2 mL of light mineral oil (Sigma, embryo tested, low viscosity) in this mineral oil bath, ensuring that there are no air bubbles under the 35-mm dish. The purpose of this mineral oil bath is to allow for even heating of the disposable 35-mm dish containing your cells, which is otherwise prevented by the plastic rims on the bottom of most 35-mm dishes (which are poor heat conductors).

The most problematic aspect of these simple approaches is the thermal drift experienced when using high magnification lenses caused by temperature fluctuations in the room the microscope is located in. Often times, it is necessary to constantly readjust the focus during image acquisition because of this thermal-drift, which can introduce artifacts during quantitative microscopy owing to the photobleaching that occurs during re-focusing. The optimal solution to this problem is to use a temperature-controlled plexiglass chamber that encloses the entire imaging area (or at least the objectives and the stage). Using such a chamber will result in highly stable temperature control and minimal thermal drift. Commercial chambers can be purchased (e.g., Zeiss, Solent Scientific), but most on-campus machine shops (or lab workers with some experience using a bandsaw) can build comparable units for less money ($10,000–$20,000 plus, vs $100–$200 for plexiglass, hinges, and hardware, and $1000–$2000 for a temperature-controlled heater). We have had great success using a home-built chamber made with 3/8 in.-thick plexiglass, an Air Theron heater purchased from World Precision Instruments, and two 80-mm computer fans purchased from a local computer store to distribute air evenly throughout the Chamber. We suggest making a cardboard box first, to ensure a snug fit around the microscope, then tracing around these pieces of cardboard on the plexiglass with a marker before cutting on a handsaw.

5. Higher quality cameras will allow for shorter exposure times, resulting in better resolution owing to less motion-induced image smearing, reduced phototoxicity, and the ability to take more pictures before the onset of photobleaching/ phototoxicity (thus resulting in higher frame rates or longer imaging periods).

6. High-quality objectives. Numerical aperture, not magnification, is the most important aspect here, as long as you are using a camera with small pixels (e.g., 7-μm pixels for a 60X 1.4 N.A. lens). The higher N.A. of the lens, the better the spatial resolution and the brighter the image. Upright microscopes are currently limited to 60X 0.9 N.A. water immersion lenses, although higher N.A. long working distance lenses are in development. The trick to using these objectives in live cell microscopy is to immerse the objective into the media, then add a layer of mineral oil to the media to prevent evaporation. As long as the oil is quickly cleaned off after imaging, no damage to the objective should occur.

   Inverted scopes are better in this respect, and lenses up to 1.65 N.A. are now available, although 1.4 N.A. lenses are most commonly used. For longer working distances, and imaging samples where the cells are not plated directly on the cover slips, water immersion cover slip corrected lenses are preferable (1.2 N.A. is currently the max for these lenses).

7. Especially if complicated multicolor fluorescence microscopy is being performed, a more sophisticated software package than what will come with your camera will be required. We are most familiar with Metamorph, which has proven to be an excellent choice, albeit an expensive one. It is relatively user-friendly, and allows for sophisticated macros to be written (called "journals") for even greater flexibility. Another good choice is Open Lab. Zeiss software is probably the most user-friendly, but experienced users may find it to be rather limited given the inability to create macros.

8. The specific fluorophores employed during imaging make a big difference in terms of signal-to-noise. Although enhanced green fluorescent protein (EGFP) is widely used to study protein dynamics, newer variants have proven to be superior for live cell microscopy. We currently use a new enhanced yellow fluorescent protein variant called "Venus" as our tag of choice, which is brighter than EGFP, and matures 20X faster than EGFP. In addition, because Venus is further red-shifted than EGFP, less scattering occurs in the sample being imaged, resulting in a greater signal-to-noise ratio. For multicolor imaging, enhanced cyan fluorescent protein-tagged proteins can easily be distinguished from Venus-tagged proteins in the same cell if the correct filter sets are used. The drawbacks to using enhanced cyan fluorescent protein are low brightness, increased scatter, and faster photobleaching compared with Venus. Alternatively, a new monomeric red fluorescent protein variant has recently been developed by Roger Tsien, known as mRFP1, that should prove to be useful for imaging protein dynamics. The main benefits of using this protein are the ability to add this protein as a third color to cyan fluorescent protein and yellow fluorescent protein imaging, as well as the fact that red-fluorescent proteins can be imaged at deeper levels in living tissue owing to less scattering. The major drawbacks for using this protein are its large size compared with GFP, as well as its relatively low brightness.

9. Specific equipment required for multicolor imaging: either a filter wheel based system (e.g., Lambda 10-2, Sutter) or a device that allows for simultaneous multicolor imaging (e.g., Dual-View, Optical Insights). The benefits of a filter wheel device are high transmission efficiency and the flexibility to use a wide range of filter sets. The filter wheels sold directly by the main microscope manufacturers are relatively slow compared with the Sutter Lambda 10-2 system. The benefits of using a dual-view device is the fact that there are no motion-related artifacts that can occur during the sequential imaging that takes place when using a filter wheel. This is especially important when doing quantitative, multicolor fluorescence experiments, such as fluorescent energy transfer, where even single-pixel motion-related registration artifacts can cause major problems. The drawbacks are the limitation to only doing two-color imaging and the fact that the effective imaging area of the CCD chip is reduced by half. It is also important to consider the light source. Although mercury arc lamps are the most prevalent lamps used in epifluorescent microscopy, they do not provide uniform illumination across the field of view and exhibit large fluctuations in light output over short periods of time. Both of these aspects become problematic when doing quantitative microscopy. To overcome the nonuniform illumination problem, one can use a liquid light guide to couple the lamphouse to the microscope. Using such a device scatters the light, and causes a uniform field of illumination. To overcome output fluctuations, either a feedback mechanism to control input power to the bulb can be used, or one can switch to a Xenon lamp, which has a much more constant output (although this is at the cost of brightness). A new light source has recently become available that combines a mercury-like spectrum and intensity with a stable output and liquid light guide (X-Cite120, εXFO) that we have been quite pleased with.

10. Important things to keep in mind when conducting live-cell fluorescent imaging.
    a. Do not use phase-contrast objectives if possible. The phase ring on the image will substantially reduce the amount of light collected.
    b. Keep light exposure to a minimum. The use of neutral density filters is encouraged when you are trying to find a good cell for imaging. Keep the shutter closed whenever you aren't looking at the cell being imaged, and minimize the time it takes to center and focus the cell being imaged.
    c. If you aren't doing sequential differential interference contrast microscopy and fluorescent microscopy, remove all prisms/analyzers from the optical path.
    d. Make sure the gain settings on the camera you are using are optimized.
    e. Consider using "binning" to increase the sensitivity of the CCD camera. A binning of 2 will increase the sensitivity 4X at the expense of a reduction in resolution by half.
    f. If one experiences high background, check to see if it is caused by reflection off of the transmitted light optics; if so, remove these from the light path.

Acknowledgments

The authors wish to thank other members of the Rao lab for establishing or helping with establishing some of the assays. Research in our lab has been supported by grants from the NIH and the National Brain Tumor Foundation.

References

1. Altman, J. and Das, G. D. (1966) Autoradiographic and histological studies of postnatal neurogenesis I: a longitudinal investigation of the kinetics, migration and transformation of cells incorporating tritiated thymidine in neonate rats, with special reference to postnatal neurogenesis in some brain regions. *J. Comp. Neurol.* **127,** 337–390.
2. Luskin, M. B. (1993) Restricted proliferation and migration of postnatally generated neurons derived from the forebrain subventricular zone. *Neuron* **11,** 173–189.
3. Lois, C. and Alvarez-Buylla, A. (1994) Long-distance neuronal migration in the adult mammalian brain. Science **264,** 1145–1148.
4. Hu, H. and Rutishauser, U. (1996) A septum-derived chemorepulsive factor for migrating olfactory interneuron precursors. *Neuron* **16,** 933–940.
5. Wu, W., Wong, K., Chen, J. H., Jiang, Z. H., Dupuis, S., Wu, J. Y., et al. (1999) Directional guidance of neuronal Migration in the olfactory system by the secreted protein Slit. *Nature* **400,** 331–336.
6. Ward, M., McCann, C., DeWulf, M., Wu, J. Y., and Rao, Y. (2003) Distinguishing between directional guidance and motility regulation in neuronal migration. *J. Neurosci.* **23,** 5170–5177.
7. Liu, G. and Rao, Y. (2003) Neuronal migration from the forebrain to the olfactory bulb requires a new attractant persistent in the olfactory bulb. *J. Neurosci.* **23,** 6651–6659.
8. Wong, K., Ren, X.-R., Huang, Y.-Z., Xie, Y., Liu, G., Saito, H., et al. (2001) Signal transduction in neuronal migration: roles of GTPase activating proteins and the small GTPase Cdc42 in the Slit-Robo pathway. *Cell* **107,** 209–221.

IV

ANALYSIS OF CELL MIGRATION IN MODEL ORGANISMS

13

Analysis of Cell Migration in *Caenorhabditis elegans*

M. Afaq Shakir and Erik A. Lundquist

Summary

This chapter is concerned with a method of analysis and quantification of cell migration defects in mutants of the nematode worm *Caenorhabditis elegans*. The method takes advantage of transgenic expression of the green fluorescent protein to visualize migrating cells. By following these protocols, one will be able to analyze cell migration defects in new mutant strains for comparison to wild-type and to other mutants. Techniques described include obtaining wild-type and mutant worm strains as well as strains harboring green fluorescent protein transgenes; maintenance and manipulation of *C. elegans* in the laboratory; introducing transgenes into different genetic backgrounds; mounting worms for fluorescence microscopy; and scoring and analysis of cell migration defects.

Key Words: Cell migration; *C. elegans*; green fluorescent protein; cell-specific promoter; mutant analysis; transgene.

1. Introduction

The nematode worm *C. elegans* has become an excellent model system in which to study the genetic and molecular control of cell migration, primarily because of the ability to combine genetics and mutant analysis with anatomical imaging *(1–4)*. By comparing cell migration defects of wild-type and mutant strains as well as single mutants to double mutants, the roles of molecules in cell migration and their interactions with other molecules can be very well defined. Information about the biology and experimentation of *C. elegans* can be found in published sources *(5,6)* and on the Internet (*Caenorhabditis* World Wide Web Server; http://elegans.swmed.edu/ and at Wormbase; http://www. wormbase.org). Analysis of cell migration in *C. elegans* is facilitated by the fact that *C. elegans* are transparent and that cells can be visualized with standard light microscopic techniques. Furthermore, *C. elegans* have relatively few

From: *Methods in Molecular Biology, vol. 294: Cell Migration: Developmental Methods and Protocols*
Edited by: J-L. Guan © Humana Press Inc., Totowa, NJ

Table 1
**Cells in *C. elegans* That Undergo Long-Range Migrations
and Genes Whose Promoters Are Active in Those Cells**

Migrating cell(s)	Promoter active in the cell(s)
Posteriorly directed migrations	
ALM neurons (2)	*mec-4 (13)*
CAN neurons (2)	*ceh-23 (10)*
Embryonic co-elomocytes (ccs) (4)	*hlh-8 (14)*
M cell (1)	*hlh-8 (14)*
muIntR (1) (intestinal muscle cell)	*R107.4 (15)*
QL	*unc-73 (16)*
PQR neuron[a]	*osm-6 (17)*
Z1 and Z4 (somatic gonad precursors)	*unc-39* (EAL, unpublished data)
Gonadal distal tip cells[b]	*him-4 (19)*
Anteriorly directed migrations	
HSN neurons (2)	*unc-86 (20)*
QR	*unc-73 (16)*
AQR neuron[c]	*osm-6 (21)*
AVM neuron[c]	*mec-4 (13)*
SDQR neuron[c]	*unc-119 (18)*
Gonadal distal tip cells[b]	*him-4 (19)*
Sex myoblasts (SMs)[d]	*hlh-8 (14)*

[a] PQR is a descendant of the QL cell. Defects in PQR migration could be the result of a Q-cell migration defect or a defect in the migration of PQR itself.

[b] The gonadal distal tip cells are descendants of the somatic gonad precursors Z1 and Z4, which also execute long-range migrations.

[c] AQR, AVM, and SDQR are descendants of the QR cell. Defects in their migration could be the result of a Q cell migration defect or a defect in the migration of the Q-descendants themselves.

[d] The SMs are descendants of the M cell, and defects in their migration could be the result of an M migration defect or a defect in the migration of the SMs themselves.

cells (959 in the soma of hermaphrodites) and fewer still undergo long-range migrations during development (*see* **Table 1**). This relatively simple organization allows a very detailed analysis of cell migration in *C. elegans*. Additionally, genetic screens for cell migration mutants in *C. elegans* have identified many genes involved in this process *(7)*.

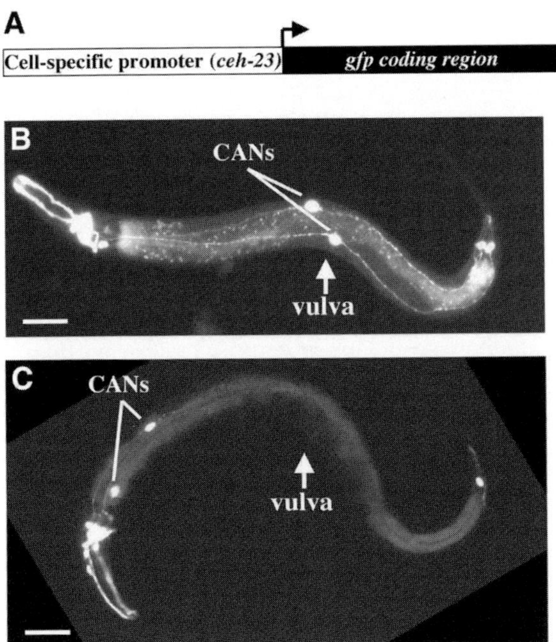

Fig. 1. (**A**) A schematic diagram of a *promoter::gfp* transgene constructed by recombinant deoxyribonucleic acid techniques. The transgene consists of the promoter that drives expression in a cell-specific manner upstream of the green fluorescent protein (*gfp*) coding region. In transgenic animals, *gfp* will be expressed in a cell-specific manner. (**B,C**) Fluorescent micrographs of animals harboring a *ceh-23::gfp* transgene that is expressed in the CAN neurons. Anterior is to the left and dorsal is up. *ceh-23::gfp* is expressed in the CAN neurons (labeled "CANs") as well as some amphid and phasmid neurons (fluorescent cells in the head and tail, respectively). The scale bars represent 50 μm. (**B**) The CAN neuron cell bodies of a wild-type animal are located in the midbody region near the vulva. (**C**) The CAN neuron cell bodies of an *unc-39(e257)* mutant animal have failed in their complete migration and are displaced anterior to the vulva.

Methods used to visualize *C. elegans* cells that undergo migration include differential interference contrast (DIC) microscopy and fixation and staining with antibodies that label particular cell types. The molecular cloning of the green fluorescent protein gene (GFP) from the fluorescent jellyfish *Aquorea victoria* (*8*) has given rise to an additional method to visualize cells in *C. elegans*: transgenic expression of GFP in migrating cells. Transgenes are constructed that consist of a promoter active in the migrating cell of interest driving the expression of GFP (**Fig. 1A**). Cell migration is then scored in transgenic animals. This technique has several notable advantages over the other techniques mentioned.

First, cell migration can be assayed in living animals because GFP visualization does not require fixation. Second, GFP allows a more detailed analysis of cell migration than does DIC microscopy because cells and their nuclei can often be difficult to identify and to locate with DIC microscopy. Third, a large number of well-defined *promoter::gfp* transgenes have been generated in *C. elegans* that are expressed in most migrating cells, whereas antibody staining is limited to a small number of available antisera. In many cases, *promoter::gfp* transgenes drive GFP expression in one cell or in a small subset of cells, whereas antisera often label many cell types, increasing experimental background. An additional advantage of the transgenic GFP approach is that transgenes, once generated, can easily be introduced into different genetic backgrounds to assay the effects of different mutations on cell migration.

Transgenes in *C. elegans* are maintained in two distinct forms: extrachromosomal arrays and integrated arrays *(9)*. Extrachromosomal arrays (designated *Ex* in *C. elegans* nomenclature) consist of repeated copies of the transgenes that are maintained extrachromosomally in a contiguous array. Although extrachromosomal arrays can be transmitted both mitotically and meiotically, their transmission is unstable, and the brood of an animal harboring an extrachromosomal array will contain both array-bearing animals and animals without the array. Therefore, to maintain a strain bearing an extrachromosomal array, it is necessary to select animals that harbor the array in each generation. Selection of GFP-expressing animals is the most common method to maintain a strain harboring a *promoter::gfp* extrachromosomal array. Transgenes can also be maintained as chromosomal integrants (designated *Is* in *C. elegans* nomenclature). Integrants can consist of either extrachromosomal arrays that have been integrated into the genome or of direct chromosomal integrants of one or more tandem copies of the transgene. When homozygous, an integrant strain need not be maintained by selection of GFP-expressing progeny. Methods used to generate novel mutant strains harboring *promoter::gfp* transgenes and to score cell migration in these strains will be presented.

2. Materials

1. *C. elegans* wild-type (N2) and mutant strains.
2. *C. elegans* transgenic *promoter::gfp* strains.
3. Nematode growth medium plates (for 1 L): add 2.5 g of tryptone, 3 g of NaCl, and 15 g of agar to 995 mL of water. Add 1 mL of each of the following sterile solutions: 5 mg/mL cholesterol in ethanol; 1 M CaCl$_2$; 1 M MgSO$_4$; 2 M Tris base; and 3.5 M Tris-HCl. Sterilize by autoclaving and aliquot medium into 15-mm plastic Petri plates.
4. LB bacterial growth medium.
5. M9 buffer with and without 5 mM NaN$_3$ (For 1 L of M9): add 3 g of KH$_2$PO$_4$, 6 g of anhydrous Na$_2$HPO$_4$, and 5 g of NaCl to 1 L of ddH$_2$O. Autocalve. After the solution has cooled, add 1 mL of sterile 1 M MgSO$_4$.

6. 2% Agarose in M9 with 5 m*M* NaN$_3$.
7. A "worm pick" (a 1-inch length of 0.3-mm diameter platinum wire melted into the tip of a borosilicate glass pipet).
8. Glass microscopy slides and cover slips.
9. An "eyedropper" pipet (a borosilicate glass pipet attached to small pipet bulb).
10. A 65°C incubator or temperature block.
11. A compound microscope with fluorescence optics (fluorescein isothiocyanate long-pass or band-pass filter set) and at least a 10X and a 40X magnification objectives.
12. A dissecting microscope.
13. A fluorescence dissecting microscope (fluorescein isothiocyanate long pass or band pass filter set).
14. *Escherichia coli* strain OP50.
15. 20°C and 30°C Incubators.
16. 95% Ethanol.
17. A Bunsen burner or an alcohol burner.
18. A small chemical spatula.
19. A bacterial cell spreader.

3. Methods

The methods discussed describe 1) how to obtain *C. elegans* wild-type and mutant strains as well as transgenic *promoter::gfp* strains, 2) methods of culturing *C. elegans* in the laboratory, 3) strategies by which *promoter::gfp* transgenes are introduced into different novel mutant backgrounds, 4) how to mount *C. elegans* for inspection with fluorescence microscopy, and 5) strategies to score and to statistically analyze defects in cell migration.

3.1. Obtaining C. elegans *Strains, Including Wild-Type and Mutants, as Well as Transgenic* promoter::gfp *Strains*

3.1.1. Obtaining C. elegans *Strains*

Wild-type and mutant strains of *C. elegans* can be requested from the *Caenorhabditis elegans* Genetics Center (Website: http://biosci.umn.edu/CGC/CGChomepage.htm) or from individual *C. elegans* laboratories (Website: http://elegans.swmed.edu/Worm_labs/).

3.1.2. Obtaining Transgenic C. elegans *That Harbor* Promoter::gfp *Transgenes*

Many strains that harbor *promoter::gfp* transgenes also can be obtained from the *Caenorhabditis elegans* Genetics Center or from individual labs. A searchable list of cells and the promoters active in given cells can be found on the Wormbase website (http://www.wormbase.org/db/searches/expr_search). Cells that undergo long-range migrations during *C. elegans* development and the promoters that are active in those cells are listed in **Table 1**.

3.2. Nematode Growth and Maintenance

3.2.1. Seeding Nematode Growth Medium
(NGM) Plates With E. coli *Strain OP50*

Nematodes are cultured on NGM plates on which has been grown a lawn of *E. coli* strain OP50, which can be obtained from the CGC.

1. Innoculate LB liquid medium without antibiotics with OP50.
2. Grow the OP50 culture overnight at 37°C with shaking.
3. From the overnight culture, place a small drop of OP50 (approx 20 μL) in the middle of an NGM plate.
4. Spread the bacteria over the plate using a sterilized cell spreader.
5. Incubate the seeded NGM plates at 37°C or at room temperature until a thin lawn of OP50 has grown.
6. Store the plates at room temperature or at 4°C.

3.2.2. Growth and Maintenance of C. elegans *on NGM Plates*

Strains of *C. elegans* can be maintained by transferring hermaphrodites to seeded NGM plates. Hermaphrodites are self-fertile, and a single *C. elegans* hermaphrodite will give rise to approx 250 progeny (self-fertilization, or "selfing"). Stock cultures of *C. elegans* should be transferred when the OP50 bacterium have all been consumed and the worms begin to starve. *C. elegans* should be cultured at 20°C unless temperature-sensitive mutations call for a different temperature (15°C or 25°C).

C. elegans can be transferred from plate to plate by two means: using a platinum wire "worm pick" or by "chunking," which is removing a portion of the NGM agar containing worms and placing the agar chunk on a seeded NGM plate. Picking is generally used to transfer single animals to plates or to set up matings. Chunking is generally used for routine stock transfer.

3.2.3. "Picking" Worms

1. Sterilize the platinum wire of the worm pick by briefly placing it in a flame (Bunsen burner or ethanol burner). Wait several seconds for the pick to cool before transferring animals.
2. Working on a dissecting microscope, gently place the platinum wire underneath a worm and lift it off the plate (*see* **Note 1**).
3. While observing through the dissecting microscope, quickly transfer the worm to a seeded NGM plate (*see* **Note 2**).
4. Repeat this procedure for each worm picked, sterilizing the pick each time to ensure that the plates do not become contaminated with bacteria or mold from the environment (*see* **Note 3**).
5. Incubate animals at 20°C.

3.2.4. "Chunking" Worms

1. Sterilize the tip of a small chemical spatula by dipping in ethanol and exposing to a flame to burn off the ethanol.
2. Use the spatula end to remove a small chunk of agar from the NGM plate.
3. Quickly transfer the agar chunk to a seeded NGM plate.
4. Repeat for each worm strain, sterilizing the spatula tip between each transfer to prevent cross-contamination of worm strains and to prevent bacterial or fungal contamination.
5. Incubate animals at 20°C.

3.3. Introducing Promoter::gfp *Transgenes* Into Different Mutant Backgrounds

In many cases, it is necessary to introduce the *promoter::gfp* transgene into different mutant backgrounds so that the mutants' effects on cell migration can be determined. This requires a genetic cross, in which a *trans*-heterozygote of the mutation and the transgene is constructed, followed by hermaphrodite self-fertilization and selection of double homozygote of the mutation and the transgene. A genetic cross requires that male worms be mated to hermaphrodite worms to produce cross progeny. Details of this genetic cross are described subsequently.

3.3.1. Heat-Shocking to Generate C. elegans *Males for Crosses*

Males arise in populations of *C. elegans* at a frequency of about 0.2%. This frequency can be increased to 2–5% males by heat-shocking the hermaphrodites during germ cell development *(6)*.

1. Place 25 L4 or young adult hermaphrodites on five seeded NGM plates, five animals per plate.
2. Incubate the plates at 30°C for 6 h.
3. Allow the hermaphrodites to self-fertilize at 20°C.
4. Inspect the progeny for the presence of male worms.
5. Pick the male worms to a seeded NGM plate (*see* **Note 4**).

3.3.2. C. elegans *Matings*

1. On a seeded NGM plate, place the 5–10 males generated in **Subheading 3.3.1.** (*see* **Note 5**). Ensure that no eggs or larvae have been accidentally transferred from the male stock. If so, remove the males to a new seeded plate or remove the eggs and larvae with a worm pick.
2. To the plate containing males, transfer 5 L4 hermaphrodites. Using L4 hermaphrodites ensures that self-fertilization has not yet begun, thus maximizing mating efficiency.
3. Incubate at 20°C.

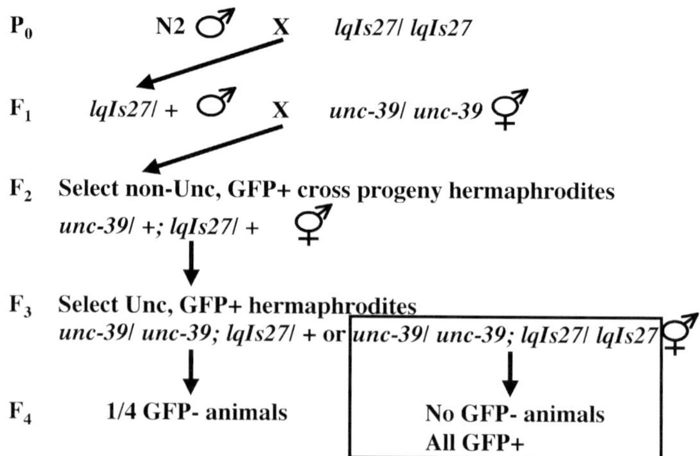

Fig. 2. A scheme to introduce *promoter::gfp* transgenes into different mutant backgrounds. *lqIs27* is the integrated *ceh-23::gfp* transgene and *unc-39* is a recessive mutant. The generations are noted as parental (P₀) and first, second, and third filial (F₁–F₃). The P₀ and F₁ generations involve crosses between males and hermaphrodites. The F₂ and F₃ generations involve hermaphrodite self-fertilization. The double homozygote *unc-39/unc-39; lqIs27/lqIs27* is boxed.

4. On each of the next 2 d, transfer the males and hermaphrodites to a new seeded plate.
5. On d 3 and 4, inspect the plates for the presence of cross progeny (*see* **Note 6**).

3.3.3. Strategy to Introduce promoter::gfp Transgenes Into Different Mutant Backgrounds

Fig. 2 shows a genetic crossing scheme to introduce a *promoter::gfp* transgene into different mutant backgrounds to score cell migration defects. In this example, *unc-39* is the mutation of interest and the *ceh-23 promoter::gfp*-integrated transgene *lqIs27* is the transgene expressed in the canal-associated neurons (the CAN neurons) (*10*). This scheme can be generalized for use with any combination of mutant strain and *promoter::gfp* transgene (*see* **Note 7**). This scheme requires three *C. elegans* strains: wild-type (N2), *unc-39* homozygotes, and *lqIs27* homozygotes.

1. Isolate N2 males as described in **Subheading 3.3.1.**
2. P₀ cross: mate N2 males to *lqIs27* hermaphrodites as described in **Subheading 3.3.2.**
3. F₁ cross: mate *lqIs27/ +* F₁ males (GFP-positive, or GFP+) to *unc-39* homozygous hermaphrodites. All F₁ males from the P₀ cross will be cross progeny, and

those that show GFP fluorescence on a fluorescent dissecting microscope have inherited the *lqIs27* transgene.

4. F_2 generation: using a fluorescence dissecting microscope, pick 5 GFP+ F_2 cross progeny L4 hermaphrodites to seeded NGM plates, one animal per plate. Picking L4 stage hermaphrodites ensures that they have not been crossed by males on the plate. These F_2 hermaphrodites are heterozygous for *unc-39* and heterozygous for the *lqIs27* transgene (*unc-39/ +*; *lqIs27/ +*). Allow the F_2 hermaphrodites to self-fertilize.

5. F_3 generation: approximately one-fourth of the F_3 animals will be homozygous for *unc-39* and approximately three-fourths will be GFP+. However, only one fourth of the animals will be *lqIs27* homozygotes. From one plate of the progeny of the F_2 hermaphrodites, pick 16 F_3 homozygous *unc-39* hermaphrodites (Unc phenotype; *see* **Note 8**) that are GFP+ to seeded NGM plates, one animal per plate. Allow the F_3 animals to self-fertilize. The Unc GFP+ F_3 will be of two genotypes, *unc-39/unc-39*; *lqIs27/ +* and *lqIs27/lqIs27*.

6. F_4 generation: inspect F_4 animals from each of the 16 F_3 plates for GFP fluorescence using a fluorescence dissecting microscope. Those F_3 that were homozygous for *lqIs27* will give rise to all GFP+ animals and no GFP- animals. Those that were heterozygous for *lqIs27* will give rise to approximately one-quarter GFP- animals. Save for analysis a strain that is homozygous for both *unc-39* and *lqIs27*.

3.4. Mounting Animals for Inspection With Fluorescence Microscopy

Once the *promoter::gfp* transgenes have been introduced into a genetic background of interest, cell migration can be assayed. Animals are first mounted on an agarose pad and then observed with fluorescence microscopy.

3.4.1. Prepare an Agarose Pad

1. Melt 2% agarose in M9 buffer with 5 m*M* NaN$_3$ and maintain molten agarose at 65°C. (NaN$_3$ is used to anesthetize the worms).
2. Place two strips of laboratory tape on the undersides of each of two slides as a spacer that determines the thickness of the agarose pad.
3. Align the two spacer slides in parallel and place a third slide in parallel between the two-spacer slides.
4. Place a drop of molten agarose (approx 100 µL) on the middle slide and immediately place a fourth slide on the top of agarose perpendicular to and supported by the spacer slides.
5. After allowing the agarose to solidify for 1 min, slowly remove the bottom slide by gently sliding it from beneath the top slide such that the agarose pad remains on the bottom slide. The agarose pad is ready to use, and can be stored in a humidified chamber for several hours.

3.4.2. Mounting Animals on an Agarose Pad

1. Place 10 µL of M9 with 5 m*M* of NaN$_3$ on the agarose pad.

2. Transfer animals into the M9 drop on the agarose pad using a worm pick. Pick animals from an NGM plate while observing under a dissecting microscope. Complete the transfer of worms onto the agarose pad before the liquid M9 evaporates from the pad (*see* **Note 9**). For most studies young adult animals should be mounted. Different experimental conditions might require that earlier larval stages be analyzed.

3. Place a cover slip slowly and gently on the top of the agarose pad containing the animals (*see* **Note 10**).

3.4.3. Recovering Animals From the Agarose Pad

In case the observed animals need to be saved, they can be recovered from the agarose pad after observation under the compound microscope, even if NaN$_3$ was used to mount them. However, recovery should be done within 30 min to avoid death of the animals by the NaN$_3$ or by desiccation of the slide. M9 without NaN$_3$ also can be used in mounting to avoid NaN$_3$-induced lethality, but the animals will not be anesthetized and will be more difficult to score. Furthermore, the slides can be stored in a humidified chamber to prevent death by desiccation.

1. Under the compound microscope, observe the position of the animal of interest relative to other animals on the slide (a "landmark").

2. Transfer the slide to a dissecting microscope. While observing the animal of interest, place 5 µL of M9 beneath the edge of the cover slip.

3. Slowly slide the cover slip to one side and off of the agarose pad without putting pressure on the agarose pad. While sliding the cover slip, continuously observe the animal of interest under a dissecting microscope.

4. Remove the animal of interest from the agarose pad using a worm pick, and place the animal on a seeded NGM plate for recovery. Often, recovery of the animal can take up to 1 h (the animal might not move for up to 1 h). Addition of 5 µL of M9 over the animal on the NGM plate aids in recovery.

3.5. Scoring Cell Migration Defects

Once the animals have been mounted on an agarose pad, they are ready for observation under a fluorescence compound microscope.

3.5.1. Observing Animals Under the Compound Microscope

1. Place the slide on the stage of a compound microscope.
2. Locate the animals under low magnification and bright field.
3. Turn off the bright field light source and turn on the fluorescence source to observe GFP fluorescence.
4. Switch to higher magnification to observe the cell of interest (usually a 40X objective is powerful enough to visualize GFP-expressing cells).
5. Once the position of the cell of interest has been noted, repeat **steps 2–4** to locate another animal on the slide (*see* **Note 11**).

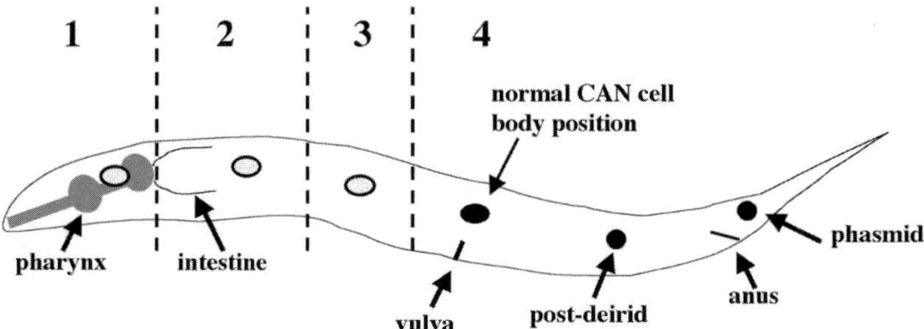

Fig. 3. A schematic diagram of a *C. elegans* adult hermaphrodite is shown. Below the diagram are listed the anterior-to-posterior anatomical landmarks used to judge cell migration defects. The normal position of the CAN cell body near the vulva is indicated; the stippled ovals represent the positions of CAN cell bodies that have failed in their migrations from the anterior to the posterior. The dashed lines demarcate the segments used to score CAN axon pathfinding: 1, the region anterior to the pharyngeal-intestinal connection; 2, posterior to the pharyngeal-intestinal connection but less than half the distance from the pharyngeal-intestinal connection to the vulva; 3, greater than half the distance from the pharyngeal-intestinal connection to the vulva but still anterior to the wild-type position; and 4, the wild-type CAN cell position near the vulva.

3.5.2. Strategies for Scoring Cell Migration Defects

The techniques described here deal with an end-point analysis of cell migration (i.e., scoring the final position of the cell body after migration has ceased). Before scoring cell migration defects, it is important to have a strategy that will be used to judge the extent and severity of cell migration defects. The following considerations should be used in devising a strategy to score cell migration defects:

1. Before observing cell migration in mutant backgrounds, the behavior of a cell of interest in a wild-type background should always be scored first in order to gage the normal variability of the cell position.
2. The position of a cell body of interest should be judged relative to other landmarks of the animals' anatomy. Anatomical landmarks in the anterior–posterior axis of *C. elegans* that can be used as gauges for cell migration defects include, from anterior to posterior: the pharyngeal-intestinal boundary; the vulva, the post-deirid sensilla, the anus, and the phasmid sensilla (**Fig. 3**). For example, the CAN neurons are a symmetrical pair of neurons born in the anterior of the embryo *(11)*. The CAN neurons execute long-range posterior migrations past the pharyngeal-intestinal connection to a final position near the vulva. In some mutant animals, such as *unc-39*, the CAN neurons fail to complete their migrations and remain in position anterior to the vulva (**ref.** *12*; **Fig. 1B,C** and **Fig. 3**).

3. Scoring both the penetrance and the expressivity of a cell migration defect can yield important information. The penetrance of a cell migration defect reflects the proportion of cells that exhibit any defect in migration, whereas expressivity relates to the severity of the cell migration defect. Depending on the situation, it might be necessary to score either one or both of these features. For example, two different mutants might display the same proportion of CAN cells that fail in their migration (equivalent penetrance), but the migrations might fail earlier in one mutant than in the other, resulting in a more severe CAN migration defect and a difference in expressivity between the two mutants. When scoring expressivity of a defect, it is useful to divide the distance of the cell migration into segments and note the position of each individually observed cell relative to this segment system (**Fig. 3**). For example, the CAN neurons migrate from the anterior pharyngeal region posteriorly to a position near the vulva. This distance can be divided into four segments to score CAN migration: the region anterior to the pharyngeal-intestinal connection; less than one half the distance between the pharyngeal intestinal connection and the vulva; greater than one-half this distance but still not the wild-type position; and wild-type cell position (**Fig. 3**).
4. When scoring cell position, it is useful to record the position of the cell relative to a segment system on a rough drawing of a worm containing the anatomical landmarks. This provides a permanent record of the experiment that can be utilized on future occasions.

3.5.3. Statistical Analysis of Cell Migration Defects

When comparing cell migrations of wild-type to mutant strains and mutant strains to one another, it is often important to demonstrate the statistical significance of differences between the strains. The following descriptive statistics are useful in analyzing cell migration defects.

1. Standard error of a proportion. This statistic is useful when comparing the penetrances of cell migration defects (the proportion of cells with defective migration). The standard error of a proportion is based solely on the proportion and the sample size and does not reflect experimental variation. Therefore, it is necessary to score the penetrance of a cell migration defect on at least three different occasions to ensure that similar results are obtained at each occasion and that experimental variation is low.
2. Standard deviation from the mean. This statistic can be used to describe both penetrance and expressivity. For penetrance, the standard deviation of means of experiments done on multiple occasions can be used to reflect variation in experimental conditions. For expressivity, standard deviation can be used to describe the variation in position of cells on segment scale described in **Subheading 3.5.2**. The mean segment position can be calculated by assigning each segment an ordered numerical score (e.g., 1 to 4 in the example in **Subheading 3.5.2.**) and calculating a mean. The cardinal segment position can be determined by rounding the mean segment position to the nearest whole number.

A standard deviation can then be calculated by determining if each individually observed cell was 0, 1, 2, or 3 segments away from the cardinal segment position. For example, mutant cells might display a mean segment position of 2.2. Each cell can then be scored as being 0, 1, or 2 segments away from the cardinal segment position of 2 in this case. A standard deviation can be calculated from these scores (e.g., 2.2 ± 0.5).

4. Notes

1. When fashioning a worm pick, it is useful to use the flat end of a razor blade or another flat object to flatten the end of the platinum wire into a shovel shape. This aids in inserting the pick beneath the animal during transfers. Also helpful in many cases is to pick up a globule of OP50 bacteria on the end of the pick and use it as a "glue" to pick up animals. By touching an animal with the globule of bacteria, surface tension in the drop will trap the animal and facilitate transfer.

2. Animals can be removed from the pick by gently touching the pick to the agar surface and allowing the animal to crawl off. Alternatively, the pick can be gently dragged across the agar surface to displace the animal from the pick to the plate. Avoid disturbing the surface of the agar, as animals will burrow into any break in the agar substrate.

3. When first learning how to pick worms, transfer one animal at a time. Later, when proficiency increases, it is possible to transfer multiple animals at once.

4. Once males have been isolated by heat shocking, it is often useful to start a male stock of the strain, especially of wild-type N2. Using the mating protocol described in **Subheading 3.3.2.**, mate the generated males to hermaphrodites of the same genotype. In this case, one-half of all cross progeny will be male and the males can be used in matings. By setting up a cross of each generation, a male stock can be maintained.

5. To increase mating efficiency, special mating NGM plates can be used. Instead of spreading a lawn of OP50 on the NGM plate, place a small drop (1 μL) of OP50 on the plate and do not spread it. This creates a small bacterial spot in which the animals will stay. For immediate use, a small globule of OP50 can be transferred to an unseeded NGM plate and gently spread in a small spot with a worm pick. The small area of the spots created by these methods increases the likelihood of mating.

6. Whenever designing a *C. elegans* mating, it is important to have a strategy to discriminate cross progeny from self-progeny. Male progeny result from cross-fertilization. If using a male strain harboring a *promoter::gfp* transgene, all progeny harboring that transgene will be cross progeny. Furthermore, if crossing males into a mutant strain with a visible recessive phenotype, all non-mutant, wild-type progeny will be cross progeny.

7. This scheme uses an integrated transgene. If an extrachromosomal array is used, simply select animals at each generation that harbor the transgene with a fluorescence dissecting microscope. The step involving isolation of a homozygous transgene is not required.

8. The phenotype of homozygous *unc-39* animals is visible uncoordinated locomotion. Many mutations affecting cell migration display a visible phenotype that will vary from mutation to mutation.

9. Animals can be removed from the pick by allowing them to swim off the pick into the M9 buffer. Gently dragging the pick through the M9 hastens this process. Multiple animals should be transferred to the pad as judged by evaporation of liquid from the pad.

10. First, put one side of the cover slip at one side of the pad and then, with help of a worm pick, slowly drop the rest of the cover slip over the agarose pad. If the M9 evaporates, there might not be enough liquid to completely fill the space between the agar pad and cover slip. This results in air bubbles under the cover slip, which can obscure observation with a compound microscope.

11. To ensure that all animals on the slide have been scored and that double scoring is avoided, start at one end of the pad and scan up or down in the *x*-axis until the entire length of the slide has been covered. Move the field of view left or right in the *y*-axis and repeat scanning in the *x*-axis. Repeat until the entire slide has been analyzed.

Acknowledgments

We thank E. Struckhoff, Y. Yang, J. Lu, K. Jiang, and S. Mariano for assistance and comments on this chapter and the National Institutes of Health (R01-NS40945) and the National Science Foundation (IBN0093192) for support.

References

1. Lehmann, R. (2001) Cell migration in invertebrates: clues from border and distal tip cells. *Curr. Opin. Genet. Dev.* **11,** 457–463.

2. Blelloch, R., Newman, C., and Kimble, J. (1999) Control of cell migration during *Caenorhabditis elegans* development. *Curr. Opin. Cell Biol.* **11,** 608–613.

3. Montell, D. J. (1999) The genetics of cell migration in *Drosophila melanogaster* and *Caenorhabditis elegans* development. *Development* **126,** 3035–3046.

4. Garriga, G. and Stern, M. J. (1994) Hams and Egls: genetic analysis of cell migration in *Caenorhabditis elegans. Curr. Opin. Genet. Dev.* **4,** 575–580.

5. Riddle, D., Blumenthal, T., Meyer, B., and Priess, J., eds. (1997) *C. elegans II.* Cold Spring Harbor Laboratory Press, Cold Spring Harbor, NY.

6. Wood, W., ed. (1988) *The Nematode Caenorhabditis elegans.* Cold Spring Harbor Laboratory Press, Cold Spring Harbor, NY.

7. Forrester, W. C. and Garriga, G. (1997) Genes necessary for *C. elegans* cell and growth cone migrations. *Development* **124,** 1831–1843.

8. Chalfie, M., Tu, Y., Euskirchen, G., Ward, W. W., and Prasher, D. C. (1994) Green fluorescent protein as a marker for gene expression. *Science* **263,** 802–805.

9. Mello, C. and Fire, A. (1995) DNA transformation. *Methods Cell Biol.* **48,** 451–482.

10. Forrester, W. C., Perens, E., Zallen, J. A., and Garriga, G. (1998) Identification of *Caenorhabditis elegans* genes required for neuronal differentiation and migration. *Genetics* **148,** 151–165.

11. White, J. G., Southgate, E., Thomson, J. N., and Brenner, S. (1986) The structure of the nervous system of the nematode *Caenorhabditis elegans. Philos. Trans. R. Soc. Lond.* **314,** 1–340.
12. Manser, J. and Wood, W. B. (1990) Mutations affecting embryonic cell migrations in *Caenorhabditis elegans. Dev. Genet.* **11,** 49–64.
13. Driscoll, M. and Chalfie, M. (1991) The *mec-4* gene is a member of a family of *Caenorhabditis elegans* genes that can mutate to induce neuronal degeneration. *Nature* **349,** 588–593.
14. Harfe, B. D., Vaz Gomes, A., Kenyon, C., Liu, J., Krause, M., and Fire, A. (1998) Analysis of a *Caenorhabditis elegans* Twist homolog identifies conserved and divergent aspects of mesodermal patterning. *Genes Dev.* **12,** 2623–2635.
15. Hope, I. A., Albertson, D. G., Martinelli, S. D., Lynch, A. S., Sonnhammer, E., and Durbin, R. (1996) The *C. elegans* expression pattern database: a beginning. *Trends Genet.* **12,** 370–371.
16. Honigberg, L. and Kenyon, C. (2000) Establishment of left/right asymmetry in neuroblast migration by UNC-40/DCC, UNC-73/Trio and DPY-19 proteins in *C. elegans. Development* **127,** 4655–4668.
17. Collet, J., Spike, C. A., Lundquist, E. A., Shaw, J. E., and Herman, R. K. (1998) Analysis of *osm-6*, a gene that affects sensory cilium structure and sensory neuron function in *Caenorhabditis elegans. Genetics* **148,** 187–200.
18. Maduro, M. and Pilgrim, D. (1995) Identification and cloning of *unc-119*, a gene expressed in the *Caenorhabditis elegans* nervous system. *Genetics* **141,** 977–988.
19. Vogel, B. E., and Hedgecock, E. M. (2001) Hemicentin, a conserved extracellular member of the immunoglobulin superfamily, organizes epithelial and other cell attachments into oriented line-shaped junctions. *Development* **128,** 883–894.
20. Finney, M. and Ruvkun, G. (1990) The *unc-86* gene product couples cell lineage and cell identity in *C. elegans. Cell* **63,** 895–905.
21. Collet, J., Spike, C. A., Lundquist, E. A., Shaw, J. E., and Herman, R. K. (1998) Analysis of *osm-6*, a gene that affects sensory cilium structure and sensory neuron function in *Caenorhabditis elegans. Genetics* **148,** 187–200.

14

Analysis of Cell Migration
Using *Drosophila* as a Model System

Jocelyn A. McDonald and Denise J. Montell

Summary

There are a number of reasons to use *Drosophila* as a model system to study cell migration. First and foremost is the availability of an arsenal of powerful genetic techniques that can be deployed, permitting the study of cell migration in vivo, in the context of the entire organism. This is especially important for the study of a complex behavior that can be dramatically affected by small changes in environmental conditions. Several different types of cell migrations occur during *Drosophila* development. In this chapter, we focus on cell migrations that have been subjected to the most intense scrutiny. We describe each of the cell types and their trajectories and provide information regarding markers that are useful for the study of each cell type and mutations that affect their migrations. In addition, we provide protocols for staining embryos and manipulating gene function in each of the migratory populations. Finally, we offer some advice concerning the analysis and interpretation of mutant phenotypes.

Key Words: *Drosophila*; cell migration; primordial germ cells; hemocytes; tracheal cells; border cells; antibody staining; mutant; overexpression.

1. Introduction

The cell migrations described here are, in chronological order, primordial germ cell (PGC) migration, hemocyte migration, and tracheal cell migration, all of which occur in the embryo, and border cell migration in the adult ovary. Cell migrations that are also of interest, but not included here because of space limitations are the movements of midgut cells, salivary gland cells, and glial cells in the embryo and other types of follicle cells in the ovary.

From: *Methods in Molecular Biology, vol. 294: Cell Migration: Developmental Methods and Protocols*
Edited by: J-L. Guan © Humana Press Inc., Totowa, NJ

1.1. PGC Migration

PGC migration is of special interest in part because PGCs in many organisms, including humans, undertake long journeys to meet up with and settle down within the somatic gonad. In *Drosophila*, PGCs are the first cells to form in the developing embryo, and they do so when the rest of the embryo is a syncytium of hundreds of nuclei within a common cytoplasm. In the early stages of PGC development, they are swept into the embryo passively as they hitch a ride on the cells that will form the posterior midgut. This occurs as the rest of the embryo elongates along the anterior–posterior axis and folds over on itself in a process known as germ band extension. Subsequently, they invade the cells of the posterior midgut, clinging to the dorsal side and avoiding the ventral side of this otherwise seemingly symmetrical tissue (**Fig. 1**). They then travel directionally to the mesoderm to contact the specific subset of somatic cells that constitute the gonadal mesoderm (**Fig. 1**). Subsequent gonad development does not really involve motility so much as differential adhesion and compaction of the cells, a process in which the PGCs become centralized within the gonad as a consequence of being surrounded by gonadal mesodermal cells (**Fig. 1**). During their migrations, the PGCs are actively repelled from many areas of the embryo and actively attracted to the cells of the gonadal mesoderm.

1.2. Hemocyte Migration

Hemocytes are macrophage cells responsible for engulfing foreign bodies, including bacteria and apoptotic cells. These cells originate within a patch of ectoderm in the head region of the embryo at stage 8 (**Fig. 2**). They then migrate away from this site in two distinct streams. One stream crosses the amnioserosa, a layer of extraembryonic cells on the dorsal side, and heads into the very posterior end of the germ band extended embryo while the other stream migrates along the ventral side of the embryo, along the midline of the developing ventral nerve cord (**Fig. 2A**). Eventually, the hemocytes spread throughout the entire embryo as the two streams meet in the middle and then cells migrate dorsally up both sides of the embryo (**Fig. 2A**).

1.3. Tracheal Cell Migration

The formation of the tracheal system is of interest because it is representative of tube formation and branching morphogenesis, processes that are important in the development of the vasculature, lung, and breast in mammals. Tracheal cells perform a series of intricate maneuvers to generate the elaborate tracheal tree from a series of segmentally repeated groups of cells known as the tracheal placodes. The cells are first specified as an oval collection of cells in a specific location within each segment of the embryonic ectoderm. As development proceeds, these cells invaginate to form a simple sac. Then, leading cells extend

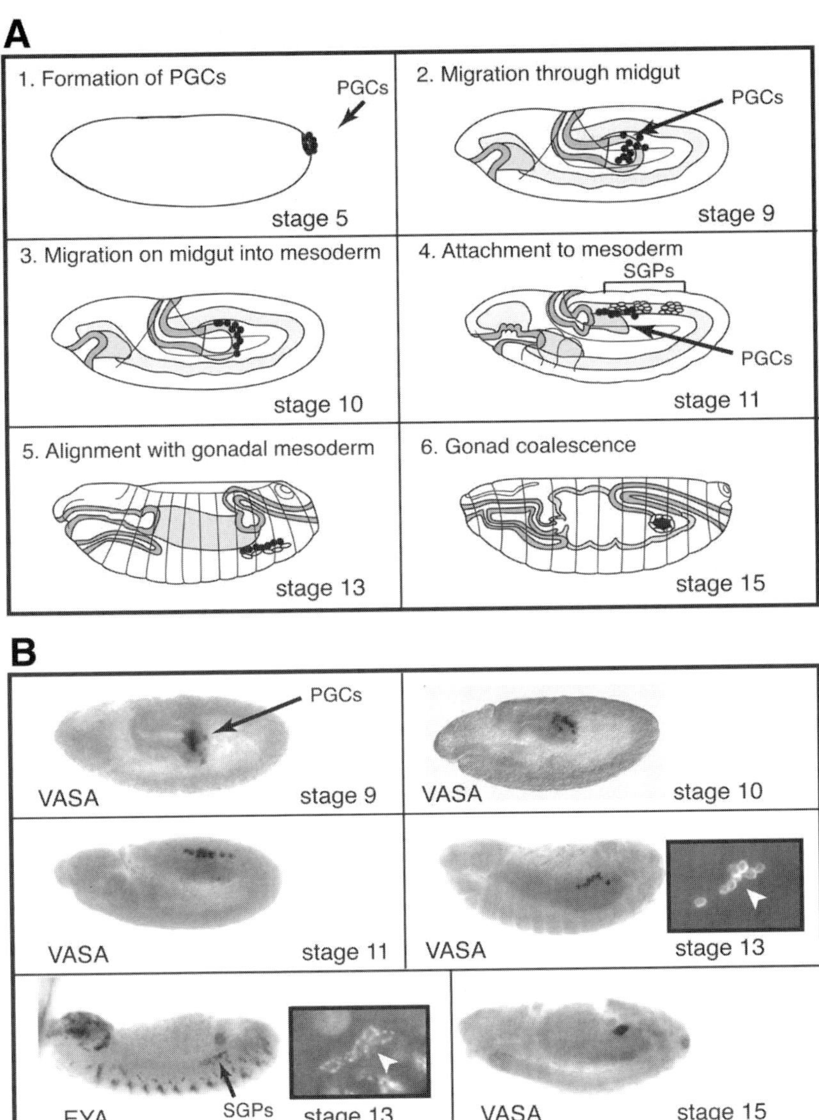

Fig. 1. Migration of the PGCs during embryogenesis. (**A**) Schematic of six stages of PGC migration; the PGCs are in black and the somatic gonadal precursors (SGPs) are in white. Reprinted with permission from **ref. *11***. Between the stages of PGC formation (stage 5) and migration through the midgut (stage 9), the posterior midgut invaginates and the PGCs are passively carried inside the embryo (not shown). (**B**) VASA staining of embryos at embryonic stages 9, 10, 11, 13, and 15, and EYA staining of SGPs at stage 13. Insets show higher magnification of fluorescently stained embryos showing the PGCs or SGPs (arrowheads). Anterior is to the left and ventral is down in all panels.

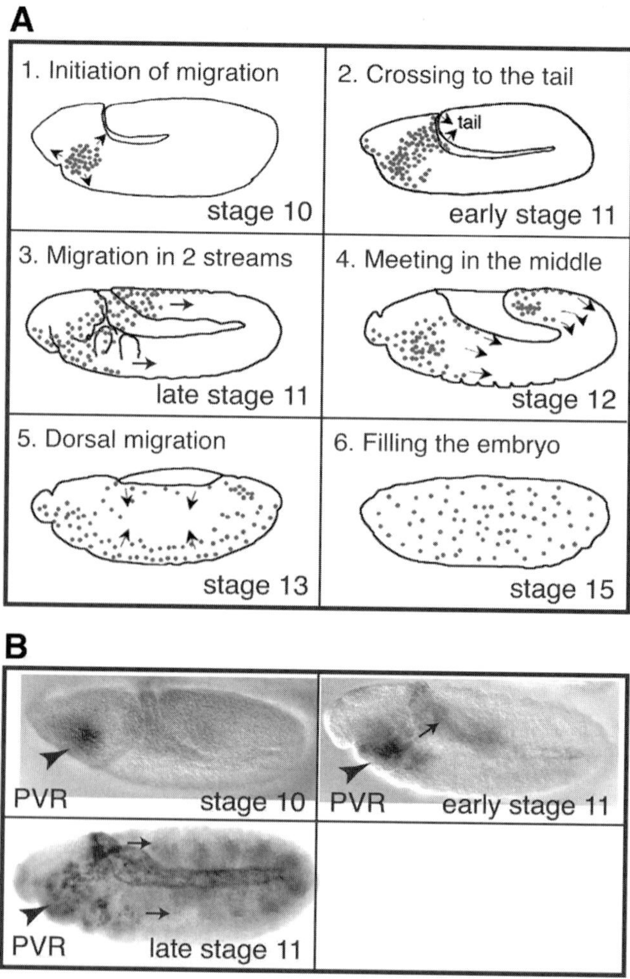

Fig. 2. Migration of the hemocytes during embryogenesis. (**A**) Schematic of six stages of hemocyte migration (*see* text for details). Arrows indicate the direction of the migration. Reprinted with permission from **ref. *16***. (**B**) PVR staining of stage 10 and 11 embryos, showing the hemocytes during the early part of their migration. Arrowheads point to the origin of hemocytes in the head and arrows indicate the direction of migration. Anterior is to the left and ventral is down in all panels.

filopodial protrusions in stereotypical directions and patterns (**Fig. 3**). The cells organize themselves into tubes and generate an interconnected network of branches referred to as ventral, lateral, visceral and dorsal branches as well as the dorsal trunk (**Fig. 3**).

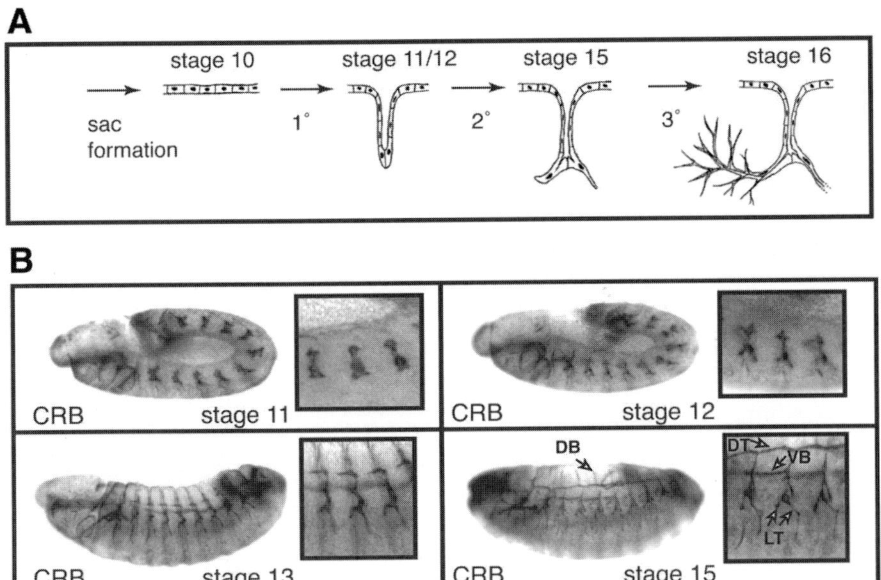

Fig. 3. Migration of the tracheal cells during embryogenesis. (**A**) Schematic of tracheal development. A portion of one segment showing the formation of the tracheal tube (sac formation) and subsequent branchings, with primary (1°), secondary (2°), and terminal (3°) branches forming off the original sac. Individual tracheal cell nuclei are shown as a dot in each cell. Reprinted with permission from **ref. 51**. (**B**) Crumbs (CRB) staining in the tracheal lumen at stages 11, 12, 13, and 15, showing the progressive branching of the tracheal cell tubes. Tracheal network formation is completed by stage 16. Insets show higher magnification of three segments. In the lower right panel (stage 15), the dorsal branch (DB), dorsal trunk (DT), visceral branch (VB) and lateral trunk (LT) branches are labeled. Anterior is to the left and ventral is down in all panels.

1.4. Border Cell Migration

Migration of the small subset of epithelial follicle cells in the ovary, known as the border cells, is useful for studying the mechanisms by which epithelial cells acquire invasive characteristics. Border cells detach from the follicular epithelium and migrate away in a process that may be similar to the behavior of invasive cancer cells of epithelial origin. Border cells are specified and initiate migration at stage 9. During the course of approx 6 h, they migrate to the anterior border of the oocyte from which they derive their name (**Fig. 4**). Border cells migrate as a coherent cluster of cells. In the center of the cluster are two non-migratory cells known as polar cells. These cells secrete a cytokine signal that stimulates the surrounding cells to migrate and carry the polar cells to their final destination.

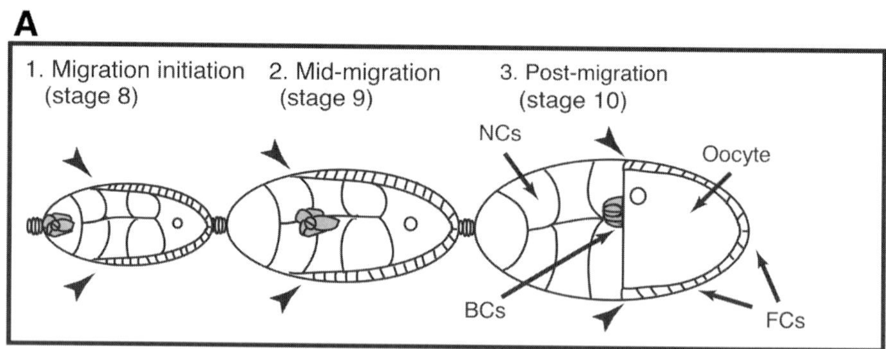

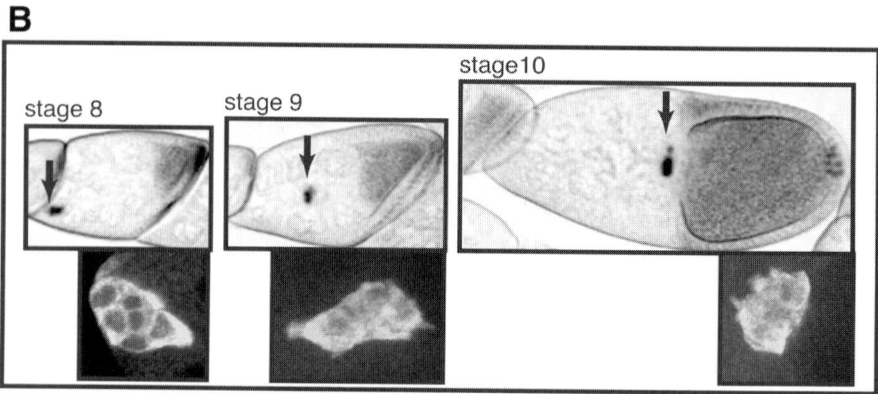

Fig. 4. Border cell migration during oogenesis. (**A**) Schematic of an ovariole with egg chambers at stage 8 (initiation of migration), stage 9 (mid-migration), and stage 10 (post-migration). The arrowheads indicate the rearrangement of the follicle cells; the border cells migrate in parallel to this rearrangement. Abbreviations used: NC, nurse cell; BC, border cell; FC, follicle cell. (**B**) Top panels: egg chambers stained for β-galactosidase activity from the enhancer trap *PZ8685-lacZ*, which is expressed in the border cells (arrows). Bottom panels: higher magnification of border cells stained for Singed (Sn; fluorescent detection) at the equivalent stages to the top panels. Sn is detected primarily in the border cell cytoplasm and shows the morphology of the cluster during migration. Anterior is to the left in all panels.

The techniques used for studying cell migration in *Drosophila* fall into three broad categories: cell-labeling procedures, genetic manipulations, and live imaging. Cell-labeling procedures include using antibodies and *lacZ* enhancer traps to stain specific migratory subsets of cells. Genetic approaches include generating, identifying, and analyzing mutant organisms, as well as the use of transgenic animals. Live imaging is a relatively new addition to the study of cell migration in *Drosophila* and will be discussed in the following chapter (*see* Chapter 15). Here, we provide protocols for labeling migratory cells in embryos and in the ovary.

2. Materials

2.1. Labeling Cells in the Embryo

2.1.1. Embryo Collection

1. Empty 6-oz plastic *Drosophila* stock bottles (Doc Frugal). Punch a few holes in the bottom with an 18-gage needle to prevent water condensation.
2. Egg collection plates (makes approx 40 plates): mix together 36 mL of molasses, 222 mL of water, and 8.8 g of bacto-agar (Difco). Autoclave on liquid cycle and when cool (approx 55°C) pour into 35-mm tissue culture dish lids (Falcon, cat. no. 35-3001; *see* **Note 1**).
3. Yeast paste: add enough water to dry baker's yeast and stir to make a spreadable paste. Store at 4°C.

2.1.2. Embryo Fixation

1. Fix solution: 100 mM PIPES, pH 6.95, 2 mM ethylenebis(oxyethylenenitrilo)-tetraacetic acid, pH 8.0, 1 mM MgSO$_4$, 4% formaldehyde (*see* **Note 2**). Make fresh daily.
2. Heptane.
3. Test tube, microcentrifuge tube, or scintillation vial.
4. 50% Bleach diluted in water. Make fresh.
5. Dechorionating baskets with Nitex membrane (Doc Frugal, cat. no. 46-101) or cell strainer (Falcon; 35-2350).
6. Paintbrush.
7. Small tissue culture dish or beaker.
8. Methanol.

2.1.3. Antibody Staining Embryos

1. PBT: 1X phosphate-buffered saline (PBS, 137 mM NaCl, 2.68 mM KCl, 10.14 mM Na$_2$HPO$_4$, 1.76 mM KH$_2$PO$_4$), 0.1% Triton X-100 (*see* **Note 2**).
2. PBT block: 1X PBS, 0.1% Triton X-100, 1% bovine serum albumin (BSA; *see* **Note 3**).
3. For fluorescent staining: fluorescent secondary antibodies, for example, Alexa Fluor 488, Alexa Fluor 568, Alexa Fluor 647 (Molecular Probes).
4. For histochemical staining with horseradish peroxidase (HRP):
   a. Biotin-conjugated secondary antibodies (Vector Labs).
   b. Vectastain Elite ABC kit (Vector Labs, cat. no. PK-6200).
   c. DAB (3,3' diaminobenzidine) Substrate kit (Pierce, cat. no. 34002) or metal enhanced DAB Substrate kit (Pierce; cat. no. 34065). Store at –20°C.
5. For histochemical staining with alkaline phosphatase (AP):
   a. AP-conjugated secondary antibodies (Southern Biotech.).

b. AP buffer: 100 mM Tris-HCl, pH 9.5, 100 mM NaCl, 50 mM MgCl$_2$, 0.1% Tween-20.

c. 50 mg/mL Nitro-blue tetrazolium chloride (NBT) in 70% dimethylformamide. Store at −20°C.

d. 50 mg/mL 5-Bromo-4-chloro-3' indolyphosphate p-toluidine salt (BCIP) in 70% dimethylformamide. Store at −20°C.

e. AP staining solution: add 4.5 µL of NBT and 3.5 µL of BCIP to 1 mL of AP buffer. Make fresh.

2.1.4. Mounting Embryos on Microscope Slides

1. Option 1: vectashield mounting medium (Vector Labs, cat. no. H-1000) or other type of antifade mounting medium (*see* **Note 4**). Option 2: series of 30%, 50%, and 85% (v/v) glycerol in 50 mM Tris-HCl, pH 7.5.
2. Cover glasses that are 22 mm × 22 mm, no. 1 thickness and 22 mm × 40 mm, no. 1 thickness.
3. Microscope slides.
4. Clear nail polish.

2.2. Labeling Cells in the Ovary

2.2.1. Ovary Dissection

1. Yeast paste (*see* **Subheading 2.1.1.**). Store at 4°C.
2. Forceps, Dumont no. 5 standard tip, Dumoxel (Fine Science Tools, cat. no. 11251-30).
3. Depression microscope slide with two depression wells (VWR International; cat. no. 48336-001).
4. Schneider's or Grace's media (Sigma) supplemented with 10% fetal bovine serum (FBS), heat-inactivated (Sigma) or modified Ringer's solution (EBR; 10 mM HEPES buffer, pH 6.9, 130 mM NaCl, 4.7 mM KCl, 1.9 mM CaCl$_2$). Store at 4°C.
5. Dissecting microscope with fiber optic light source.
6. Small bulb transfer pipets (e.g., PGC Scientific, cat. no. 313-001).

2.2.2. Antibody Staining Ovarioles

1. Fix solution: 0.1 M potassium phosphate buffer (KH$_2$PO$_4$/K$_2$HPO$_4$), pH 7.4, 4% formaldehyde (*see* **Notes 2**, **5**, and **6**). Make fresh daily.
2. NP40 Wash: 50 mM Tris-HCl, pH 7.4, 150 mM NaCl, 0.5% NP40, 1 mg/mL BSA (*see* **Note 2**).
3. NP40 block: 50 mM Tris-HCl, pH 7.4, 150 mM NaCl, 0.5% NP40, 5 mg/mL BSA (*see* **Note 3**).
4. Fluorescent secondary antibodies, for example, Alexa Fluor 488, Alexa Fluor 568, Alexa Fluor 647 (Molecular Probes).
5. 4',6-Diamidino-2-Phenylindole (DAPI, Sigma).

6. Phalloidin conjugated to rhodamine, Oregon Green, or Alexa Fluor dyes (Molecular Probes).
7. Vectashield mounting medium (Vector Labs, cat. no. H-1000; *see* **Note 4**).

2.2.3. Staining for β-Galactosidase Activity in Ovaries

1. Fix solution: 0.1 *M* potassium phosphate buffer (KH_2PO_4/K_2HPO_4), pH 7.4, 4% formaldehyde or 0.2% glutaraldehyde (*see* **Notes 5–7**). Make fresh daily.
2. PBT: 1X PBS, 0.1% Triton X-100 (*see* **Note 2**).
3. Staining solution: 10 m*M* $NaH_2PO_4 \cdot Na_2HPO_4$, pH 7.2, 150 m*M* NaCl, 1 m*M* $MgCl_2$, 3.1 m*M* $K_4[Fe^{II}(CN)_6]$, 3.1 m*M* $K_3[Fe^{III}(CN)_6]$, 0.3% Triton X-100. Store at 4°C in the dark.
4. 8% (w/v) X-gal (5-bromo-4-chloro-3-indolyl-β-D-galactopyranoside, Sigma) in dimethylformamide or dimethyl sulfoxide. Store at –20°C in the dark. Discard solution if it turns yellow.
5. 50% (v/v) glycerol.

2.2.4. Mounting Ovaries on Microscope Slides

1. Forceps, Dumont no. 5 standard tip, Dumoxel (Fine Science Tools, cat. no. 11251-30).
2. Cover glasses: 22 mm × 22 mm, no. 1 thickness.
3. 3-mL Syringe loaded with grease or petroleum jelly.
4. Microscope slides.
5. Clear nail polish.

3. Methods

3.1. Labeling Cells in the Embryo

Antibody staining is a powerful tool for identifying migrating cells and the tissues through which they migrate. Most commonly, cell migrations in *Drosophila* are studied in fixed tissue, although methods have been developed to look at several types of migrating cells in live embryos (*see* Chapter 15). Some of the commonly used antibodies and enhancer trap lines for identifying migrating cells in the embryo are listed in **Tables 1** and **2** and shown in **Figs. 1–3**. Enhancer trap lines expressing *LacZ* should be visualized using an anti-β-galactosidase antibody (**Table 1**). The first four methods (**Subheadings 3.1.1.–3.1.4.**) together are required for antibody staining of whole *Drosophila* embryos.

3.1.1. Embryo Collection

1. Place 50 to 100 anesthetized male and female adult flies in an empty *Drosophila* stock bottle.
2. Place a small dab of yeast paste onto an egg collection plate, place on mouth of bottle, and affix with a piece of labeling tape.
3. Invert bottle (plate down) and let flies lay eggs on the egg collection plate at 25°C (*see* **Note 8**). Change plate daily.

Table 1
Antibodies Used to Examine Cell Migration in *Drosophila*

Anitbody	Antibody type and name	Stage expressed	Tissue expressed	Dilution	Source	References
VASA	Polyclonal	Embryo, all stages	PGC cytoplasm	1:5000	*See* **ref.**	*(22)*
Eyes absent	Mouse monoclonal, 10H6	Embryo, stage 11 to 16 SGPs	SGPs, other cells	1:25	DSHB	*(23)*
PVR	Polyclonal	Embryo, stages 9 to 16	Hemocytes	1:0000–1:2000	*See* **refs.**	*(24,25)*
Croquemort (CRQ)	Polyclonal	Embryo, stage 11 on	Hemocytes	1:0000	*See* **ref.**	*(26)*
Peroxidasin (PXN)	Polyclonal	Embryo, stage 10 on	Hemocytes	1:500–1:2000	*See* **ref.**	*(27)*
2A12	Mouse monoclonal (IgM)	Embryo, stage 14 on	Tracheal lumen	1:5	DSHB	*(28)*
Trachealess (TRH)	Polyclonal	Embryo, stage 11 on	Tracheal nuclei	1:1000	*See* **ref.**	*(29)*
Crumbs (CRB)	Mouse monoclonal, Cq4	Embryo, stage 11 on	Tracheal lumen, other cells	1:100	DSHB	*(30)*
Punch	Polyclonal, TL1	Embryo, stages 11 to 14	Tracheal lumen	1:2000–1:4000	*See* **ref.**	*(17)*
β-galactosidase	Polyclonal		Visualize enhancer trap lines	1:1000–1:4000	Cappel, Promega	
Singed (SN)	Mouse monoclonal, Sn7C	Ovary, stages 8 to 14	BCs	1:25	DSHB	*(31)*
Phalloidin	Dye	All stages ovary, embryo	Labels actin, all cell membranes	1:400–1:1000	Molecular Probes	
Armadillo (ARM)	Mouse monoclonal, N27A1	All stages ovary, in all FCs and germline; all stages embryo	All cell membranes, upregulated in BCs	1:75	DSHB	*(32)*
Fasciclin III (FAS III)	Mouse monclonal, 7G10	Ovary, all FCs until stage 6 and PCs stages 2 to 14	All FCs early, membrane between PCs	1:10	DSHB	*(33)*

BC, border cell; DSHB, Developmental Studies Hybridoma Bank; FC, follicle cell; PC, polar cell; PGC, primordial germ cell; SGP, somatic gonadal precursor.

Table 2
Enhancer Trap Lines to Label Specific Migrating Cells

Enhancer trap line	Stage expressed	Tissue expressed	Source	References
faf (fat facets)-lacZ	Embryo, all stages	PGCs	Bloomington	*(34)*
1-eve-1-lacZ	Embryo, stage 11 on	Tracheal cell nuclei	*See* **refs.**	*(35,36)*
slbo^{1310} (slow border cells)-lacZ[a]	Ovary, stages 8 to 14	BCs, subset of FCs	Bloomington	*(37)*
PZ6356-lacZ	Ovary, stages 8 to 14	BCs, oocyte nucleus	Montell Lab	D.J.M., unpublished
PZ8685-lacZ	Ovary, stages 8 to 14	BCs, dorsal-anterior FCs	Bloomington	D.J.M., unpublished

BC, border cell; Bloomington, Bloomington Stock Center; FC, follicle cell; PGC, primordial germ cell.
[a] Homozygotes display migration defects; use this line as a heterozygote.

3.1.2. Embryo Fixation

1. Collect embryos on an egg collection plate (*see* **Subheading 3.1.1.**). An overnight collection (0–16 h) is typical, but shorter collections can be used to enrich for specific embryonic stages (*see* **ref.** *1* for estimating the timing of different stages).
2. In a test tube or microcentrifuge tube, add 1:1 heptane:fix solution (*see* **Note 9**). Set aside.
3. Meanwhile, place a basket or cell strainer into a Petri dish. Use a squirt bottle with water or a paintbrush to wash or transfer embryos from the egg collection plate into the basket.
4. To dechorionate embryos, submerge the basket into a tissue culture dish or a small beaker filled with 50% bleach. Incubate for 3 min, swirling the basket every 30 s to ensure all of the embryos have access to bleach.
5. Rinse the embryos in the basket under running tap water to remove bleach, which will take about 2 min or until the bleach smell is no longer detected.
6. Blot the basket onto a Kim wipe and use a paintbrush to transfer the dechorionated embryos to the tube with the heptane:fix solution.
7. Fix embryos for 20 min with agitation at room temperature.
8. Remove the lower fix phase, leaving the upper heptane phase and embryos in the tube (*see* **Note 9**).
9. Add enough methanol to the tube to make a final dilution of 1:1 heptane:methanol (a visible interface will form). Vigorously shake or vortex the tube for 30 s to break open the vitelline membrane. Embryos that have settled to the bottom of the tube are free of the vitelline membrane (*see* **Note 10**).
10. Remove the embryos from the bottom of the tube using a transfer pipet and transfer them to a new microcentrifuge tube.
11. Fill the tube with methanol and allow the embryos to settle. Wash the embryos three times with methanol, letting the embryos settle each time, to remove all traces of heptane (*see* **Notes 9** and **11**).

3.1.3. Antibody Staining Embryos

Methods to visualize staining patterns by fluorescence and colorimetric detection of HRP- and AP-conjugated secondary antibodies are listed as follows: Because the yolk is autofluorescent, fluorescent antibody detection is best examined with a confocal microscope. Multiple antibody detection can also be performed (*see* **Note 12**).

1. Collect and fix embryos (*see* **Subheading 3.1.2.**). Approximately 10–20 µL of embryos should be sufficient for each staining.
2. Remove methanol and add PBT to embryos to wash off methanol. Rinse three times, letting embryos settle to the bottom of the tube each time, then incubate in PBT for an additional 30 min with agitation at room temperature.
3. Block nonspecific staining by incubating in PBT block for 30 min with agitation at room temperature.

4. Add primary antibody at the working dilution in PBT block solution. Incubate for several hours at room temperature or overnight at 4°C, with or without agitation (*see* **Notes 12–14**).
5. Rinse embryos three times in PBT, letting embryos settle each time. Wash for an additional 30 min with agitation at room temperature.
6. Incubate in PBT block for 30 min with agitation at room temperature.
7. For fluorescent staining:
   a. Add fluorescent secondary, raised against the same species as the primary antibody, at 1:400 dilution in PBT block. Incubate for 1 to 2 h at room temperature. To prevent photobleaching of the fluorescent label, wrap tubes in aluminum foil in this and subsequent steps.
   b. Rinse embryos three times in PBT, letting embryos settle each time. Wash for an additional 30 min with agitation at room temperature.
   c. Mount embryos according to **Subheading 3.1.4.**
8. For histochemical staining with HRP:
   a. Add biotin-conjugated secondary antibody, raised against the same species as the primary antibody, at 1:200 to 1:500 dilution in PBT block (*see* **Note 15**). Incubate for 1 to 2 h at room temperature.
   b. Rinse embryos three times in PBT, letting embryos settle each time. Wash for an additional 30 min with agitation at room temperature.
   c. Incubate in PBT block for 30 min with agitation at room temperature.
   d. While embryos are incubating in PBT block solution, make the ABC solution from the Vectastain kit (10 µL A, 10 µL B, 980 µL of PBT block). Incubate this solution for at least 30 min at room temperature before adding to the embryos.
   e. Add the ABC solution to embryos and incubate for 30 min.
   f. Rinse embryos three times in PBT, letting embryos settle each time. Wash for an additional 30 min with agitation at room temperature.
   g. Develop the HRP reaction using the DAB Substrate kit. Add 1:10 DAB: Peroxidase buffer to embryos (*see* **Notes 9**, **16**, and **17**). The metal enhanced DAB will give a dark brown-black precipitate.
   h. Stop the reaction by rinsing embryos three times in PBT, letting embryos settle each time. Wash for an additional 30 min with agitation at room temperature.
   i. Mount embryos according to **Subheading 3.1.4.**
9. For histochemical staining with AP:
   a. Add AP-conjugated secondary antibody, raised against the same species as the primary antibody, at 1:200 to 1:500 dilution in PBT block. Incubate for 1 to 2 h at room temperature.
   b. Rinse embryos three times in PBT, letting embryos settle each time. Wash for an additional 30 min with agitation at room temperature.
   c. Rinse embryos three times in AP buffer, letting embryos settle each time. Wash for an additional 15 min with agitation at room temperature.

   d. Remove AP buffer and develop the AP reaction by adding AP staining solution.

   e. Incubate 5 min or longer in the dark until the purple signal is sufficiently developed (*see* **Notes 16** and **17**).

   f. Stop the AP reaction by rinsing embryos three times in PBT, letting embryos settle each time. Wash for an additional 30 min with agitation at room temperature.

   g. Mount embryos according to **Subheading 3.1.4.**

3.1.4. Mounting Embryos on Microscope Slides

Although embryos can be mounted in many types of mounting media (e.g., glycerol, methyl salicylate, Epon), the following method uses glycerol-based mounting media. Embryos mounted in glycerol retain good cell morphology and are easily handled.

1. Remove all PBT from the embryos stained according to **Subheading 3.1.3.** Add mounting media to embryos: to fluorescently stained embryos add several drops (approx 100 µL) of Vectashield or other antifade mounting medium to prevent photobleaching of samples; to histochemically stained embryos add 200 to 300 µL of glycerol in a series of 30%, 50%, and 85% dilutions, letting the embryos settle each time. Embryos are ready to mount on slides when they settle to the bottom of the tube, which can take from several hours to overnight.

2. Use either a pipetman or a transfer pipet to transfer the embryos in a small volume (30 to 50 µL) of mounting medium onto the center of a glass slide.

3. Place a 22 × 22-mm cover glass on either side of the pool of embryos; this provides a "bridge" to prevent the embryos from being crushed.

4. Gently place a 22 × 40-mm cover glass on top of embryos and overlap the cover glasses on either side of the embryos. Add additional media to fill the space under the cover glasses. If needed, wick away excess media with a Kim wipe.

5. Make a semipermanent mount by sealing the edges of the cover glasses with clear nail polish. Store slides at 4°C.

3.2. Labeling Cells in the Ovary

Just as with embryos, migrating cells in the ovary are easily visualized with various antibodies and enhancer trap lines, which are listed in **Tables 1** and **2** and shown in **Fig. 4**. The ovaries are first removed from the female (**Subheading 3.2.1.**) and then are fixed and stained to detect the migrating border cells, either by antibody staining (**Subheading 3.2.2.**) or by β-galactosidase activity staining of various enhancer trap lines (**Subheading 3.2.3.**).

3.2.1. Ovary Dissection

Ovaries from well-fed females swell to fill the abdomen and are easily removed by dissection (**Fig. 5**). Ovaries can either be dissected further into ovarioles for antibody staining (**Subheading 3.2.2.**) or can be kept whole for β-galactosidase activity staining (**Subheading 3.2.3.**).

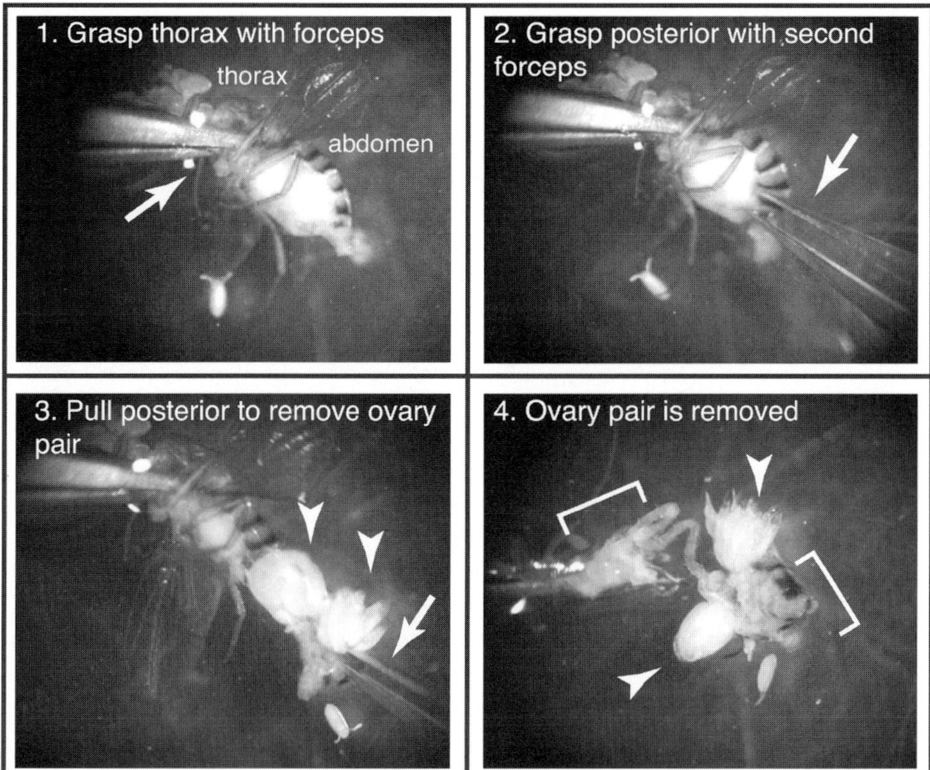

Fig. 5. Dissection of ovaries from adult *Drosophila* females. **(1)** Grasp the fly with one pair of forceps (arrow) between the thorax and the abdomen. **(2)** Next, using a second pair of forceps (arrow), grasp the posterior. **(3)** Pull with the second pair of forceps (arrow) to remove the ovary pair (arrowheads). **(4)** Once the ovary pair (arrowheads) is free, remove the excess tissue and cuticle (brackets) using forceps.

1. Fatten several 3- to 5-d-old females by adding a small amount of wet yeast paste (about the size of a thumbnail) to the side of a fly food vial. Allow the females to feed for 16 to 24 h at 25°C. If left to fatten longer, the ovaries will mainly consist of mature eggs.
2. Place a depression slide filled with ice-cold media or EBR on the stage of a dissecting microscope, preferably on a black background (*see* **Note 18**).
3. Anesthetize the flies with either CO_2 or ether.
4. With one forceps (in right hand, if right-handed), grab a fly by the wing and place the fly into the media with its back/dorsal side down while using the other forceps (in left hand) to grasp the thorax and hold the fly down at the bottom of the depression well (**Fig. 5**).

5. While still grasping the fly, use the other forceps (in right hand) to grasp the posterior end of the abdomen and pull; if done properly the ovary pair will pop out (**Fig. 5**). With forceps, tease away any cuticle or tissue clinging to the ovary (**Fig. 5**).
6. Dissect the remaining flies into the same well. Add more media if the well dries out. When finished dissecting out all the ovaries, remove extra debris from the well with a transfer pipet or forceps.
7. If the ovaries are to be left whole, use a transfer pipet to transfer the dissected ovaries along with a small amount of media to a microcentrifuge tube for fixation and proceed to **Subheading 3.2.2.** or **3.2.3.** When dissecting several lines, place the tube with ovaries on ice until ready to be fixed (*see* **Note 19**).
8. If ovaries will be needed for antibody staining, the outer muscle sheath should be removed and the ovaries further dissected into ovarioles. To dissect into ovarioles, hold the ovary pair down at the bottom of the depression well with one forceps by grasping the larger end of the ovary (where the older egg chambers are located). With the other forceps, grasp the smaller, anterior tip of one ovariole and slowly pull. A string of egg chambers (ovarioles) will emerge from the sheath, although older egg chambers (stage 10–14) tend to stay within the ovary. To liberate these older egg chambers, squeeze them out by pushing with a forceps from posterior (larger end) to anterior (smaller end), much like squeezing toothpaste from a tube.
9. Transfer dissected ovarioles using a transfer pipet to a microcentrifuge tube for fixation and antibody staining. Place the tube with ovarioles on ice until ready to be fixed (*see* **Note 19**).

3.2.2. Antibody Staining Ovarioles

Many primary antibodies do not penetrate the ovary or the surrounding muscle sheath very well. For this reason, it is best to remove the sheath and dissect the ovaries into individual ovarioles (*see* **Note 20** and **Subheading 3.2.1.**). Antibody detection in ovarioles is best done with fluorescent secondary antibodies, possibly because enzyme-linked antibodies are larger and do not penetrate the egg chambers as well. Although the oocyte has slight autofluorescence, this should not cause too much trouble for analyzing border cell migration, especially when the ovarioles are analyzed on a confocal microscope. As with embryos, labeling with multiple antibodies or dyes is easily performed (*see* **Note 12**). The following method works well for the antibodies listed in **Table 1**.

1. Dissect ovaries into ovarioles (*see* **Subheading 3.2.1.**).
2. Remove excess buffer from dissected ovarioles and add 100 µL of fix solution (*see* **Note 9**). Agitate for 10 to 20 min at room temperature.
3. Rinse ovarioles three times with NP40 wash buffer, letting ovarioles settle to the bottom of the tube each time, then wash for an additional 30 min with agitation at room temperature.

4. Block nonspecific staining by incubating in NP40 block solution for 30 min with agitation at room temperature.
5. Add the primary antibody at working dilution in NP40 block solution (*see* **Notes 12** and **13**). Incubate in primary antibody for several hours at room temperature or overnight at 4°C, with or without agitation (*see* **Note 14**).
6. Rinse ovarioles three times with NP40 Wash buffer, letting ovarioles settle to the bottom of the tube each time, then wash for an additional 30 min with agitation at room temperature.
7. Incubate in NP40 block for 30 min with agitation at room temperature.
8. Add fluorescent secondary antibody, against the appropriate species of the primary antibody, at 1:200 to 1:500 dilution in NP40 block solution. Incubate for 1 to 2 h at room temperature with agitation. To prevent photobleaching of the fluorescent label, wrap tubes in aluminum foil in this and subsequent steps.
9. To label nuclei, add 0.5 µg DAPI during secondary antibody incubation and incubate for 10 min to 2 h (*see* **Note 21**).
10. To label actin, add fluorescent-conjugated phalloidin at 1:400 dilution during secondary antibody incubation and incubate for 10 min to 2 h.
11. Rinse ovarioles three times with NP40 Wash buffer, letting ovarioles settle to the bottom of the tube each time, and then wash for an additional 30 min to several hours with agitation at room temperature.
12. Remove all NP40 Wash and add two drops (approx 100 µL) of Vectashield or other antifade mounting media. Mount ovarioles according to **Subheading 3.2.4.**

3.2.3. Staining for β-Galactosidase Activity in Ovaries

Many screens have identified enhancer trap lines that express β-galactosidase in the nuclei of specific cells in the ovary. Some of the enhancer trap lines that are useful for identifying the border cells are listed in **Table 2**. β-galactosidase is visualized either by activity staining using X-gal as a substrate (this **Subheading 3.2.3.**) or using an anti-β-galactosidase antibody (**Subheading 3.2.2., Table 1**). In the ovary, X-gal staining is more sensitive and quicker than antibody staining.

1. Dissect ovaries and leave whole (*see* **Subheading 3.2.1.**).
2. Remove excess buffer from dissected ovaries and add 100 µL of fix solution (*see* **Note 9**). Agitate for 10 min at room temperature.
3. Remove fix and rinse with PBT three times, letting ovaries settle to the bottom of the tube each time. Wash for an additional 10 to 30 min with agitation at room temperature.
4. Prewarm the staining solution to 37°C and add 2.5 µL of 8% X-gal for every 100 µL of staining solution.
5. Remove PBT from ovaries and add warm staining solution.
6. Incubate ovaries in staining solution at 37°C in the dark for 10 min to 24 h (*see* **Note 22**).
7. Remove staining solution and rinse three times with PBT, letting ovaries settle each time.

8. Add 50–100 µL of 50% glycerol to the ovaries and let them settle to the bottom of the tube. Mount ovaries according to **Subheading 3.2.4.**

3.2.4. Mounting Ovaries on Microscope Slides

1. To mount stained ovaries, use a transfer pipet to transfer ovaries or ovarioles in mounting media (*see* **Subheadings 3.2.2.** and **3.2.3.**) onto a microscope slide.
2. If mounting whole ovaries, use forceps to tease apart the ovaries into individual ovarioles and egg chambers.
3. To prevent flattening of the egg chambers, use a syringe to place small dabs of grease or petroleum jelly onto the four corners of a cover glass and place on top of the stained ovarioles. Alternatively, place small shards of a broken cover glass under the main cover glass to provide support.
4. Make a semipermanent mount by sealing the edges of the cover glass with clear nail polish. Store slides at 4°C.

3.3. Genetic Techniques I: Loss of Function Analysis

Basic genetic techniques, including generating mutant chromosomes, the use of balancer chromosomes and identification of homozygous mutant embryos, are described elsewhere *(2)*. When examining loss of function phenotypes in the embryo and the adult ovary, it is usually best to use strong loss of function or null alleles and analyze multiple alleles. Oregon-R, Canton-S, *white*, and *yellow* flies are commonly used as wild-type controls and can be obtained from the Bloomington Stock Center at http://fly.bio.indiana.edu/. Cell migration phenotypes are most commonly analyzed in homozygous mutant embryos or adults. However, it is important to note that some genes may have a strong maternal contribution of transcript in the early embryo, including the PGCs. Messenger ribonucleic acid (mRNA) deposited in the egg by a heterozygous mother may produce a functional protein, even in a homozygous mutant embryo, which may mask zygotic embryonic phenotypes. In such cases, it may be necessary to make germline clones using the dominant female sterile technique in order to generate embryos that lack the maternally contributed mRNA. (For a description of this technique *see* **refs. *3,4*.**)

3.3.1. Mosaic Clones to Study Lethal Genes in the Adult

Although zygotic mutant phenotypes can be studied easily in the embryo, it is more difficult to study the effects of such mutations in the adult since many mutations that affect cell motility cause lethality and therefore homozygous mutant adults do not survive. The FLP (FLPase)-FRT (FLPase recombination target) technique was developed to selectively induce patches of homozygous mutant tissue, called mosaic clones, in an otherwise heterozygous, and therefore phenotypically normal, adult organism *(5)*. Recombination normally occurs only during meiosis, whereas the FLP-FRT technique induces mitotic

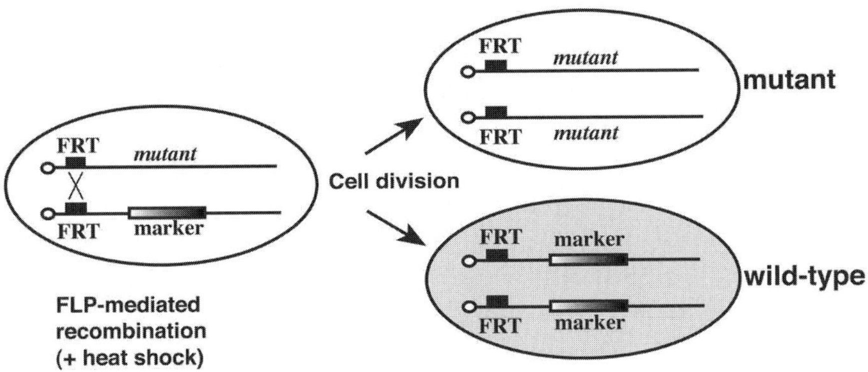

Fig. 6. Diagram of FLP-FRT mosaic clones. The FLP enzyme catalyzes mitotic recombination at the FRT sites between homologous chromosomes; in this example, one chromosome carries a mutant allele distal to the FRT site, the other chromosome carries a visible marker (e.g., GFP) distal to the FRT. After recombination and cell division, one daughter cell is homozygous mutant (top right), the other cell is homozygous wild-type (bottom right, gray cell) and carries the marker.

recombination at specific FRT sites. Fly stocks are available that have been engineered so that they contain FRT sites at useful locations in the fly genome (i.e., near each centromere). In short, after induced expression of the FLP enzyme, mitotic recombination occurs between two homologous chromosomes, one of which bears the mutant allele distal to an FRT site, and the other of which bears a wild-type allele distal to the same FRT site (**Fig. 6**). The result is one daughter cell that is homozygous mutant and another cell that is homozygous wild-type (**Fig. 6**). As these cells divide, clones are created consisting of multiple cells that are homozygous mutant or homozygous wild-type. To identify which cells are mutant, it is important to mark either the mutant chromosome or the wild-type chromosome (*see* **Notes 23** and **24**). The source of FLP enzyme to catalyze recombination can either come from heat inducible FLP, *hs-FLP*, or by control of the GAL4-upstream activator sequence (UAS) method, in which FLP is produced by GAL4-driven expression of UAS-FLP *(6)* (*see* **Note 25**). One method for inducing mosaic clones in follicle cells and border cells of the adult ovary by *hs-FLP* is described here.

1. Use meiotic recombination to recombine the appropriate FRT with your mutant allele (*see* **Note 26**; **ref. 2**).
2. Cross the *FRT, mutant* stock to a stock containing *hs-FLP* and the same *FRT* that is proximal to a ubiquitous cell marker on the same chromosome (*see* **Notes 23** and **24**). Select 5 to 10 adult females (3 to 5 d old is best) from the progeny of this cross.

3. To induce clones in the adult, place these females in a food vial. Submerge the vial with flies into a 37°C water bath for 1 h. Induce heat shocks one to three times a day for 1 to 4 d (*see* **Note 27**).
4. Wait 3 to 10 d after the last heat shock to dissect ovaries.

3.4. Genetic Techniques II: GAL4/UAS System to Overexpress Genes

The ability to over- or mis-express genes using the GAL4/UAS system in *Drosophila* is a powerful tool and complements loss-of-function analysis. The GAL4/UAS system was adapted from yeast and consists of the gene to be mis-expressed, called the UAS line, and the driver line, which expresses the GAL4 transcription factor in a tissue-specific pattern *(7)*. The UAS line contains several GAL4 binding sites upstream of the gene of interest and transcription of this gene is inactive until GAL4 binds to these sites (reviewed in **ref. 8**). This technique is useful for assessing the effects of over-expressing a gene, or the effects of expressing a dominant negative or constitutively active version of a protein. For example, the roles of the Rac, Cdc42, and Rho small GTPases in border cell migration have been studied using dominant-negative and constitutively active versions of these proteins *(9)*. Because the GAL4/UAS system is temperature sensitive, the level and timing of expression can be varied to some extent. GAL4 is less active at 18°C, more active at 25°C, and most active at 29°C. **Table 3** lists several useful GAL4 lines expressed in migrating cells in the embryo and the ovary. To test over- or mis-expression of a particular gene, clone the gene into the pUAST vector *(7)*, make a transgenic fly line by injecting the construct into embryos *(10)*, and cross the subsequent UAS line to a relevant GAL4 line. Alternatively, a variety of UAS lines are listed as alleles in FlyBase (http://flybase. bio.indiana.edu/) and are available from the Bloomington Stock Center (http:// flystocks.bio.indiana.edu/) or from the laboratories that generated them.

3.5. Testing Mutants for Cell Migration Defects

There are some general rules for analyzing mutants for cell migration defects. First, it is important to become familiar with the wild-type process by staining wild-type embryos or egg chambers for antibodies specific to the migrating cell type (**Table 1**) and examining the normal pattern of staining in the tissue (**Figs. 1–4**). Second, when examining mutant embryos, make sure that there are no gross abnormalities in the development of the embryo, particularly in pattern formation (e.g., segmentation), gut development (which can push against the epidermis) or germ band retraction. These abnormalities can cause secondary defects in the migration of many cell types in the embryo. This is also true in the ovary, where loss of some genes can disrupt the follicle cell layer and cause secondary defects in border cell migration, in which case it is useful to make smaller mosaic clones that only encompass the border cells (*see* **Subheading 3.3.1.**). Third, make sure that you are analyzing the phenotype at the right stage(s)

Table 3
GAL4 Lines Used to Mis-Express Genes in Migrating Cells

Gal4 line	Stage expressed	Tissue expressed	Source	References
nanos-GAL::VP16	Maternally contibuted to embryo[a]	PGCs	Bloomington	(*34*)
twist-GAL4	Embryo	Mesoderm (test PGC migration to mesoderm)	Bloomington	(*39*)
hs-GAL4	GAL4 under control of heat shock promoter	Ubiquitous when turned on by 37°C heat shock	Bloomington	(*38*)
breathless-GAL4	Embryo, stage 11 on	Trachea	*See* **ref.**	(*40*)
slbo-GAL4	Ovary, stages 8 to 14	BCs, subset of FCs	Bloomington	(*41*)
c306-GAL4	All stages ovary	Subset of anterior and posterior FCs	Bloomington	(*42*)
c522-GAL4	Ovary, stages 8 to 14	BCs	Bloomington	(*42*)
T155-GAL4	All stages ovary	All FCs	Bloomington	(*6*)

BC, border cell; Bloomington, Bloomington Stock Center; FC, follicle cell; PGC, primordial germ cell.

[a] Because *nos*-GAL::VP16 is maternally contributed to the embryo, GAL4 needs to come from the mother. UAS lines are only expressed in the PGCs after stage 9 owing to transcriptional repression before this stage.

of embryogenesis or oogenesis. In many cases it may be easier to look at the "final" phenotype, and then backtrack to earlier stages to determine when the first abnormality appears (*see* **Figs. 1–4**). It can also be helpful to examine mutants known to affect cell migration (**Table 4**) and compare to your mutant phenotype. (*See* **refs.** *1,11* for the correct staging of embryos and **ref.** *12* for the staging of egg chambers.) Additional help with analyzing these migrations in wild-type and mutant tissues can be found in the following references: PGC migration *(13,14)*; hemocyte migration *(15,16)*; tracheal cell migration *(17,18)*; and border cell migration *(19,20)*.

4. Notes

1. Egg collection plates should be stored at 4°C and warmed to room temperature before use; if the plates will be needed long-term, add 4 mL of Tegosept (Doc Frugal, cat. no. 20-258) solution (10% solution of *p*-hydroxy-benzoic acid methyl ester dissolved in 95% ethanol) as a preservative after autoclaved molasses agar has cooled to 55°C.
2. Different buffers/detergents can be used; however, these are the best buffer/ detergent combinations for each system based on empirical evidence.
3. BSA is used to block nonspecific staining; 5–10% normal goat serum can be substituted.
4. Other antifade mounting media for fluorescently stained embryos and ovarioles can be purchased from various suppliers (e.g., Molecular Probes). Alternatively, home-made antifade media can be made by dissolving 1% (w/v) *n*-propyl-gallate in 80% glycerol at 55°C for several hours. Store at 4°C in the dark.
5. 1X PBS can be substituted for potassium phosphate buffer in the fix solution.
6. Using methanol-free formaldehyde (16% ultrapure EM grade; Polysciences, Inc., cat. no. 18814) in the fix solution works well when staining with antibodies that are sensitive to methanol or for visualizing green fluorescent protein (GFP).
7. Either formaldehyde or glutaraldehyde fixative can be used. Glutaraldehyde fix results in stronger staining whereas formaldehyde fix results in better egg chamber morphology.
8. Flies generally prefer to lay eggs in the dark or on a 12-h light/12-h dark cycle.
9. Wear gloves and dispose of waste according to institutional guidelines.
10. Unfertilized embryos and embryos with an intact vitelline membrane will remain at the interface and should be discarded.
11. Embryos can be stored in methanol at –20°C for several months to a year. This is especially beneficial when many embryos of the same genotype need to be collected; multiple embryo collections can be pooled together in the same tube.
12. To double label with two antibodies from different species, add both antibodies together during primary incubation. If using fluorescence, choose secondary antibodies of different color; for example, anti-rabbit Alexa 488 and anti-mouse Alexa 568 if using primary antibodies made in rabbit and mouse. For histochemical staining, incubate the biotin-conjugated secondary antibody (to develop the brown HRP signal) and AP-conjugated secondary antibody of the appropriate species together.

Table 4
Examples of Mutant Alleles That Affect Cell Migration in *Drosophila* Development

Gene	Allele	Type of allele	Protein product	Migration affected	Phenotype	Source of allele	References
wunen (wun) and *wunen2 (wun2)*	wun^{CE}	EL, lof	Phosphatidic acid phosphatase	PGC	PGCs fail to migrate on midgvt	*See* **refs.**	*(43,44)*
columbus	clb^1, clb^2	EL, null	HMGCoA reductase	PGC	PGCs fails to migrate to mesoderm	*See* **ref.**	*(45)*
Pvr (Pvf receptor)	Pvr^{c2195}	EL, lof	PDGF/VEGF receptor tyro-sine kinase	Hemocyte; BC	No migration to tail (hemocyte); delayed migration (BC)	Exelixis	*(16)*
breathless (btl)	btl^{LG18} or btl^{LG19}	EL, null	FGF receptor tyrosine kinase	Tracheal cell	No migration	*See* **ref.**	*(46)*
branchless (bnl)	bnl^{P1} (bnl^{00857})	EL, null	FGF, ligand for Btl	Tracheal cell	No migration	Bloomington	*(47)*
slow border cells (slbo)	$slbo^{1310}$	FS, hypomorph	C/EBP homolog	BC	No migration	Bloomington	*(37)*
shotgun (shg) [a]	shg^{R69}	EL, null	DE-cadherin	BC	No migration	*See* **ref.**	*(48)*
Stat92E [a]	$Stat^{06346}$	EL, lof	Stat92E	BC	No migration	Bloomington	*(49,50)*

BC, border cell; Bloomington, Bloomington Stock Center; EL, embryonic lethal; FS, female sterile; FGF, fibroblast growth factor; lof, loss of function; PDGF, platelet derived growth factor; PGC, primordial germ cell; VEGF, vascular endothelial growth factor.

[a] Make mosaic clones to analyze border cell migration.

197

Develop the HRP signal first and develop the AP signal last. As a general rule, the stronger primary antibody should be developed with alkaline phosphatase.

13. Many polyclonal primary antibodies work better and have less non-specific staining when they are preadsorbed. Preincubate the antibody at a dilution of 1:10 or at the working dilution in PBT block on fixed wild-type embryos (or in NP40 block on fixed wild-type ovarioles) for approx 2 h at room temperature. Discard the embryos (or ovarioles) and store the preadsorbed antibody solution at 4°C with the addition of 0.01% (w/v) sodium azide.

14. Primary antibodies are typically incubated at 4°C overnight. However, the signal from some antibodies can be improved by instead incubating at room temperature for several hours, but this must be determined for each individual antibody. Agitation is optional.

15. Secondary antibodies directly conjugated to HRP can be used instead. After incubation in secondary antibody, skip **steps 8(b–e)** and proceed with **step 8(f)** of **Subheading 3.1.3.**

16. Develop the reaction until the background starts to appear and the signal is dark; if the signal is not developed sufficiently, the signal will appear weaker when examined at high magnification.

17. Because the signal can appear very quickly, it is best to develop the reaction in a watch glass or depression slide and observe development of the signal on a dissecting scope.

18. Dissect in EBR if only a few ovaries need to be dissected. However, when dissecting several lines, use Schneider's or Grace's supplemented with serum in order to keep the ovaries alive for several hours on ice.

19. Ovaries should be fixed within one hour of dissection.

20. Some antibodies do not require dissection of ovaries into ovarioles, but this must be determined for each individual antibody. Ovaries must be dissected into ovarioles if staining for phalloidin, since phalloidin stains the outer muscle sheath surrounding each ovariole.

21. If using a confocal microscope without UV fluorescence, use propidium iodide, which fluoresces in the red channel, to stain the nucleus instead of DAPI. After **step 3** of **Subheading 3.2.2.**, remove RNA (and subsequent diffuse staining) by treating ovarioles with 400 µg/mL RNAse A in 1X PBS for 2 to 3 h at room temperature. Repeat **step 3** and proceed with the rest of the protocol. At **step 9**, add 10 µg/mL of propidium iodide instead of DAPI to the ovarioles and incubate for 20 min. Proceed with **step 11** of the staining protocol.

22. Some enhancer trap strains develop signal in a few minutes (e.g., *slbo*[1310] takes 10 min) whereas others take longer to develop (e.g., PZ6356 takes 6 to 12 h to develop fully).

23. Ubiquitously expressed markers are generally used to mark the wild-type cells. We use a GFP marker that is both nuclear and expressed under the ubiquitin promoter (e.g., P[w + mC = Ubi – GFPnls]) and has been recombined onto the same chromosome arm as each of the commonly used FRT insertions. *See* Website: http://fly.bio.indiana.edu/frt.htm for useful FRT lines with markers.

24. The Mosaic Analysis with a Repressible Cell Marker system is used to mark only the mutant cells. (*See* **refs.** *8,21* and Website: http://fly.bio.indiana.edu/gal80. htm for details).
25. e22c-GAL4, UAS-FLP and T155-GAL4, UAS-FLP drive expression of FLP without heat shock in most follicle cells in the ovary *(6,20)*. These lines are available from the Bloomington Stock Center, but will need to be combined with the appropriate FRT *(2)*.
26. FRTs exist on all chromosome arms close to the centromere, allowing most mutant alleles to be recombined distal to an FRT. FRT lines can be obtained from the Bloomington Stock Center (Website: http://fly.bio.indiana.edu/frt.htm).
27. The number of heat shocks and length of time after heat shock needed before examining the mutant phenotype can vary depending on the individual mutant allele. Because of the rate of cell division, it can take from 3 to 10 d after heat shock to make large enough mutant clones that encompass most or all of the border cells. If no mutant clones are detected, then the mutant allele is likely to be cell lethal.

References

1. Campos-Ortega, J. A. and Hartenstein, V. (1997) *The Embryonic Development of* Drosophila *Melanogaster,* 2nd Ed., Springer-Verlag, Berlin, Germany.
2. Greenspan, R. J. (1997) *Fly Pushing: The Theory and Practice of* Drosophila *Genetics*, Cold Spring Harbor Laboratory Press, Cold Spring Harbor, NY.
3. Chou, T-B. and Perrimon, N. (1996) The autosomal FLP-DFS technique for generating germline mosaics in *Drosophila melanogaster*. *Genetics* **144,** 1673–1679.
4. Chou, T-B. and Perrimon, N. (1992) Use of a yeast site-specific recombinase to produce female germline chimeras in *Drosophila*. *Genetics* **131,** 643–653.
5. Xu, T. and Rubin, G. M. (1993) Analysis of genetic mosaics in developing and adult *Drosophila* tissues. *Development* **117,** 1223–1237.
6. Duffy, J. B., Harrison, D. A., and Perrimon, N. (1998) Identifying loci required for follicular patterning using directed mosaics. *Development* **125,** 2263–2271.
7. Brand, A. and Perrimon, N. (1993) Targeted gene expression as a means of altering cell fates and generating dominant phenotypes. *Development* **118,** 401–415.
8. Duffy, J. B. (2002) GAL4 system in *Drosophila*: a fly geneticist's swiss army knife. *Genesis* **34,** 1–15.
9. Murphy, A. M. and Montell, D. J. (1996) Cell type-specific roles for Cdc42, Rac, and RhoL in *Drosophila* oogenesis. *J. Cell Biol.* **133,** 617–630.
10. Spradling, A. C. (1986) *Pelement-mediated germline transformation, in* Drosophila *a Practical Approach* (Roberts, D. B., ed.), IRL Press, Oxford, UK, pp. 175–196.
11. Hartenstein, V. (1993) *Atlas of* Drosophila *Development*, Cold Spring Harbor Laboratory Press, Cold Spring Harbor, NY.
12. Spradling, A. C. (1993) Developmental Genetics of Oogenesis, in *The Development of* Drosophila melanogaster, 1st Ed. (Bate, M. and Martinez-Arias, A., eds.), Cold Spring Harbor Laboratory Press, Cold Spring Harbor, NY, pp. 1–70.

13. Starz-Gaiano, M. and Lehmann, R. (2001) Moving towards the next generation. *Mech. Dev.* **105,** 5–18.
14. Moore, L. A., Broihier, H. T., Van Doren, M., Lunsford, L. B., and Lehmann, R. (1998) Identification of genes controlling germ cell migration and embryonic gonad formation in *Drosophila. Development* **125,** 667–678.
15. Tepass, U., Fessler, L. I., Aziz, A., and Hartenstein, V. (1994) Embryonic origin of hemocytes and their relationship to cell death in *Drosophila. Development* **120,** 1829–1837.
16. Cho, N. K., Keyes, L., Johnson, E., Heller, J., Ryner, L., Karim, F., and Krasnow, M. A. (2002) Developmental control of blood cell migration by the *Drosophila* VEGF pathway. *Cell* **108,** 865–876.
17. Samakovlis, C., Manning, G., Steneberg, P., Hacohen, N., Cantera, R., and Krasnow, M. A. (1996) Genetic control of epithelial tube fusion during *Drosophila* tracheal development. *Development* **122,** 3531–3536.
18. Uv, A., Cantera, R., and Samakovlis, C. (2003) *Drosophila* tracheal morphogenesis: intricate cellular solutions to basic plumbing problems. *Trends Cell. Biol.* **13,** 301–309.
19. Montell, D. J. (2003) Border-cell migration: The race is on. *Nat. Rev. Mol. Cell Biol.* **4,** 13–24.
20. Liu, Y. and Montell, D. J. (1999) Identification of mutations that cause cell migration defects in mosaic clones. *Development* **126,** 1869–1878.
21. Lee, T. and Luo, L. (1999) Mosaic analysis with a repressible neurotechnique cell marker for studies of gene function in neuronal morphogenesis. *Neuron* **22,** 451–461.
22. Lasko, P. F. and Ashburner, M. (1988) The product of the *Drosophila* gene vasa is very similar to eukaryotic initiation factor-4A. *Nature* **335,** 611–617.
23. Bonini, N. M., Leiserson, W. M., and Benzer, S. (1993) The eyes absent gene: genetic control of cell survival and differentiation in the developing *Drosophila* eye. *Cell* **72,** 379–395.
24. Duchek, P., Somogyi, K., Jekely, G., Beccari, S., and Rørth, P. (2001) Guidance of cell migration by the *Drosophila* PDGF/VEGF receptor. *Cell* **107,** 17–26.
25. McDonald, J. A., Pinheiro, E. M., and Montell, D. J. (2003) Pvf1, a PDGF/VEGF homolog, is sufficient to guide border cells and interacts genetically with Taiman. *Development* **130,** 3469–3478.
26. Franc, N. C., Dimarcq, J. L., Lagueux, M., Hoffmann, J., and Ezekowitz, R. A. (1996) Croquemort, a novel *Drosophila* hemocyte/macrophage receptor that recognizes apoptotic cells. *Immunity* **4,** 431–443.
27. Nelson, R. E., Fessler, L. I., Takagi, Y., Blumberg, B., Keene, D. R., Olson, P. F., et al. (1994) Peroxidasin: a novel enzyme-matrix protein of *Drosophila* development. *EMBO J.* **13,** 3438–3447.
28. Patel, N. H. (1994) Imaging neuronal subsets and other cell types in whole-mount *Drosophila* embryos and larvae using antibody probes, in *Drosophila melanogaster: Practical Uses in Cell and Molecular Biology* (Goldstein, L. S. B. and Fyrberg, E. A., eds.), Vol. 44, Academic Press, San Diego, CA, pp. 445–487.

29. Henderson, K. D., Isaac, D. D., and Andrew, D. J. (1999) Cell fate specification in the *Drosophila* salivary gland: the integration of homeotic gene function with the DPP signaling cascade. *Dev. Biol.* **205,** 10–21.

30. Wodarz, A., Grawe, F., and Knust, E. (1993) CRUMBS is involved in the control of apical protein targeting during *Drosophila* epithelial development. *Mech. Dev.* **44,** 175–187.

31. Cant, K., Knowles, B. A., Mooseker, M. S., and Cooley, L. (1994) *Drosophila* Singed, a Fascin homolog, is required for actin bundle formation during oogenesis and bristle extension. *J. Cell Biol.* **125,** 369–380.

32. Riggleman, B., Schedl, P., and Wieschaus, E. (1990) Spatial expression of the *Drosophila* segment polarity gene armadillo is posttranscriptionally regulated by wingless. *Cell* **63,** 549–560.

33. Patel, N. H., Snow, P. M., and Goodman, C. S. (1987) Characterization and cloning of fasciclin III: a glycoprotein expressed on a subset of neurons and axon pathways in *Drosophila. Cell* **48,** 975–988.

34. Fischer-Vize, J. A., Rubin, G. M., and Lehmann, R. (1992) The fat facets gene is required for *Drosophila* eye and embryo development. *Development* **116,** 985–1000.

35. Perrimon, N., Noll, E., McCall, K., and Brand, A. (1991) Generating lineage-specific markers to study *Drosophila* development. *Dev. Genet.* **12,** 238–252.

36. Wilk, R., Weizman, I., and Shilo, B. Z. (1996) Trachealess encodes a bHLH-PAS protein that is an inducer of tracheal cell fates in *Drosophila. Genes Dev.* **10,** 93–102.

37. Montell, D. J., Rørth, P., and Spradling, A. C. (1992) *Slow border cells,* a locus required for a developmentally regulated cell migration during oogenesis, encodes *Drosophila* C/EBP. *Cell* **71,** 51–62.

38. Van Doren, M., Williamson, A. L., and Lehmann, R. (1998) Regulation of zygotic gene expression in *Drosophila* primordial germ cells. *Curr. Biol.* **8,** 243–246.

39. Greig, S. and Akam, M. (1995) The role of homeotic genes in the specification of the *Drosophila* gonad. *Curr. Biol.* **5,** 1057–1062.

40. Shiga, Y., Tanaka-Matakatsu, M., and Hayashi, S. (1996) A nuclear GFP/beta-galactosidase fusion protein as a marker for morphogenesis in living *Drosophila. Develop. Growth Differ.* **38,** 99–106.

41. Rørth, P., Szabo, K., Bailey, A., Laverty, T., Rehm, J., Rubin, G. M., et al. (1998) Systematic gain-of-function genetics in *Drosophila. Development* **125,** 1049–1057.

42. Manseau, L., Baradaran, A., Brower, D., Budhu, A., Elefant, F., Phan, H., et al. (1997) GAL4 enhancer traps expressed in the embryo, larval brain, imaginal discs, and ovary of *Drosophila. Dev. Dyn.* **209,** 310–322.

43. Zhang, N., Zhang, J., Purcell, K. J., Cheng, Y., and Howard, K. (1997) The *Drosophila* protein Wunen repels migrating germ cells. *Nature* **385,** 64–67.

44. Starz-Gaiano, M., Cho, N. K., Forbes, A., and Lehmann, R. (2001) Spatially restricted activity of a *Drosophila* lipid phosphatase guides migrating germ cells. *Development* **128,** 983–991.

45. Van Doren, M., Broihier, H. T., Moore, L. A., and Lehmann, R. (1998) HMG-CoA reductase guides migrating primordial germ cells. *Nature* **396,** 466–469.

46. Klambt, C., Glazer, L., and Shilo, B. Z. (1992) Breathless, a *Drosophila* FGF receptor homolog, is essential for migration of tracheal and specific midline glial cells. *Genes Dev.* **6,** 1668–1678.

47. Sutherland, D., Samakovlis, C., and Krasnow, M. A. (1996) Branchless encodes a *Drosophila* FGF homolog that controls tracheal cell migration and the pattern of branching. *Cell* **87,** 1091–1101.

48. Niewiadomska, P., Godt, D., and Tepass, U. (1999) DE-cadherin is required for intercellular motility during *Drosophila* oogenesis. *J. Cell Biol.* **144,** 533–547.

49. Beccari, S., Teixeira, L., and Rørth, P. (2002) The JAK/STAT pathway is required for border cell migration during *Drosophila* oogenesis. *Mech. Dev.* **111,** 115–123.

50. Silver, D. L. and Montell, D. J. (2001) Paracrine signaling through the JAK/STAT pathway activates invasive behavior of ovarian epithelial cells in *Drosophila*. *Cell* **107,** 831–841.

51. Metzger, R. J. and Krasnow, M. A. (1999) Genetic control of branching morphogenesis. *Science* **284,** 1635–1639.

15

Imaging Cell Movement During Dorsal Closure in *Drosophila* Embryos

William Wood and Antonio Jacinto

Summary

During *Drosophila* embryogenesis, many cells and tissues undergo complex morphogenetic movements, such as ventral furrow formation, germ band extension and retraction, and dorsal closure. The best way to study and understand the cell behaviors during such tissue movements is to image them live using time course analysis. The *Drosophila* embryo lends itself perfectly to live imaging for several reasons: powerful genetics allow transgenic embryos expressing green fluorescent protein–fusion proteins to be quickly and easily generated and the expression domains of these reporters can be efficiently controlled using the GAL4-upstream activator sequence system. Embryos will survive for several hours mounted on a slide in a gas-permeable oil, such as Voltalef or Halocarbon, during which time they will undergo normal development and can be easily imaged using a confocal laser scanning microscope. Here, we describe in detail a protocol for live imaging of *Drosophila* embryos which, in our hands, is routinely used to study dorsal closure but is suitable for the live study of any process involving cell motility, be it a coordinated tissue movement such as dorsal closure or the movement of individual cells such as hemocytes within the embryo.

Key Words: *Drosophila*; embryogenesis; dorsal closure; epithelia; cell movement; live imaging; green fluorescent protein (GFP); GAL4–UAS system.

1. Introduction

The process of dorsal closure in *Drosophila* provides a useful model to study the problem of how epithelial sheets migrate and fuse during the shaping of the embryo in development. It is the last major morphogenetic movement to occur during *Drosophila* embryogenesis and exhibits many similarities to wound healing in mammalian epithelia *(1–3)*. For many years genetic analysis has been used to elucidate the signaling cascades that regulate dorsal closure. More

From: *Methods in Molecular Biology, vol. 294: Cell Migration: Developmental Methods and Protocols*
Edited by: J-L. Guan © Humana Press Inc., Totowa, NJ

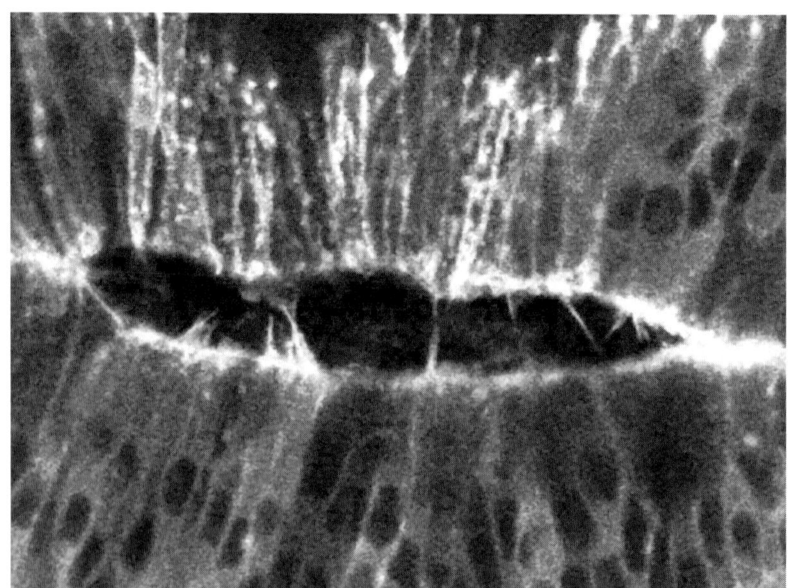

Fig. 1. *Drosophila* embryonic epithelia during the final phases of dorsal closure. The actin cytoskeleton is labeled with an actin–GFP fusion protein. Adapted from **ref. *1***.

recently, live imaging using green fluorescent protein (GFP) fusion *(4)* proteins and the GAL4–upstream activator sequence (UAS) system *(5)*, has added a powerful and dynamic dimension to the study of this morphogenetic process (*see* **Fig. 1**).

Dorsal closure begins around 12 h after egg laying when germband elongation and retraction are complete (*see* Chapter 14 for a description of these processes). At this stage of embryogenesis, the epidermis does not cover the dorsal side of the embryo; instead, the dorsal most tissue comprises a single layer of extraembryonic cells, known as the amnioserosa. During dorsal closure, two lateral epithelial sheets migrate dorsally, meet in the dorsal midline, and fuse to one another to seal the dorsal side of the embryo. Initially, this fusion occurs at the anterior and posterior ends of the embryo, where opposing epithelial cells first come into contact. Subsequently, the dorsal hole "zips" closed from both ends via the interactions of actin rich filopodia, which extend from the dorsal most epithelial cells and are essential for the "knitting together" of these cells along the dorsal midline. In this chapter, we describe a method for visualizing this process in real time in living embryos.

2. Materials

2.1. Flies

In our studies of *Drosophila* dorsal closure, we routinely use the following fly transgenics. (For a detailed account of the use of the GAL4-UAS system and GFP in *Drosophila, see* **refs. *6–8*.**)

2.1.1. GAL4 Lines

1. e22cGAL4: GAL4 is expressed in epidermal cells during embryonic development *(5)*.
2. enGAL4: GAL4 is driven by the engrailed promoter in epidermal stripes *(5)*.
3. GAL4$^{332.3}$: expression of GAL4 in the amnioserosa during dorsal closure *(9)*.

2.1.2. GFP Lines

1. UAS-GFP: enhanced GFP under the control of UAS *(6)*.
2. UAS-GFP-actin: GFP-actin fusion under UAS control *(10)*.
3. UAS-αcatenin-GFP: αcatenin-GFP fusion under UAS control *(11)*.
4. ubi-ε-cadherin-GFP: ubiquitin promoter driving *Drosophila* ε-cadherin-GFP *(12)*.
5. sqh-GFP: GFP fused to spaghetti squash (encodes Myosin II regulatory light chain) driven by its own promoter *(13)*.
6. sGMCA: GFP–moesin fusion under the control of the spaghetti squash promoter *(14)*.

2.2. Supplies

2.2.1. Embryo Collection

1. Egg-laying cages.
2. Apple juice agar plates.

2.2.2. Dechorination

1. Egg-collection basket.
2. Commercial bleach (e.g., Clorox) diluted 50% in water (final concentration should be 2–2.5% hypochlorite).
3. Thin forceps (FST, no. 5).
4. Fine paintbrushes.
5. Watch glass.

2.2.3. Mounting

1. Microscope slides.
2. Cover slips (square and rectangular with standard thickness: 0.17 mm).
3. Double-sided sticky tape (Scotch 3MM).
4. Voltalef oil 10S (Atofina; Website: www.atofina.com).
5. Halocarbon oil: mixture of 50% Halocarbon 27 and 50% Halocarbon 700, from Sigma or Halocarbon (Website: www.halocarbon.com).

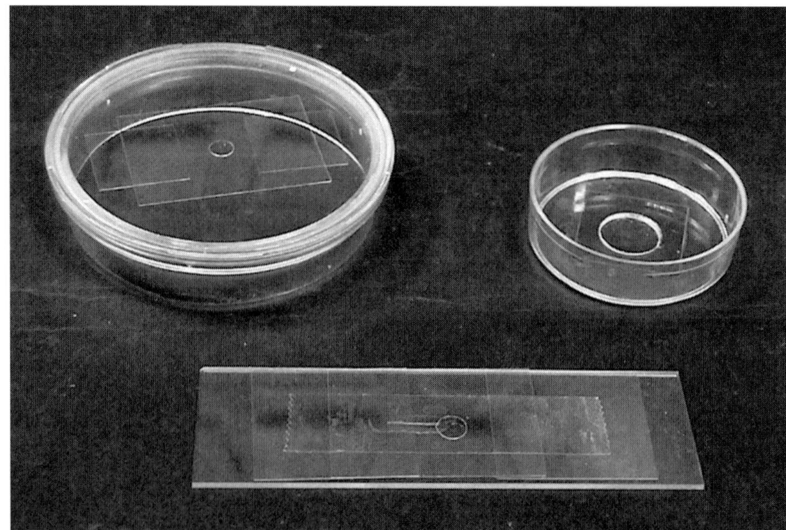

Fig. 2. Mounting supports for *Drosophila* embryos. Top left is an inverted petriPERM dish supporting an embryo mounted in Voltalef oil (method C). In the top right a Petri dish with glass bottom is used to mount an embryo in oil (method B). The bottom slide shows an embryo mounted on double-sided sticky tape covered with oil and a cover slip (method A).

6. petriPERM culture dishes (Vivascience, www.vivascience.com), to use as gas permeable embryo culture support (*see* **Note 1**).
7. Glass bottom Petri dishes (MatTek; www.glass-bottom-dishes.com).

3. Methods

We provide three different methods for both the dechorination and the mounting of the embryos, adapted from **refs. *15*** and ***16***. Although the overall approach is the same in each method, different methods will be more suitable depending on the length of imaging time and the equipment available to the researcher. (For pictures of the different mounting methods described, *see* **Fig. 2**.)

3.1. Embryo Collection

1. Incubate flies in a laying cage with a base comprising a Petri dish filled with apple juice agar, supplemented with a smear of yeast paste for optimal laying conditions.
2. Change the apple juice agar plates every 1–2 h and age appropriately to obtain embryos at the required developmental stage.

3.2. Dechorination

3.2.1. Method A

1. Using water and a paintbrush, wash the embryos into a collection basket placed on a watch glass.
2. Fill basket with 50% commercial bleach in water and leave for 5 min to dechorinate before washing thoroughly again in water.
3. Using the paintbrush, transfer the embryos to a fresh agar plate.
4. Select the correct stage embryos under a fluorescent dissecting microscope and, using forceps, transfer them to a slide, permeable membrane or a Petri dish depending on your chosen method of mounting (*see* **Subheading 3.3.**).

3.2.2. Method B

1. Using a paintbrush, transfer the embryos from the apple juice agar plates to a watch glass filled with 50% commercial bleach in water.
2. Observe the embryos under a fluorescent dissecting microscope and, once the chorion begins to dissolve, transfer the correct stage embryos to a drop of water on a clean Petri dish using forceps. (This stage must be carried out rapidly since care must be taken to ensure that the embryos do not stay in the bleach solution for more than 5 min).
3. Using forceps, transfer the embryos to your chosen method of mounting.

3.2.3. Method C

1. Using forceps transfer the embryos from the apple juice agar plate to double sided sticky tape stuck to a slide.
2. Gently roll the embryo on the sticky tape to remove the chorion (the embryo is now ready to be mounted as described in **Subheading 3.3.**).

3.3. Mounting

3.3.1. Method A

Suitable for shorter time-lapse studies and both upright and inverted microscopes:

1. Cover a glass slide with 3 cm of double-sided sticky tape.
2. Place the embryo on the tape and roll gently with forceps until the embryo becomes appropriately positioned for live imaging (i.e., dorsal side up for dorsal closure studies). This positioning should be performed quickly because the embryo rapidly dries out if left on the tape for more than a few minutes.
3. Cover with a small drop of Voltalef or Halocarbon oil (*see* **Note 2**).
4. Place a small cover slip on either side of the embryo to prevent them from being squashed and cover with a rectangular cover slip (*see* **Note 3**). This cover slip arrangement creates a small oil chamber, inside which the embryo will develop normally (*see* **Fig. 2**).

3.3.2. Method B

Suitable for upright microscopes and longer time-lapse studies:

1. Put a drop of Voltalef or Halocarbon oil onto an air-permeable membrane stretched over a supporting frame. This can be custom made, but petriPERM culture dishes can be adapted for this purpose and are available ready made (*see* **Fig. 2**).
2. Using forceps, transfer the embryos to the oil and orientate the embryos into the appropriate position for imaging.
3. Place a cover slip on either side of the embryo approx 1 cm apart and cover with a clean rectangular cover slip.

3.3.3. Method C

Suitable for inverted microscopes:

1. In a small Petri dish, cut out a circle with a diameter of approx 1 cm and glue a cover slip on the underside so that it covers the hole (these can be bought ready made from several suppliers, e.g., MatTek; *see* **Fig. 2**).
2. Place the embryo on the cover slip and cover with a drop of Voltalef or Halocarbon oil.
3. Image on an inverted microscope.

3.4. Imaging

The mounted embryos can be imaged on any fluorescence microscope equipped with a digital acquisition system. However, confocal microscopes are especially suitable for this application because embryos are relatively thick specimens and basic software of most confocal systems include time-lapse capabilities. One key aspect of imaging live embryos is to make sure that the imaging settings do not damage cells or affect embryo viability. The chosen imaging conditions are therefore always a compromise between maximum image quality and minimal phototoxicity. We normally acquire an image stack every 1–2 min, and each stack comprises four to eight confocal sections separated by 1–2 μm. Using these time-lapse settings and reduced laser excitation, we routinely image developing embryos for several hours.

Image processing of the time-lapse series acquired can be performed using the software available in the acquisition system or ImageJ software (Website: http://rsb.info.nih.gov/ij/), a public domain image processing program written in Java.

4. Notes

1. Embryo culture supports can also be custom made by stretching and attaching a gas permeable membrane (e.g., Teflon) over a small plastic frame (e.g., Perspex).
2. To reduce anoxia during longer time-lapse imaging, the Voltalef or Halocarbon oil can be saturated with oxygen. To achieve this transfer 1 mL of oil into a

microcentrifuge tube and gently bubble oxygen into the oil for about 1 min. Keep the tube tightly capped until the oil is used.

3. For mounting using methods A and B the cover slip is generally not sealed with nail varnish as it is often advantageous to use the cover slip to roll the embryos if the original orientation during the mounting was not ideal for imaging. Large adjustments cannot be made if embryos are mounted using method A however, since the embryos are held to the slide by sticky tape.

Acknowledgments

We thank Paul Martin for all his support and inspiration that led us to use live imaging and Sérgio Simões and Beatriz Garcia Fernandez for comments on the manuscript. The authors are supported by Fundação para a Ciência e a Tecnologia.

References

1. Jacinto, A., Wood, W., Balayo, T., Turmaine, M., Martinez-Arias, A., and Martin, P. (2000) Dynamic actin-based epithelial adhesion and cell matching during *Drosophila* dorsal closure. *Curr. Biol.* **10,** 1420–1426.
2. Wood, W., Jacinto, A., Grose, R., Woolner, S., Gale, J., Wilson, C., and Martin, P. (2002) Wound healing recapitulates morphogenesis in *Drosophila* embryos. *Nat. Cell Biol.* **4,** 907–912.
3. Jacinto, A., Woolner, S., and Martin, P. (2002) Dynamic analysis of dorsal closure in *Drosophila*: from genetics to cell biology. *Dev. Cell* **3,** 9–19.
4. Prasher, D. C. (1995) Using GFP to see the light. *Trends Genet.* **11,** 320–323.
5. Brand, A. H. and Perrimon, N. (1993) Targeted gene expression as a means of altering cell fates and generating dominant phenotypes. *Development* **118,** 401–415.
6. Brand, A. (1995) GFP in *Drosophila. Trends Genet.* **11,** 324–325.
7. Brand, A. (1999) GFP as a cell and developmental marker in the *Drosophila* nervous system. *Methods Cell Biol.* **58,** 165–181.
8. van Roessel, P. and Brand, A. H. (2002) Imaging into the future: visualizing gene expression and protein interactions with fluorescent proteins. *Nat. Cell Biol.* **4,** E15–E20.
9. Wodarz, A., Hinz, U., Engelbert, M., and Knust, E. (1995) Expression of crumbs confers apical character on plasma membrane domains of ectodermal epithelia of *Drosophila Cell* **82,** 67–76.
10. Verkhusha, V. V., Tsukita, S., and Oda, H. (1999) Actin dynamics in lamellipodia of migrating border cells in the *Drosophila* ovary revealed by a GFP-actin fusion protein. *FEBS Lett.* **445,** 395–401.
11. Oda, H. and Tsukita, S. (1999) Dynamic features of adherens junctions during *Drosophila* embryonic epithelial morphogenesis revealed by a Dalpha-catenin-GFP fusion protein. *Dev. Genes Evol.* **209,** 218–225.
12. Oda, H. and Tsukita, S. (2001) Real-time imaging of cell-cell adherens junctions reveals that *Drosophila* mesoderm invagination begins with two phases of apical constriction of cells. *J. Cell. Sci.* **114,** 493–501.

13. Royou, A., Sullivan, W., and Karess, R. (2002) Cortical recruitment of nonmuscle myosin II in early syncytial *Drosophila* embryos: its role in nuclear axial expansion and its regulation by Cdc2 activity. *J. Cell Biol.* **158,** 127–137.

14. Kiehart, D. P., Galbraith, C. G., Edwards, K. A., Rickoll, W. L., and Montague, R. A. (2000) Multiple forces contribute to cell sheet morphogenesis for dorsal closure in *Drosophila. J. Cell Biol.* **149,** 471–490.

15. Endow, S. A. and Komma, D. J. (1996) Centrosome and spindle function of the *Drosophila* Ncd microtubule motor visualized in live embryos using Ncd-GFP fusion proteins. *J. Cell Sci.* **109,** 2429–2442.

16. Edwards, K. A., Demsky, M., Montague, R. A., Weymouth, N., and Kiehart, D. P. (1997) GFP-moesin illuminates actin cytoskeleton dynamics in living tissue and demonstrates cell shape changes during morphogenesis in *Drosophila Dev. Biol.* **191,** 103–117.

16

Analysis of Cell Movements in Zebrafish Embryos

Morphometrics and Measuring Movement
of Labeled Cell Populations In Vivo

Diane S. Sepich and Lilianna Solnica-Krezel

Summary

Cell movements occur in all phases of animal life from embryogenesis, to maintaining adult organs, to comprising a critical component of pathology. During gastrulation, cells demonstrate a repertoire of morphogenetic movements coordinated with fate inductions to sculpt the embryonic body. The morphogenetic behaviors, underlying mechanisms, and their control, are the subject of much current study. External development of the transparent zebrafish embryo, the abundance of mutations influencing cell movements, as well as a range of observation and manipulation methods, make the zebrafish valuable for cell movement studies. This chapter offers a conceptual background for analysis of gastrulation cell movements by reviewing how region specific cell movements shape the wild-type zebrafish embryo, and how defective morphogenetic movements alone or in combination with altered cell fate specification distort the body plans of known zebrafish mutants. We furnish methods for the morphometric analysis of embryonic shape and organ rudiments in live and fixed embryos, and present data collected from live wild-type, dorsoventral patterning (*somitabun* and *chordino*) and convergence and extension (*knypek* and *trilobite*) classes of mutants. We provide a method for quantitative assessment of the movements of cell populations in vivo, and a method for determining whether cell fate and/or movement are disturbed.

Key Words: Photo-activatable dye; uncaged dye; time-lapse; live measurement; morphometry; epiboly; gastrulation; convergence and extension; segmentation.

1. Introduction

Cell movement is a recurrent theme in animal ontogeny, contributing to early embryo morphogenesis, organogenesis, adult homeostasis, and to the development and progression of disease states. Cell motility is never as preva-

From: *Methods in Molecular Biology, vol. 294: Cell Migration: Developmental Methods and Protocols*
Edited by: J-L. Guan © Humana Press Inc., Totowa, NJ

lent, however, as during the process of gastrulation, when essentially all cells of a developing embryo engage in sweeping morphogenetic movements. These gastrulation movements create the germ layers of mesoderm, endoderm, and ectoderm, sculpt them into an invertebrate or vertebrate body plan, and place organ rudiments in proper positions. Normal gastrulation requires that motility behaviors of individual cells, or of cells within epithelial sheets, are precisely coordinated in space and time. The behaviors themselves, as well as their control, are just beginning to be understood. The zebrafish is well suited to studies of gastrulation cell movements. The optical transparency and rapid external development of the embryo and the availability of mutants, allow easy genetic and embryological experimentation and observation. The aim of this chapter is to provide a framework for the analysis of gastrulation cell movements in normal embryos and embryos suspected of having abnormal gastrulation movements. We start by discussing cell movements during normal zebrafish embryogenesis with emphasis on gastrulation and how specific gastrulation cell movements influence embryonic morphology. Next, we provide an overview of known zebrafish mutants with altered body plan caused by defective morphogenetic movements alone or by defects in cell movements and cell fate specification. We also describe methods for the morphometric analysis of normal and altered embryonic shape and organ rudiments, both in vivo and in embryos with visualized ribonucleic expression domains. We provide simple cell tracing methods for quantitative monitoring of movements of cell populations in vivo. The chapter ends with a description of cell tracing methods that afford simultaneous assessment of whether cell fate and/or movement are affected.

1.1. Cell Movements in Vertebrate Embryos: An Embarrassment of Riches

Cell movements occur during nearly every stage of zebrafish embryogenesis. During the 12 h after fertilization (or hours post-fertilization, hpf), the single-cell zebrafish zygote transforms into an embryo with a clearly defined body axis, including head and trunk rudiments, and tailbud, curled along the dorsal side of the yolk cell (for detailed stage by stage description and images of zebrafish development, *see* website: http://zfin.org/zf_info/zfbook/stages/stages.html). The embryo initially is partitioned into a large non-cleaving yolk cell and a cytosolic blastodisc, which undergoes a series of synchronous cleavages *(1)*. The resulting blastula subdivides into the enveloping layer (EVL), a superficial epithelial sheet of cells, the deep cells that give rise to the embryo proper, and a layer of marginal blastomeres. Subsequent fusion of the marginal blastomeres with the underlying yolk cell forms the yolk syncytial layer (YSL) *(2)*.

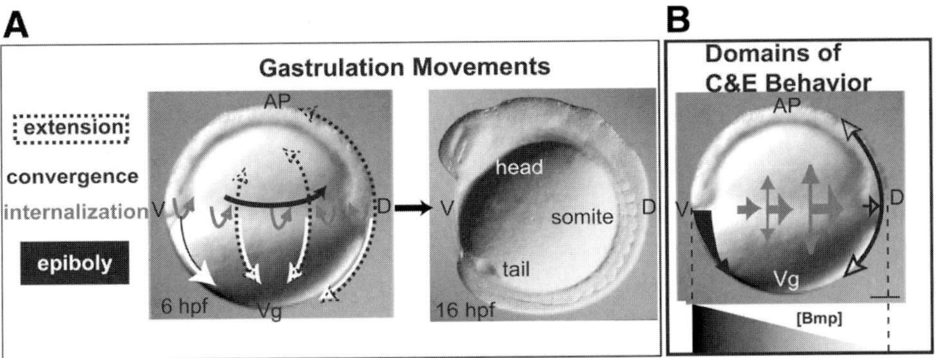

Fig. 1. The morphogenetic movements that transform the zebrafish embryo are represented in (**A**). The intent is to depict the major tissue rearrangements without implying underlying cellular behaviors. Epiboly (white) spreads the blastoderm over and encloses the yolk cell. Internalization (gray) moves cells at the blastoderm margin deeper, creating a layer between the yolk cell and the ectoderm. Convergence (black) moves cells to the dorsal midline, and depletes them from the ventral side of the embryo. Extension (dotted line) is meant to include all cell behaviors that lengthen tissues in the anterior–posterior direction. In 10 h, these movements change a cap of cells into an embryo with an easily recognizable body plan. (**B**), The morphogenetic movements of convergence and extension vary over the embryo along the high ventral to low dorsal gradient of Bmp signaling. In the ventral domain (black line with solid arrowhead, no convergence no extension), cells do not converge, and their complex set of cell behaviors results in no extension. In the lateral domain (gray), cells undergo increasing convergence and extension as they progress dorsally during gastrulation. In the dorsal domain (black line with open arrowhead), cell groups engage in generous extension and limited convergence.

Several movements begin during the blastula stage. Shortly after the onset of zygotic transcription, cells develop a blebbing motility, revealed by simple observation or time-lapse recording using differential interference contrast or Nomarski optics, and move along short random paths that mix cells locally without reshaping the embryo (*3*). During the late blastula stage, the marginal blastomeres shift dorsally, whereas the dorsal blastomeres shift toward the margin, such that blastomeres in the dorsal region become more densely packed (*4*). The dorsalward shift was revealed by randomly injecting single blastomeres with a fluorescent dye at midblastula and determining their locations at late blastula. These changes may account for the thinner and more pointed profile of the dorsal margin observed in late blastula stages (*5*). At about the same time, the blastoderm, including the EVL and deep cells, together with the YSL, begin to thin and spread over the yolk cell, in the movement termed epiboly (**Fig. 1A**). The first sign of epiboly is the doming up of the yolk cell into the blastoderm *(6)*, a movement that is hypothesized to push deep blastomeres on a course of radial intercalation between the more superficial ones. Mixing of cells is more prevalent in the center of the blastoderm than near the margin and correlates with the most dramatic thinning *(7)*. Correspondingly, clones of cells labeled with fluorescent dextran scatter more in the center of the blastoderm than clones at the margin. Epiboly of the EVL and YSL does not require participation of deep cells as demonstrated in mutants *half-baked* and *volcano*, where epiboly of deep cells is blocked, whereas the vegetalward spreading of EVL and YSL continues *(8,9)*.

The defining step in the initiation of the gastrula period is the formation of the internal mesodermal and endodermal germ layers relative to the superficial ectoderm, also known as mesendoderm internalization or emboly (**Fig. 1A; ref. 10**). The previously loose blastomeres pack into a more sheet-like structure at the blastoderm margin *(11)*, which thickens into a "germ ring" as cells move internally, piling up against the yolk cell. Concurrently, cells gather dorsally, forming the embryonic shield *(6)*. The internalization movement at the margin gives the appearance of a flowing sheet. However, time-lapse analysis revealed that cells make this passage as individuals in the company of similarly behaving neighbors *(12)*. During the gastrula period, the germ layers both move dorsally (convergence), and spread from anterior to posterior (extension), to congregate along the dorsal midline as a mediolaterally narrow and anteroposteriorly lengthened body rudiment (**Fig. 1A**). In zebrafish, convergence and extension movements within the mesendoderm vary in extent and employ distinct cell behaviors in different gastrula regions (**Fig. 1B**). Labeling of groups of cells at the onset of gastrulation at different dorsoventral positions of the blastoderm margin by photo-activation of caged fluorescein revealed three domains of movement *(13)*. In a narrow ventral region, cell behaviors produce no convergence and no extension of the tissue. In the lateral region, cell behaviors drive increasing convergence and extension as they move dorsally. In the dorsal region of the mesendoderm, cell behaviors result in minor convergence with tremendous extension of the tissue. These movement domains within the mesendoderm reside along a high ventral to low dorsal gradient of bone morphogenetic protein (BMP) signaling (**Fig. 1C**). They are regulated by BMP signaling levels as demonstrated by their shift in response to manipulation of BMP signaling through overexpression of constitutively active BMP receptor or mutations in BMP signaling components *(14)*.

In the frog *Xenopus laevis*, the germ layers are composed of continuous sheets of cells. During *Xenopus* gastrulation, the narrowing of the axis is linked to its lengthening, because both are driven by one cell behavior, mediolaterally oriented intercalation of bipolar cells *(15)*. This process is called "convergent extension" to emphasize the linked nature of the movements. In contrast, several different cell behaviors contribute to convergence and extension of germ layers in zebrafish. In the dorsal region, time-lapse confocal microscopy provides evidence for mediolateral intercalation as one of several mechanisms driving rearrangement of the notochord precursors *(16)*. During early gastrulation, animalward migration of well-separated mesendodermal cells and vegetalward epiboly extend the mesendoderm at all dorsoventral levels *(6,17,18)* (D. S. S. and L. S. K., unpublished results). Animalward migration of the early internalized cells continues until midgastrulation (D. S. S. and L. S. K., unpublished results). By mid-gastrulation, cells of lateral mesendoderm take meandering paths that result in net dorsal movement (net dorsal speed 81 ± 21 µm/h) *(19)*, and contribute to tissue convergence. Also at midgastrulation, the ventral-most mesendodermal cells change direction and migrate toward the vegetal pole, eventually forming the posterior portion of the tailbud (D. S. S. and L. S. K., unpublished results). By late gastrula stages, mediolaterally elongated mesendodermal cells migrate dorsally along straighter paths as a closely packed cohort with increased net dorsal speed (105 ± 22 µm/h) *(17,19)*. Over the course of gastrulation, cells leave the ventral-anterior region of the embryo by epiboly and convergence until the region is noticeably thinner and lacking mesoderm. The trailing edge of the departing mesendoderm delineates the limits of this evacuation zone.

The ectoderm overlying the mesendoderm also undergoes epiboly, convergence, and extension. However, ectodermal cells move as a coherent sheet *(11)*; using an underlying cellular mechanism that is not understood. Time-lapse observations using differential interference contrast optics show that cells in the sheet move predominantly vegetally (epiboly), with dorsalward bias that increases in later stages and in the dorsal half of the embryo (convergence). Most cells depart the ventral side, leaving a layer one-cell thick, which stretches, but maintains continuity. The dorsal midline extends, with cells approx 200 µm above the equator moving animally while simultaneously cells below this point move vegetally *(11,17)*.

Posterior body morphogenesis during segmentation stages involves both continuation of gastrulation cell movements and unique morphogenetic processes. The tailbud is formed from cells at the margin; dorsal marginal cells make up the anterior whereas ventral cells make up the posterior portions of the tailbud *(20)*. The anterior tailbud generates all cell types found in tail *(21,22)*; whereas the posterior tailbud can contribute all cell types but notochord *(22)*. Deep cells in the anterior tailbud (notochord precursors) extend the midline, whereas the more superficial layer deposits cells at the end of the axis and rolls the tailbud distally. Posterior tailbud cells subduct below the anterior tailbud and turn laterally to flow anteriorly beside it *(20)*.

In summary, gastrulation movements remodel a radially symmetric cap of cells into a long and narrow body rudiment. The initially spherical wild-type embryo takes on an ovoid shape, longer on the animal-vegetal axis relative to the dorsoventral axis. The axial mesendoderm increases more than ninefold in length from the start of gastrulation to the closing of the blastopore at the end of epiboly (**Table 1**). The processes of convergence and extension continue from the end of epiboly into segmentation stages, as the embryo lengthens nearly 60% more (**Table 1**) and the paraxial mesoderm narrows by about 40% (**Table 1**) and subsequently contribute to tail formation.

The migration of endodermal cells *(23,24)*, nuclei within the yolk cell *(25)*, forerunner cells *(26)*, muscle cells and somitogenesis *(27)*, neurons and their growth cones *(28)*, neural crest cells *(29,30)*, and primordial germ cells *(31,32)* are discussed in several recent reviews.

1.2. Understanding Altered Embryos

The shape of an early segmentation stage embryo is a product of multiple morphogenetic movements and inductive processes during gastrulation. Any new gastrulation phenotype presents us with several questions. Which morphogenetic processes are affected: gastrulation cell movements, cell proliferation, or cell death? Do these morphogenetic defects originate in altered embryonic patterning, such as germ layer induction, dorsoventral (DV), or anteroposterior patterning (AP)?

Simple morphometric comparison of an altered experimental embryo to wild-type, as well as to previously characterized gastrulation mutants can suggest which morphogenetic processes are changed and when alterations begin (**Fig. 2**). Similarly, comparison between experimental and wild-type embryos of the expression patterns of genes marking tissue boundaries can provide insights into which morphogenetic processes are affected. Examination of the expression of patterning genes and cell-type specific genes can address whether patterning and inductive processes are compromised as well.

In our previous work, we measured various embryonic structures through gastrulation and early segmentation stages that we anticipated would reflect the ongoing gastrulation cell rearrangements (**Table 1; refs.** *13,19,33,34*). We measured the length of the dorsal mesendodermal hypoblast (hypoblast length, **Fig. 2C,F**) to study the elongation of the body axis driven by cell behaviors in axial mesendoderm. Elongation of the body axis resulting from both lengthening of axial mesendoderm and from epiboly was analyzed by measuring the length of the body rudiment, from its anterior boundary at the animal pole to the edge of the blastoderm margin (**Fig. 2K**). Narrowing of the body axis was tracked by measuring the mediolateral width of the somites and notochord (**Fig. 2X**).

Table I
Results of Morphological Analyses of Live Embryos During Gastrulation and Early Somitogenesis [a]

Morphology Measures (in μm)

Stage	Number	Genotype	Embryo length	Hypoblast length	Thickness at ventral margin	Dorsal margin	Animal pole margin	Animal view Dorsoventral diameter	Mediolateral diameter	Lateral view Animal-vegetal diameter
Shield	$n = 11$	WT	489 ± 29	95 ± 21	63 ± 5	75 ± 8	60 ± 6	616 ± 31	600 ± 38	609 ± 27
	23	*sbn*	**512 ± 30**	**113 ± 18**	64 ± 6	**87 ± 11**	61 ± 6	627 ± 32	619 ± 28	624 ± 29
60% Epiboly	18	WT	562 ± 42	227 ± 40	53 ± 10	66 ± 8	39 ± 7	620 ± 27	619 ± 27	613 ± 42
	12	*sbn*	564 ± 21	222 ± 31	52 ± 6	61 ± 7	37 ± 2	**639 ± 10**	631 ± 8	627 ± 20
80% Epiboly	16	WT	727 ± 56	583 ± 57	33 ± 6	53 ± 8	19 ± 5	610 ± 41	603 ± 46	616 ± 45
	10	*sbn*	708 ± 48	**509 ± 65**	**35 ± 5**	**43 ± 9**	19 ± 4	599 ± 20	599 ± 25	619 ± 45
	10	*din*	692 ± 51	**516 ± 81**	**39 ± 5**	**64 ± 9**	18 ± 4	604 ± 36	575 ± 66	590 ± 39
	11	*tri*	697 ± 46	**534 ± 57**	34 ± 6	51 ± 8	**16 ± 2**	595 ± 15	601 ± 22	601 ± 47
	10	*kny*	**676 ± 34**	450 ± 40	**24 ± 3**	**46 ± 6**	**15 ± 4**	612 ± 35	599 ± 34	**583 ± 21**
YPC	20	WT	1005 ± 69	923 ± 66	15 ± 4	38 ± 7	40 ± 12	583 ± 34	577 ± 34	592 ± 33
	20	*sbn*	**1115 ± 74**	**1027 ± 73**	**24 ± 4**	**29 ± 5**	46 ± 8	605 ± 37	595 ± 38	**727 ± 24**
	10	*din*	**922 ± 23**	**813 ± 34**	**19 ± 4**	**46 ± 8**	**27 ± 6**	586 ± 39	579 ± 46	**561 ± 21**
	15	*tri*	960 ± 65	883 ± 57	13 ± 4	40 ± 5	**11 ± 15**	588 ± 27	589 ± 41	589 ± 34
	14	*kny*	**896 ± 92**	**822 ± 91**	15 ± 4	43 ± 11	**7 ± 11**	**632 ± 51**	**624 ± 57**	613 ± 50
2 somite	10	WT	1387 ± 57			44 ± 5	58 ± 6			620 ± 20
	13	*sbn*	**1235 ± 61**			**31 ± 5**	**52 ± 8**			**825 ± 57**
	9	*din*	**1175 ± 86**			47 ± 4	53 ± 7			**582 ± 37**
	11	*tri*	**1115 ± 49**			40 ± 5	**34 ± 7**			611 ± 23
	10	*kny*	**1084 ± 75**			45 ± 6	**19 ± 21**			**597 ± 11**

Values that are statistically different from wild-type are in boldface type ($p \leq 0.05$).

[a] Collected from previous publications (**refs. *14,19,33,34***) and D. S. S. and L. S. K., unpublished data.

Continued on next page

Table 1 *(Continued)*
Results of Morphological Analyses of Live Embryos During Gastrulation and Early Somitogenesis [a]

Morphology Measures (in µm)

								Animal view		Lateral view
Stage	Number	Genotype	Embryo length	Hypoblast length	Thickness at ventral margin	Dorsal margin	Animal pole margin	Dorsoventral diameter	Mediolateral diameter	Animal-vegetal diameter
5 somite	10	WT	1604 ± 59			60 ± 13	64 ± 11			630 ± 43
	9	*din*	**1253 ± 75**			56 ± 5	61 ± 6			589 ± 33
	13	*tri*	**1310 ± 66**			65 ± 13	74 ± 13			653 ± 43
	11	*kny*	**1274 ± 45**			**72 ± 8**	44 ± 30			626 ± 31

Somite Measurements (in µm)

			Ant-post length		Mediolateral width		Notochord width	Mediolateral width		Ant-post length
Stage	Number	Genotype	L somite #1	R somite #1	L somite #1	R somite #1	Somite #1	L somite #5	R somite #5	R somite #5
2 somite	10	WT	38 ± 5	37 ± 6	106 ± 15	105 ± 17	28 ± 4			
	13	*sbn*	**55 ± 17**	**57 ± 18**			**54 ± 18**			
	9	*din*	29 ± 8	31 ± 9	**86 ± 22**	86 ± 24	32 ± 7			
	11	*tri*	32 ± 3	29 ± 7	**131 ± 20**	**133 ± 25**	**35 ± 6**			
	10	*kny*	27 ± 5	25 ± 5	**152 ± 8**	**155 ± 7**	**38 ± 7**			
5 somite	10	WT			68 ± 16	65 ± 15	27 ± 4	77 ± 14	77 ± 14	38 ± 6
	9	*din*			**30 ± 7**	**32 ± 5**	29 ± 4	**49 ± 13**	**48 ± 12**	**27 ± 3**
	13	*tri*			**108 ± 19**	**108 ± 17**	**35 ± 10**	**136 ± 31**	**136 ± 28**	**26 ± 7**
	11	*kny*			**108 ± 15**	**110 ± 14**	**34 ± 7**	**144 ± 11**	**145 ± 14**	**25 ± 5**

Values that are statistically different from wild-type are in boldface type ($p \leq 0.05$). L, Left; R, right.

[a] Collected from previous publications (**refs. *14,19,33,34***) and D. S. S. and L. S. K., unpublished data.

The embryo length increases substantially after the end of epiboly, as the somites and the presomitic mesoderm narrow. The anterior–posterior length of each somite pair remains constant between the 2 and 5 somite stages, suggesting that somites increased in size along the unmeasured dimension, the dorsoventral axis. The width of the notochord does not change, suggesting its convergence was complete. As an early gauge of internalization, anterior migration of mesendoderm and epiboly movements, we determined the thickness of the blastoderm at the equator on the ventral and dorsal sides and at the animal pole (**Fig. 2A**). The ventral margin becomes thinner losing cells to epiboly and convergence. Changes at the dorsal margin are more complex: thinning from epiboly, anterior mesendoderm migration, but then thickening from convergence. The blastoderm thins at the animal pole reflecting epibolic movement and only thickens again with the arrival of the prechordal mes-endoderm.

Mutations in genes encoding components of the non-canonical Wnt signaling pathway, (*trilobite/strabismus* [*tri*; **refs. 9,35**], *knypek/glypican4/6* [*kny*; **ref. 9**], *silberblick* [*slb/wnt11*; **ref. 36**]), or interference with the transcriptional regulator, STAT3 *(37)* result in simple reduction of convergence and extension movements without alteration in embryonic patterning. Mutations in genes that cause the misspecification of cell fates, such as *spadetail* (*spt*; **ref. 38**), and the dorsoventral patterning genes *chordino* (*din*; **ref. 39**) and *somitabun* (*sbn*; **ref. 40**), demonstrate more complex phenotypes, including altered cell fates and expanded or ectopic movement domains.

After gastrulation, the *trilobite* mutants (**Fig. 2G, L, Q, U, W,** and **Y**) *(19)*, exhibit all organ primordia, notochord, somites, neural tube, dorsal and ventral portions of the tailbud in a form that is anteroposteriorly shorter and mediolaterally broader compared with the wild-type (**Fig. 2F, K, P, T, V,** and **X**). The pre-gastrulation expression domains of DV patterning genes are indistinguishable from those in wild-type, while the changes observed after the end of the gastrula period are consistent with impaired movements alone. Measurement of the morphology of *tri* mutants suggests that convergence and extension of the body rudiment is relatively normal (the *p* values for statistical significance are just above 0.05) until the two-somite stage. At that time, the body is clearly shorter, falling to 80% of normal length (**Table 1**). In addition, the notochord is wider and somites are wider and shorter in AP length. Groups of labeled cells in the lateral mesoderm of *tri* converge and extend normally until after the one-somite stage when they fall behind *(33)*. Similarly, time-lapse analysis shows normal cell behaviors and morphology in the lateral mesendoderm of midgastulae, but reduced dorsalward speed and rounder, less mediolaterally aligned cells at the end of epiboly when morphological defects become striking *(19)*.

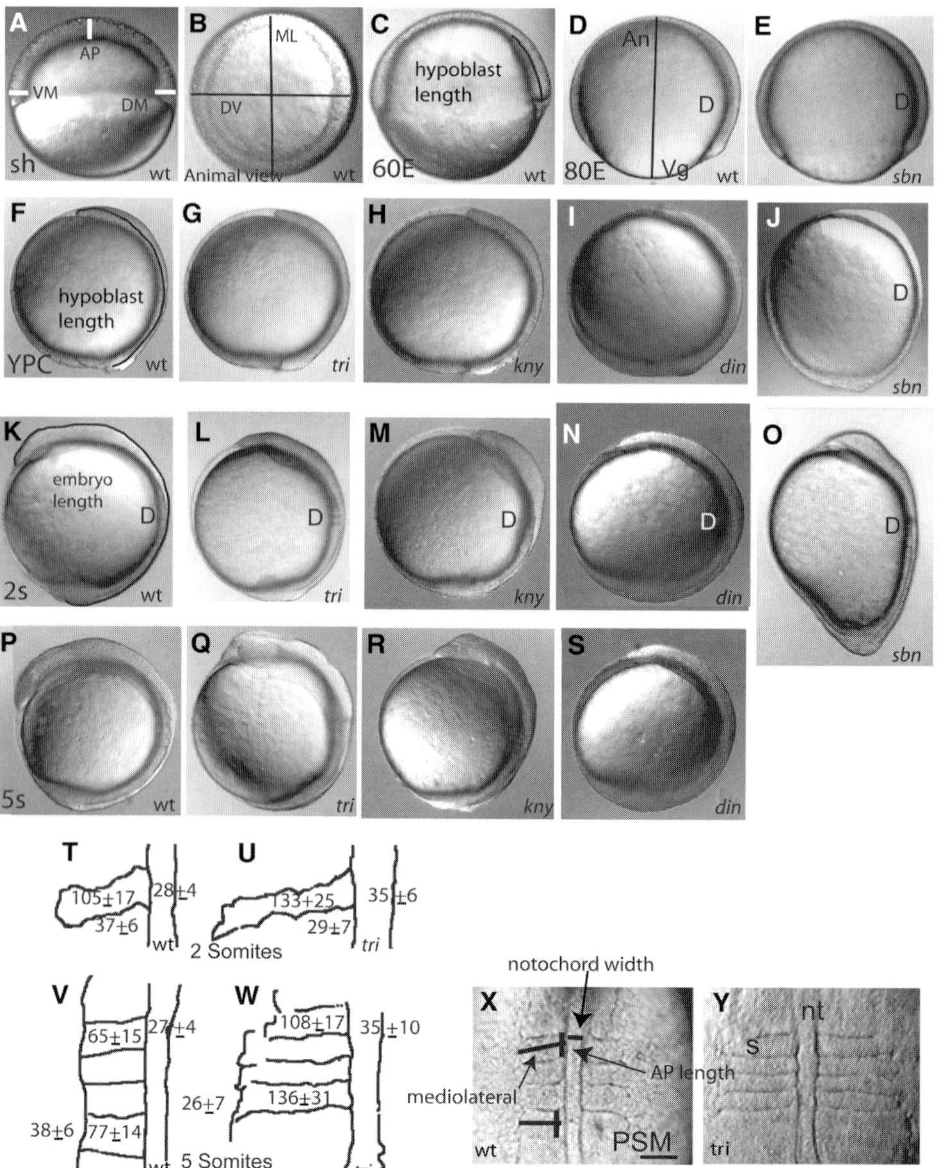

Fig. 2. Morphological changes of wild-type and mutant embryos during gastrula and early segmentation periods. Embryos are in lateral views, dorsal to right, anterior at top, except as noted. The structures measured are illustrated with black or white lines. Please *see* text for details. Top row shows wild-type embryos at (**A**) shield stage (sh, 6 hpf); (**B**) shield stage, animal pole view; (**C**) 60% epiboly (60E, 7 hpf); (**D**) 80% epiboly (80E, 8.5 hpf), and (**E**) *sbn* at 80% epiboly. (**F**) Wild-type, (**G**) *tri*, (**H**) *kny*,

(Continued)

The *knypek* mutations, which act on a transmembrane protein of the Glypican 4/6 class, were proposed to promote non-canonical Wnt signaling *(34)*. As in *tri*, all organ primordia are present in *kny* but are shorter and broader compared to wild-type (**Fig. 2H**, **M**, and **R**). Early DV patterning appears to be intact, and changes in RNA expression patterns by the end of epiboly suggest impaired movements alone. Morphometric analysis of *kny* mutants shows reduced axial length from midgastrula stages (**Table 1**). Accordingly tracing labeled cell populations in the dorsal margin shows reduced extension from the earliest time measured, end of epiboly *(34)*. Also at this time, labeled cells in the lateral mesoderm demonstrate reduced convergence movement, while labeled cells in the paraxial region are rounder and less mediolaterally aligned.

The genes *silberblick* and *pipetail* encode Wnt 11 *(41)* and Wnt 5 *(42)*, respectively, which function as ligands in non-canonical Wnt signaling. At 24 hpf, the phenotypes of *slb* and *ppt* mutants manifest in different regions of the embryo, with frequent cyclopia in *slb* and a bent tail in *ppt*. Both *slb* and *ppt* mutants were shown to have shorter body axes at the end of epiboly, a study aided by visualizing the RNA expression domains for cathepsin L also known as *(hgg1)* in the prechordal plate and no tail (*ntl*) in the notochord *(43)*. The 24 hpf phenotypes of *slb* and *ppt* develop as a consequence of region specific movement defects. The *ppt* embryo, unlike *slb*, has deficits in the movements of tail development *(44)*, whereas the *slb* embryo has a more severe delay in the migration of the anterior axial mesoderm *(43)*. In confocal time-lapse images, labeled prechordal plate cells in *slb* fail to separate well from the overlying ectoderm and fail to migrate efficiently anteriorly *(45)*.

The JAK/STAT signaling pathway is involved in a wide range of developmental processes. Depletion of STAT3, a transcriptional regulator, using an antisense morpholine-modified oligonucleotide results in an anterior-posteriorly shortened and mediolaterally broadened zebrafish embryo with an enlarged posterior tailbud by the end of epiboly *(37)*. The expression domains of tissue-specific genes are consistent with reduced convergence and extension without influence on patterning. In support, labeling of lateral cells demonstrated that

---

Fig. 2. *(Continued)* (**I**) *din*, and (**J**) *sbn* embryos are shown at yolk plug closure (YPC, 9.5 hpf). (**K–O**) At two-somite stage (2s, 10.7 hpf). (**P–S**) At five-somite stage (5s, 11.7 hpf). Dorsal views of somites and notochord in wild-type and *tri* embryos (**T–Y**). Outlines of somites and notochord with average measures for two-somite stage (**T,U**) and five-somite stage embryos (**V,W**). The regions measured are illustrated.

AP, animal pole margin; D, dorsal; DM, dorsal margin; E, percent epiboly; nt, notochord; PSM presomitic mesoderm; s, somite; VM, ventral margin; YPC, yolk plug closure stage.

convergence, but not animalward migration or epiboly, is inhibited from midgastrulation. Labeling of dorsal margin demonstrated a similar early inhibition of extension.

Inappropriate specification of cell fates can alter the movements of the misspecified cells and disturb embryonic morphology. The *spadetail* (*spt*) mutations block the functioning of a mesodermally expressed T-box gene necessary for paraxial mesoderm specification *(46)*. At early segmentation stages, the *spt* embryo has relatively normal body length and notochord but has a reduced number of cells in the somites and an enlarged posterior tailbud. At 24 hpf, the excess tailbud cells contribute to normal tail structures and a bolus of cells at the end of the tail, suggesting its name *(38)*. Transplantation experiments placing both wild-type and *spt* cells together into either type of host showed that *spt* paraxial mesoderm autonomously failed to converge, and instead migrated vegetally similar to ventral mesoderm *(38)*. Subsequently, the excess cells migrate correctly for their new location into the tail.

Mutations in dorsoventral patterning genes cause alterations in early patterning and have complex movement phenotypes. *somitabun* (*sbn*) mutations disturb Smad5, a downstream component of Bmp signaling, resulting in a dorsalized embryo *(47)*. At the end of epiboly, the *sbn* embryo acquires an elongated shape resembling an ice cream cone (**Table 1, Fig. 2E, J**, and **O**); notochord cells are not dramatically increased in number, but somites expand laterally *(40)*. Both somitic and neural markers show moderate expansion of dorsal structures in the anterior region, whereas posterior structures, such as hindbrain and posterior somites, expand to encircle the embryo *(48,49)*. The elongated shape of *sbn* led us to measure the dorsoventral, animal-vegetal, and mediolateral diameters *(14)* (**Fig. 2B,C**) as a summary of overall "shape." Measurement of the embryonic diameters demonstrated that mutant mid-gastrulae are the same size as wild-type ones. By the end of epiboly, morphogenesis has gone awry, with *sbn* embryos displaying an elongated shape (animal-vegetal axis of 727 ± 24 vs 592 ± 33 for wt) and delayed formation of the evacuation zone. Measurements show the ventral margin of the blastoderm fails to thin properly. Tracing labeled groups of mesodermal cells demonstrates reduced convergence of lateral and blocked normal vegetal migration of ventral midline cells at the end of epiboly when the elongated shape becomes apparent. Furthermore, cells at the ventral midline elongate and orient mediolaterally, suggesting they adopt the cellular morphology and behavior of paraxial mesodermal cells, which may drive the formation of the elongated shape.

Morphometric analysis can be used to test a specific hypothesis. During the characterization of *sbn*, it became clear that the embryo length falls behind wild-type between YPC and the two-somite stage and the anterior edge of the head rudiment appears stalled at the animal pole. We tested if the *sbn* embryo length was primarily caused by the lack of extension of the region between the first somite and the leading edge of the head. We found that two thirds of the difference in embryo length was contributed by defective extension of this region *(14)*.

Mutations in *chordino* (*din*) inactivate *Chordin*, a dorsally expressed antagonist of Bmp signaling *(50)*. At 24 hpf, *din* embryos are characterized by a reduced head, lack of notochord in the tail, ectopic cells at the ventral tail fin, and posterior to the widened yolk extension *(39)*. *din* mutants assume a rounder profile by the end of epiboly (**Table 1, Fig. 2I**, **N**, and **S**). The body length of *din* mutants is shorter than wild-type, suggesting that tissues fail to extend. Reduced extension movement was confirmed by the labeling of cell populations at the dorsal margin and quantifying the extension of the cell array. Additionally, the posterior tailbud is greatly expanded, hinting that convergence may be inhibited. In agreement, labeling of lateral cells demonstrates that convergence is strongly reduced by the end of epiboly *(14)*.

The morphometric analysis of live embryo anatomy is easily expanded to include the use of mRNA expression domains as landmarks to reveal hidden structures. For example, to measure the distance between otherwise anonymous cell populations, Marlow et al. revealed axial mesoderm with sonic hedgehog *(shh)* and the edge of the neural plate with distal-less 3 *(dlx3)* *(51)*. Similarly, the reduced axial length of ppt embryos at the end of epiboly was only revealed by measuring the ntl expression domain *(43)*. We expect that use of transgenic zebrafish lines expressing fluorescent proteins under the control of gene specific promoters will permit additional measurements in vivo.

Morphometric analyses using ribonucleic expression patterns and live embryos are critical for demonstrating overall defects and especially for focusing subsequent experiments on relevant tissues at the correct stages of development. While morphological studies are important, they cannot identify which specific cell movement is deficient. Following cells in the live embryo provides the fundamental test of cell movement defects.

1.3. Distinguishing Fate or Movement as the Primary Defect

Abnormal gastrulation movement can result from inactivation of proteins mediating motile properties of cells. However, gastrulation defects can be also a consequence of defects in embryonic patterning. In this latter case, altered positional information can change cellular fates and cell movements *(13)*. To distinguish between these two scenarios, three approaches can be employed. First, one can test whether expression domains for dorsoventral patterning genes are altered before the onset of gastrulation, for example, the ventrally expressed *bmp2* and *bmp4* and the dorsally expressed *goosecoid* and *chordino* at shield stage.

Second, at the end of gastrulation, expression domains of tissue specific genes can be examined for alterations that are consistent with disturbed movement alone. For example, in *kny* mutant embryos, expression domains of somites and notochord appear to be present, but are expanded laterally, consistent with the proposed reduced convergence. In contrast, in *din*, anterior somites and paraxial mesoderm, as labeled by paraxial protocadherin *(papc)*, are reduced, and the ventral tailbud, labeled by *bmp4*, is greatly expanded *(14,34)*.

Third, any cell population can be simultaneously examined for whether it moves normally and assumes its expected fate. The queried cell population is labeled at the beginning of gastrulation and allowed to develop until abnormal morphology appears. The embryo is then processed to reveal the location of the labeled cells and tissue specific gene expression. To test whether more lateral cells are recruited to build up the laterally expanded somites in *kny*, at the onset of gastrulation we labeled cells that were expected to contribute to somites. We found that after the somites had formed, the labeled cells had converged less than wild-type cells but gave rise to an equivalent region of the somite, indicating that movements are disturbed without additional fate changes.

2. Materials

1. Fish water: 0.3 g/L "Instant Ocean" salts in deionized water.
2. An agarose injection plate (1% agarose in 1X fish water) is prepared using a custom milled plastic mold (5×7 cm having six bars, each $1 \times 1 \times 50$ mm, spaced 3.5 mm apart).
3. Fine forceps (Fine Science Tools; Dumont, Biologie tip; cat. no. 11255-20).
4. Fine probes: glue a fine filament (e.g., eyebrow hair) into a capillary tube.
5. Eye piece reticle marked with a protractor (Klarmann Rulings, *see* website: www.reticles.com).
6. 100% Danieau's buffer: 58 mM NaCl, 0.7 mM KCl, 0.4 mM MgSO$_4$, 0.6 mM Ca(NO$_3$)$_2$, 5 mM HEPES, pH 7.6 *(52)*.
7. 1.5% and 2.0% Methylcellulose (Sigma, cat no. M-0387) in 30% Danieau's buffer. Methylcellulose requires 3–4 d to hydrate. To remove the visible particles, spin in a microfuge for 5 min at maximum speed. Remove clear solution to a new tube.
8. Dextran DMNB caged fluorescein 10,000 MW (Molecular Probes D-3310), dissolved to a final concentration of 1% in sterile filtered 120 mM KCl, 20 mM HEPES, pH 7.5. Spin for 5 min in microfuge at maximum speed; remove supernatant and aliquot for storage at –20°C.
9. Bridged slides: using cyanoacrylate-type glue, build two parallel supports of cover slips (any size, of no. 2 thickness), separated by a 1.5-cm gap on a 75×25-mm slide. Cover the embryo with a 25-mm square cover slip no. 1 1/2 thickness.
10. 1X PBS: 0.8% NaCl, 0.02% KCl, 0.02 M phosphate buffer, pH 7.3.
11. PBT: 1X PBS, 0.1% Tween-20.

12. 4% Paraformaldehyde in 1X PBT.
13. Antibody dilution and blocking solution: PBT, 2% sheep serum, 2 mg/mL bovine serum albumin.
14. Anti-fluorescein alkaline phosphatase, FAb fragment (Roche).
15. Fast red tablets (Roche).
16. Stripping buffer: 0.1 M glycine, pH 2.2, 0.1% Tween-20.
17. 0.1 M Tris-HCl, pH 8.2, 0.1% Tween-20.
18. Software to analyze data: object Image (*see* website: http://simon.bio.uva.nl/object-image.html), is a modified version of the Mac-based NIH Image (*see* website: http://rsb.info.nih.gov/nih-image/), which allows nondestructive marking of images, assisting record keeping of one's measurements.
19. Data storage: images require a great deal of computer space. We store archival images on compact discs and DVDs. Removable hard drives are very convenient for organizing these large datasets.
20. A microscope room held at 28°C.
21. Compound epifluorescence microscope (Zeiss Axiophot II).
22. Video camera, computer, and software to drive the camera (our several-year-old system is Zeiss Axiocam, Dell Dimension XPS 800T, NVIDIA GeForce2 video card, Zeiss Axiovision).
23. Micropipet puller (Flaming/Brown P-97, Sutter Instruments).
24. Pneumatic picopump (MPPI-2), backpressure unit (BP15), and footswitch (FW; Applied Scientific Instruments). Manipulator (M3301R), stand (M1), base plate (5052), and piconozzle kit (5430-10) (World Precision Instruments).
25. Dissecting microscope.

3. Methods

3.1. Measuring Embryo Shape

It is essential for morphometric analysis to compare embryos that are the same developmental age and, ideally, the same genetic background. Embryos will quickly develop beyond the desired stage, requiring data to be collected from several measurement sessions. To obtain matched developmental stages, embryo age is judged by a morphological criterion that can be measured from collected images, for example, percentage epiboly, rather than age in hours *(53)*. For example, only embryos within ± 3% of the named epiboly stage were used in our studies. Injected embryos frequently suffer from developmental delay. As convergence and extension are progressive events, a younger embryo will exhibit mediolaterally wider and anterioposteriorly shorter axis and organ rudiments. The number of somites can be used as a staging criterion during segmentation stages, however somites are often difficult to see in live embryos that exhibit mediolateral widening. *Krx20* expression in rhombomeres 3 and 5 of the hindbrain is very dynamic over the interval from 90% epiboly to two-somite stage and can be used to assure the comparison of same stage fixed experimental and wild-type sibling embryos.

The following measurements can be made on lateral view images. Animal-vegetal dimension (AnVg) is defined as a straight line that bisects the embryo from animal to vegetal pole (**Fig. 2D**). The DV dimension is measured at the equator, or perpendicular to the AnVg dimension as a line that bisected the embryo (**Fig. 2B**). Embryo length is defined for all gastrulation stages as beginning at the animal pole and following a curved path along the outside of the ectoderm to the edge of the mesendoderm cleft in the dorsal margin (**Fig. 2K**). Beginning at yolk plug closure stage, embryo length is measured starting at anterior mesendoderm, and ending at the tip of the extending tail. The cleft between dorsal mesendoderm and ectoderm at the margin serves as the starting point for measuring mesendoderm (hypoblast) length, which then continues between mesendoderm and ectoderm to the anterior end of the visible cleft. The blastoderm is measured at the thickest point of the dorsal or ventral margin (DM and VM) at shield and 60% epiboly stages (**Fig. 2A**). At 80% epiboly and later, the blastoderm is measured at the equator of the embryo. The blastoderm at the animal pole is measured parallel to the AnVg line, extending from yolk to blastoderm edge (**Fig. 2A**). Mediolateral dimension is viewed from animal pole, in an optical cross-section at the equator of the embryo (**Fig. 2B**). Animal anterio posterior pole and mediolateral dimensions of somites 1 and 5 and the width of the notochord at the level of somite 1, are measured in dorsal view (**Fig. 2T–Y**).

3.2. Quantitative Analysis of Movement of Cell Groups Using Photo-Activated Dye

This labeling method was modified from Kozlowski et al. *(54)*.

3.2.1. Inject the Dye

1. Collect embryos from breeding zebrafish within 15 min of spawning.
2. Place embryos in their chorions on injection plate (1% agarose in 1X fish water).
3. Calibrate the injection needle to deliver 1 nL of caged dye. This can be accomplished by ejecting a drop over a stage micrometer, or into a bead of oil on the micrometer and measuring the drop size (*see* **Note 1**).
4. Inject 0.5 to 1 nL of 1% dye, Dextran DMNB caged fluorescein 10,000 MW in 120 mM KCl, 20 mM HEPES, pH 7.5, into the yolk of a one to four-cell embryo.
5. Remove the embryos to fresh fish water and raise at 28.5°C in the dark until 30% epiboly stage. Move embryos to a glass dish filled with 30% Danieau's buffer to manually remove their chorions (*see* **Note 2**). Return to the incubator until shield stage.

3.2.2. Activate the Dye in the Chosen Cell Group and Follow Their Movement Over Time

1. Using a dissecting microscope, mount the shield stage embryo on a bridged slide, in a drop of 2% methylcellulose, 30% Danieau's buffer (*see* **Notes 3** and **4**).

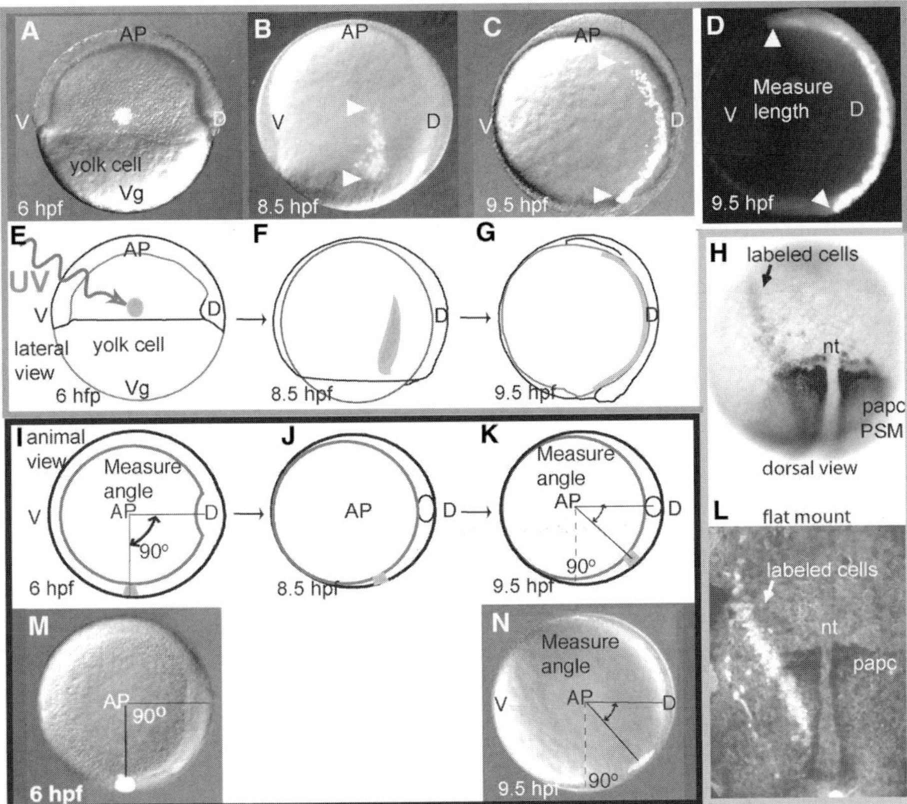

Fig. 3. Labeling cell populations by photo-activation of caged dye is illustrated with live unrelated and schematic embryo images. An embryo is labeled at the lateral blastoderm margin at the start of gastrulation; (**A,E**) shield stage. The location relative to the dorsal midline is measured in degrees using an eyepiece protractor reticle (**I,M,K,N**). The labeled cell group is followed as it elongates, (**B,F,C,G**) ends of array indicated by arrowheads; and converges toward the dorsal midline (**J,K,N**). To measure the length, the labeled cell array is placed in profile (usually not a lateral view), and the image of the array is collected (**D**) and measured in Object-Image. This example shows an embryo labeled at the dorsal margin at shield stage (**D**). Cells labeled with photo-activated dye are revealed with an anti-FITC antibody and the color substrate Fast Red (**H**). The presomitic expression domains of *paraxial protocadherin*, *papc*, provide landmarks. The labeled cells glow strongly under Rhodamine epifluorescence (**I**). nt, Notochord; PSM presomitic mesoderm.

Orient the embryo with a fine probe, and then cover the embryo with a cover slip. Make final adjustments to its position by moving the cover slip. To label lateral cells, place the embryo in a lateral view (**Fig. 3A,E**).

2. Place the protractor reticle in the eyepiece of the compound microscope. View the embryo in lateral view using the 10X objective and white light only. Using the protractor as a reference, center the cells to be labeled. Open the shutter on the fluorescence light path to view the embryo under the fluorescein isothiocyanate (FITC) filter. The embryo should be completely dark.

3. Close down the iris diaphragm on the fluorescence light path (*see* **Note 5**). This will narrow the light beam to a small portion of the visual field. Switch to the 40X objective. Close the fluorescence shutter! To label lateral cells, focus on the lateral margin. Place the DAPI filter in the fluorescence light path. Open the shutter on the fluorescence light path for 2 s to photoactive a small spot of cells. Close the shutter!

4. Switch the filter to FITC, and open the iris diaphragm and the shutter to view the photoactivated cells. The embryo should have a green spot in a dark field (**Fig. 3A**).

5. Reorient the embryo to an animal pole view on the dissecting microscope (**Fig. 3I, M**). Using the protractor reticle on the compound microscope and the 10X objective, determine the location of the spot relative to the dorsal midline. If the spot is not at the desired angle from dorsal, try again with the same embryo, using the spot as a new reference point.

6. Raise the embryos in a dark incubator until it is time to measure the location of the labeled cells (**Fig. 3I–K,M,N**; *see* **Note 6**).

7. To measure the length of a cell group (**Fig. 3A–G**), orient the embryo to place the labeled group in profile (**Fig. 3D**). Collect an image with the digital camera. Measure the length of the cell group with Object-Image. To convert the pixel-based measurement to micrometers collect an image of a stage micrometer at each of the magnifications used for acquiring embryo images and measure known distances from the micrometer.

8. Fix the embryos at the one- or two-somite stage in 4% paraformaldehyde in 1X PBT at 4°C overnight. To remove the embryos from slides, submerge the slide in 30% Danieau's buffer, and allow the cover slip and embryo to float off.

3.3. Revealing the Fate of Labeled Cells Using In Situ *Hybridization*

1. Prepare an *in situ* hybridization using 5-bromo-4-chloro-3-indolyl phosphate / Nitro blue tetrazolium or BM Purple (Roche) as the color substrate. The *in situ* hybridization method of *C. Thisse* and *B. Thisse* can be found at (website: http://zfin.org/zf_info/monitor/vol5.1/vol5.1.html; link: *High Resolution Whole-Mount* In Situ *Hybridization*).

2. Prepare 1:1000 dilution of anti-fluorescein antibody in PBT, 2% sheep serum, and 2 mg/mL bovine serum albumin. Preadsorb antibody by incubation for 2 h at room temperature (RT) with fixed, permeabilized embryos, as in a standard *in situ* hybridization.

3. Wash 3×5 min with PBT.

4. Remove the anti-digoxygenin antibody by washing 3×5 min with 0.1 *M* glycine, pH 2.2, 0.1% Tween-20.

5. Wash 3 × 5 min with PBT.

6. Incubate the embryos in anti-fluorescein antibody in PBT, 2% sheep serum, 2 mg/mL BSA at a final dilution of 1:5000 for 2 h at RT.

7. Wash 6 × 15 min in PBT.

8. Wash 2 × 15 min in 0.1 M Tris-HCl, pH 8.2, 0.1% Tween-20.

9. Prepare Fast Red solution. Disperse the tablet in 2 mL of 0.1 M Tris-HCl, pH 8.2, 0.1% Tween-20 with vigorous vortexing. Remove insoluble particles by spinning at top speed in a microfuge for 2 min.

10. Stain embryos in Fast Red solution at RT in the dark. Expect to see red color (and a yellow background) within 10 min (*see* **Note 7**, **Fig. 3H**).

11. Stop the staining reaction by washing the embryos in PBT, 25 mM ethylenediamine tetraacetic acid. Store in 10 mM ethylenediamine tetraacetic acid in the dark at 4°C.

4. Notes

1. It is important to inject no more than 1 nL of 1% dye because the embryos do not tolerate it well. Lower doses of dye (less than 0.5 nL) may give faint fluorescence that, although hard to see for tracking the progress of cells during gastrulation, will still give good labeling when revealed by anti-fluorescein antibody and Fast Red coloring. The dye may uncage if exposed too long to room light. Avoid exposure to light while resuspending and aliquoting the dye and injecting embryos. Raise the embryos in the dark. Some lots of this dye can be more toxic, causing autoactivation of dye, and deformed embryos. Reducing the dose to 0.5 nL can reduce this problem. Similar autoactivation can occur if embryos are unhealthy. Keeping a "back-up" stock of dye is not recommended.

2. To control the position of an embryo, it is necessary to remove the chorion. We place embryos in a glass dish of 30% Danieau's buffer and manually remove chorions using a pair of sharp forceps.

3. A drop of methylcellulose is placed between the supports on the slide. The methylcellulose drop should not touch the supports. The goal is for the methylcellulose to act as a hanging drop under the cover slip so movement of the cover slip will turn the embryo. The embryo is added in a minimal volume of buffer, with the tip of the pipet in the methylcellulose. Never allow an embryo younger than YPC to pass through the air; the yolk cell will burst. Orient the embryo with a fine probe, cover with a cover slip, and make final adjustments to position by moving the cover slip.

4. Adjusting the position of an embryo is much easier if one takes advantage of the optics, specifically the greater depth of field, on the dissection microscope and performs adjustments there.

5. There seem to be two styles of microscope, the type where a diaphragm is placed at the focal plane of the fluorescent light path and the type where an opening or slider is placed there. If your microscope is the former, you can narrow the light spot by closing the diaphragm and by switching to higher power objectives. If it is the latter, a pinhole device, literally a foil pierced with a pin, can be placed in the light path to narrow the beam. The higher the objective magnification, the smaller and more intense will be the light beam.

6. To avoid injuring gastrulation stage embryos, we raise them on the slides on a support in a sealed humid plastic box at 28.5°C. To remove the embryos from slides, submerge the slide in a glass dish of 30% Danieau's buffer, and allow the cover slip and embryo to float off. Prying up the cover slip will most likely kill the embryo.

7. Fast Red is soluble in methylsalicylate, making this reagent unsuitable for clearing embryos. Different lots of anti-fluorescein antibody should be tested for optimal dilution, which will be 1:2000 to 1:5000. Fast Red also glows under Rhodamine optics and this is sometimes better labeling than the red color (**Fig. 3L**). We have always performed the *in situ* hybridization technique before revealing the FITC, and do not recommend changing the order. The labeled cells may alternatively be visualized with 5-bromo-4-chloro-3-indolyl phosphate/Nitro blue tetrazolium (blue) and the *in situ* markers with Fast Red or Magenta Phosphate in place of Fast Red.

Acknowledgments

We thank our colleagues Adi Inbal, Seok-hyung Kim, and Jennifer Ray for comments on this chapter and all the members of the Solnica-Krezel lab for interest and lively discussion. D.S.S. is supported by NIH training grant 5 T32 HL07751. Work in the L.S-K lab is supported by NIH grants GM55101 and GM62283.

References

1. Kimmel, C. B. and Law, R. D. (1985) Cell lineage of zebrafish blastomeres. I. Cleavage pattern and cytoplasmic bridges between cells. *Dev. Biol.* **108,** 78–85.

2. Kimmel, C. B. and Law, R. D. (1985) Cell lineage of zebrafish blastomeres. II. Formation of the yolk syncytial layer. *Dev. Biol.* **108,** 86–93.

3. Kane, D. A. and Kimmel, C. B. (1993) The zebrafish midblastula transition. *Development* **119,** 447–456.

4. Warga, R. M. and Nusslein-Volhard, C. (1998) *Spadetail*-dependent cell compaction of the dorsal zebrafish blastula. *Dev. Biol.* **203,** 116–121.

5. Schmitz, B. and Campos-Ortega, J. A. (1994) Dorso-ventral polarity of the zebrafish embryo is distinguishable prior to the onset of gastrulation. *Roux. Arch. Dev. Biol.* **203,** 374–380.

6. Warga, R. M. and Kimmel, C. B. (1990) Cell movements during epiboly and gastrulation in zebrafish. *Development* **108,** 569–580.

7. Wilson, E. T., Helde, K. A., and Grunwald, D. J. (1993) Something's fishy here—rethinking cell movements and cell fate in the zebrafish embryo. *Trends Genet.* **9,** 348–352.

8. Kane, D. A., Hammerschmidt, M., Mullins, M. C., Maischein, H.-M., Brand, M., van Eeden, F. J., et al. (1996) The zebrafish epiboly mutants. *Development* **123,** 47–55.

9. Stemple, D. L., Solnica-Krezel, L., Zwartkruis, F., Neuhauss, S. C., Schier, A. F., Malicki, J., et al. (1996) Mutations affecting Development of the notochord in zebrafish. *Development* **123,** 117–128.

10. Schoenwolf, G. C. and Smith, J. L. (2000) Gastrulation and early mesodermal patterning in vertebrates. *Methods Mol. Biol.* **135**, 113–125.
11. Concha, M. L. and Adams, R. J. (1998) Oriented cell divisions and cellular morphogenesis in the zebrafish gastrula and neurula: a time-lapse analysis. *Development* **125**, 983–994.
12. Kane, D. and Adams, R. (2002) Life at the edge: epiboly and involution in the zebrafish. *Results Probl. Cell Differ.* **40**, 117–135.
13. Myers, D. C., Sepich, D. S., and Solnica-Krezel, L. (2002) Convergence and extension in vertebrate gastrulae: cell movements according to or in search of identity? *Trends Genet.* **18**, 447–455.
14. Myers, D. C., Sepich, D. S., and Solnica-Krezel, L. (2002) Bmp activity gradient regulates convergent extension during zebrafish gastrulation. *Dev. Biol.* **243**, 81–98.
15. Keller, R., Davidson, L., Edlund, A., Elul, T., Ezin, M., Shook, D., and Skoglund, P. (2000) Mechanisms of convergence and extension by cell intercalation. *Philos. Trans. R Soc. Lond. B Biol. Sci.* **355**, 897–922.
16. Glickman, N. S., Kimmel, C. B., Jones, M. A., and Adams, R. J. (2003) Shaping the zebrafish notochord. *Development* **130**, 873–887.
17. Kane, D. A. and Warga, R. M. (1994) Domains of movement in the zebrafish gastrula. *Semin. Dev. Biol.* **5**, 101–109.
18. Heisenberg, C. P. and Tada, M. (2002) Zebrafish gastrulation movements: bridging cell and Developmental biology. *Semin. Cell. Dev. Biol.* **13**, 471–479.
19. Jessen, J. R., Topczewski, J., Bingham, S., Sepich, D. S., Marlow, F., Chandrasekhar, A., et al. (2002) Zebrafish trilobite identifies new roles for Strabismus in gastrulation and neuronal movements. *Nat. Cell Biol.* **4**, 610–615.
20. Kanki, J. P. and Ho, R. K. (1997) The *Development* of the posterior body in zebrafish. *Development* **124**, 881–893.
21. Kanki, J. P., Chang, S., and Kuwada, J. Y. (1994) The molecular cloning and characterization of potential chick DM-GRASP homologs in zebrafish and mouse. *J. Neurobiol.* **25**, 831–845.
22. Agathon, A., Thisse, C., and Thisse, B. (2003) The molecular nature of the zebrafish tail organizer. *Nature* **424**, 448–452.
23. Warga, R. M. and Nusslein-Volhard, C. (1999) Origin and Development of the zebrafish endoderm. *Development* **126**, 827–838.
24. Warga, R. M. and Stainier, D. Y. (2002) The guts of endoderm formation. *Results Probl. Cell Differ.* **40**, 28–47.
25. D'Amico, L. A. and Cooper, M. S. (2001) Morphogenetic domains in the yolk syncytial layer of axiating zebrafish embryos. *Dev. Dyn.* **222**, 611–624.
26. D'Amico, L. A. and Cooper, M. S. (1997) Spatially distinct domains of cell behavior in the zebrafish organizer region. *Biochem. Cell Biol.* **75**, 563–577.
27. Brennan, C., Amacher, S. L., and Currie, P. D. (2002) Somitogenesis. *Results Probl. Cell Differ.* **40**, 271–297.
28. Beattie, C. E., Granato, M., and Kuwada, J. Y. (2002) Cellular, genetic and molecular mechanisms of axonal guidance in the zebrafish. *Results Probl. Cell Differ.* **40**, 252–269.

29. Raible, D. W., Wood, A., Hodsdon, W., Henion, P. D., Weston, J. A., and Eisen, J. S. (1992) Segregation and early dispersal of neural crest cells in the embryonic zebrafish. *Dev. Dyn.* **195,** 29–42.
30. Jesuthasan, S. (1996) Contact inhibition/collapse and pathfinding of neural crest cells in the zebrafish trunk. *Development* **122,** 381–389.
31. Raz, E. (2003) Primordial germ-cell Development: the zebrafish perspective. *Nat. Rev. Genet.* **4,** 690–700.
32. Raz, E. and Hopkins, N. (2002) Primordial germ-cell development in zebrafish. *Results Probl. Cell Differ.* **40,** 166–179.
33. Sepich, D. S., Myers, D. C., Short, R., Topczewski, J., Marlow, F., and Solnica-Krezel, L. (2000) Role of the zebrafish trilobite locus in gastrulation movements of convergence and extension. *Genesis* **27,** 159–173.
34. Topczewski, J., Sepich, D. S., Myers, D. C., Walker, C., Amores, A., Lele, Z., et al. (2001) The zebrafish glypican knypek controls cell polarity during gastrulation movements of convergent extension. *Dev. Cell* **1,** 251–264.
35. Hammerschmidt, M., Pelegri, F., Mullins, M. C., Kane, D. A., Brand, M., van Eden, F. J., et al. (1996) Mutations affecting morphogenesis during gastrulation and tail formation in the zebrafish, Danio rerio. *Development* **123,** 143–151.
36. Heisenberg, C-P., Brand, M., Jiang, Y-J., Warga, R. M., Beuchle, D., van Eeden, F. J. M., et al. (1996) Genes involved in forebrain development in the zebrafish, Danio rerio. *Development* **123,** 191–203.
37. Yamashita, S., Miyagi, C., Carmany-Rampey, A., Shimizu, T., Fujii, R., Schier, A. F., and et al. (2002) Stat3 controls cell movements during zebrafish gastrulation. *Dev. Cell* **2,** 363–375.
38. Ho, R. K. and Kane, D. A. (1990) Cell-autonomous action of zebrafish spt-1 mutation in specific mesodermal precursors. *Nature* **348,** 728–730.
39. Hammerschmidt, M., Pelegri, F., Mullins, M. C., Kane, D. A., van Eeden, F. J., Granato, M., et al. (1996) dino and mercedes, two genes regulating dorsal development in the zebrafish embryo. *Development* **123,** 95–102.
40. Mullins, M. C., Hammerschmidt, M., Kane, D. A., Odenthal, J., Brand, M., van Eeden, F. J., et al. (1996) Genes establishing dorsoventral pattern formation in the zebrafish embryo: the ventral specifying genes. *Development* **123,** 81–93.
41. Heisenberg, C. P., Tada, M., Rauch, G. J., Saude, L., Concha, M. L., Geisler, R., et al. (2000) Silberblick/Wnt11 mediates convergent extension movements during zebrafish gastrulation. *Nature* **405,** 76–81.
42. Rauch, G. J., Hammerschmidt, M., Blader, P., Schauerte, H. E., Strahle, U., Ingham, P. W., et al. (1997) Wnt5 is required for tail formation in the zebrafish embryo. *Cold Spring Harb. Symp. Quant. Biol.* **62,** 227–234.
43. Kilian, B., Mansukoski, H., Barbosa, F. C., Ulrich, F., Tada, M., and Heisenberg, C. P. (2003) The role of Ppt/Wnt5 in regulating cell shape and movement during zebrafish gastrulation. *Mech. Dev.* **120,** 467–476.
44. Marlow F, Gonzalez EM, Yin C, Rojo C, and Solnica-Krezel L. (2004) No tail co-operates with non-canonical Wnt signaling to regulate posterior body morphogenesis in zebrafish. *Development* **131,** 203–216.

45. Ulrich, F., Concha, M. L., Heid, P. J., Voss, E., Witzel, S., Roehl, H., et al. (2003) Slb/Wnt11 controls hypoblast cell migration and morphogenesis at the onset of zebrafish gastrulation. *Development* **130,** 5375–5384.

46. Griffin, K. J., Amacher, S. L., Kimmel, C. B., and Kimelman, D. (1998) Molecular identification of spadetail: regulation of zebrafish trunk and tail mesoderm formation by T-box genes. *Development* **125,** 3379–3388.

47. Hild, M., Dick, A., Rauch, G. J., Meier, A., Bouwmeester, T., Haffter, P., et al. (1999) The smad5 mutation somitabun blocks Bmp2b signaling during early dorsoventral patterning of the zebrafish embryo. *Development* **126,** 2149–2159.

48. Nguyen, V. H., Schmid, B., Trout, J., Connors, S. A., Ekker, M., and Mullins, M. C. (1998) Ventral and lateral regions of the zebrafish gastrula, including the neural crest progenitors, are established by a bmp2b/swirl pathway of genes. *Dev. Biol.* **199,** 93–110.

49. Nguyen, V. H., Trout, J., Connors, S. A., Andermann, P., Weinberg, E., and Mullins, M. C. (2000) Dorsal and intermediate neuronal cell types of the spinal cord are established by a BMP signaling pathway. *Development* **127,** 1209–1220.

50. Schulte-Merker, S., Lee, K. J., McMahon, A. P., and Hammerschmidt, M. (1997) The zebrafish organizer requires chordino. *Nature* **387,** 862–863.

51. Marlow, F., Zwartkruis, F., Malicki, J., Neuhauss, S. C., Abbas, L., Weaver, M., et al. (1998) Functional interactions of genes mediating convergent extension, knypek and trilobite, during the partitioning of the eye primordium in zebrafish. *Dev. Biol.* **203,** 382–399.

52. Saude, L., Woolley, K., Martin, P., Driever, W., and Stemple, D. L. (2000) Axis-inducing activities and cell fates of the zebrafish organizer. *Development* **127,** 3407–3417.

53. Kimmel, C. B., Ballard, W. W., Kimmel, S. R., Ullmann, B., and Schilling, T. F. (1995) Stages of embryonic development of the zebrafish. *Dev. Dyn.* **203,** 253–310.

54. Kozlowski, D. J., Murakami, T., Ho, R. K., and Weinberg, E. S. (1997) Regional cell movement and tissue patterning in the zebrafish embryo revealed by fate mapping with caged fluorescein. *Biochem. Cell Biol.* **75,** 551–562.

17

The *Xenopus* Embryo as a Model System for Studies of Cell Migration

Douglas W. DeSimone, Lance Davidson,
Mungo Marsden, and Dominique Alfandari

Summary

In this chapter, we describe procedures for the microsurgical removal of cells and tissues from early-stage embryos of the amphibian *Xenopus laevis*. Using simple culture conditions and artificial substrates, these preparations undergo a variety of quantifiable cellular behaviors that closely mimic cell migration in vivo. Two general methods are described. The first includes procedures for obtaining a dorsal marginal zone explant from early gastrulae in order to investigate the sheet-like extension and migration of the mesendoderm that spreads to cover the inner surface of the blastocoel roof in intact embryos. This preparation allows high-resolution analyses of cellular and subcellular events in a contiguous tissue preparation. The second describes methods for the isolation of cranial neural crest cells from tailbud stage embryos. Cranial neural crest tissue cultured in vitro on fibronectin will undergo segmentation and migrate as streams of cells as they do in the developing head. Each of these robust preparations provides an excellent example of the migratory events that are possible to observe in vitro using amphibian embryos.

Key Words: *Xenopus*; embryo; explant; cell migration; cell adhesion; gastrulation; neurulation; neural crest cells.

1. Introduction

An ideal system for the study of cell migration would be applicable to the wide variety of experimental methods that are made possible through studies of cells in culture but would also enable detailed investigations of individual cell behaviors and functions in vivo. Although no single system is likely to accommodate this scenario in its entirety, the *Xenopus* embryo affords a number of unique advantages to investigators interested in the problem of cell

From: *Methods in Molecular Biology, vol. 294: Cell Migration: Developmental Methods and Protocols*
Edited by: J-L. Guan © Humana Press Inc., Totowa, NJ

migration at multiple levels. First, the embryos are large and easy to dissect at most stages of early development. This feature makes it possible to identify and remove individual tissues and/or small numbers of defined cells from the embryo. The motility of these cell and tissue fragments can then be studied on artificial substrates as intact pieces or as dissociated single cells. Alternatively, a variety of well-characterized explant preparations have been devised that allow cell behaviors to be visualized within tissues undergoing motility-driven changes *(1)*. Tissues can also be extirpated from fluorescently labeled embryos and transferred by grafting into unlabeled host embryos or by "seeding" dissociated cells into unlabeled host tissues. In each case, the bulk movements and protrusive behaviors of individual cells can be resolved by using low-light time-lapse, digital imaging methods. Finally, one remarkable feature of embryonic *Xenopus* tissues and single cells is that they can be cultured in vitro at room temperature in simple salt solutions without CO_2 or serum supplementation, in large part, owing to their intracellular reserves of yolk protein.

This chapter focuses on two areas of research in cell migration where methods have been developed using *Xenopus* embryos. The first is the motility of cells involved in gastrulation movements, which drive the rearrangement of the blastula yielding an embryo with such recognizable landmarks as a head, tail, front, back, and left and right sides. It is arguably the most dramatic cellular rearrangement in vertebrate development, and much is now known about the cell behaviors that characterize these movements *(2,3)*. The second is the migration of cranial neural crest cells, which move from the midline of the developing embryo in several contiguous streams and come to populate regions of the head where they will give rise to the cartilaginous and skeletal elements of the head and face, among other things *(4)*. Two methods are considered, beginning with an "intact" explant preparation recently described for analyses of mesendoderm migration and bulk morphogenetic movements *(5)*. The second describes procedures used to investigate cranial neural crest cell migration *(6)*.

2. Materials

2.1. Solutions and Culture Medium

1. Modified Barth's Saline (MBS): 88 m*M* NaCl, 1 m*M* KCl, 2.5 m*M* $NaHCO_3$, 0.3 m*M* $CaNO_3$, 0.41 m*M* $CaCl_2$, 0.82 m*M* $MgSO_4$, 15 m*M* HEPES (pH 7.6) with or without antibiotic (i.e., 50 µg/mL gentamicin sulfate).
2. Danilchik's for Amy (DFA): 53 m*M* NaCl, 5 m*M* Na_2CO_3, 4.5 m*M* K gluconate, 32 m*M* Na gluconate, 1 m*M* $MgSO_4$, 1 m*M* $CaCl_2$, and 0.1% BSA *(7)*. The pH is adjusted after addition of bovine serum albumin (BSA) with 1 *M* bicine to pH 8.3. Stock DFA can be sterile filtered and stored at –20°C. DFA is

formulated to mimic the interstitial fluid found within the embryo and contributes to the capacity of cells within explants to recapitulate the cell behavior programs found in vivo.

3. Phosphate-buffered saline (PBS): standard mammalian tissue culture formulations of PBS can be used. The solution must contain 1 mM divalent cations (magnesium and calcium) for effective adhesion.

4. MOPS/EGTA/magnesium sulfate/formaldehyde buffer (MEMFA): 0.1 M MOPS (pH 7.4), 2 mM ethylenebis(oxyethylenenitrilo)tetraacetic acid, 1 mM MgSO$_4$, and 3.7% formaldehyde. A 10X stock of MEMFA salts should be prepared without adding formaldehyde until needed for use. One-tenth volume of fresh 37% formaldehyde can be added to 1X MEMFA salts to make a working fix.

2.2. Embryo Preparation

Xenopus embryos can be fertilized in vitro using standard methods *(8)*. Once fertilized eggs undergo cortical rotation, their jelly-coats are removed with 2% cysteine (pH 8.0) in one-third strength MBS (1/3XMBS). Whole embryos should remain in 1/3X MBS until the required stage. Vitelline membranes are removed manually at the necessary stage with watchmaker's forceps *(9)* to facilitate microsurgery. We use custom-designed aluminum "cold-plates" connected to a chilled water circulator (15°C) to keep embryos cold during microsurgery to retard wound healing and to slow the rate of development.

2.3. Fluorescent Labeling of Embryos and Cells

1. Intracellular labeling: lysine-fixable anionic dextrans coupled to fluorophores can be used to increase contrast in embryos and cells so that cell boundaries can be visualized by live epifluorescence and confocal imaging. Whole embryos can be injected at the one-cell stage with up to 2 nL of 25 mg/mL of fluorescent dextran (Molecular Probes). Aldehyde-based fixation of these embryos at later stages also allows visualization of cell shapes and general histology, following clearing in 2:1 benzyl benzoate to benzyl alcohol. Collection of z-series image planes by confocal microscopy is a particularly effective method for visualizing cellular details in these preparations.

2. Labeling of cell outlines: injection of up to 2 ng of synthetic capped transcripts encoding green fluorescent protein (GFP) conjugated to protein domains directing subcellular localization allow further increases in contrast in live explants and the identification of unique cells within a control tissue. The ability to use GFP localized to different parts of the cell allows a variety of tissue recombinant-type experiments where the behavior of cells expressing one set of proteins can be compared within-the-same-explant to cells expressing a second set of proteins. Membrane localized GFP (GAP43-GFP or mem-GFP *[10]*) is especially useful in studies where cell boundaries or protrusions need to be observed.

3. Methods

3.1. The Mesendoderm Explant

A variety of methods have been used to analyze the migration of the mesendoderm at gastrulation. Recently, we described a novel explant preparation derived from early gastrulae that includes a 180° swath of dorsal marginal zone tissue *(5)* centered on the midline of the embryo. It contains both superficial and deep tissues and includes neural, mesodermal and some endodermal tissues. The mesendoderm in this explant migrates as a sheet, extending away from the dorsal lip along an artificial substrate. In vivo, this tissue spreads over the inside of the fibronectin-rich blastocoel roof (its normal substrate) where it is not possible to observe cell behaviors. In contrast, individual cell behaviors, coordinated cell migration, and subcellular protein localization are readily observed in the mesendoderm explant. This assay is useful for investigating regulation of mesendoderm adhesion and motility and the responses of cells and tissues to defined substrates.

3.1.1. Preparation of Substrates and Viewing Chambers

Custom substrates can be presented to explants to investigate cell behaviors, morphogenesis, and patterns of gene expression. Our standard assays use fibronectins isolated from *Xenopus* or mammalian plasma or bacterially derived fibronectin–GST fusion proteins *(5,11)*. Careful preparation of clean glass is essential for the success of these assays because a large field of homogeneously applied substrate is required.

1. Clean glass cover slips by first washing with 75% ethanol 25% distilled water at pH 10.0 (approx one drop of 10 *N* NaOH per 20 mL). Cover slips should be washed extensively with distilled water followed by a 100% ethanol rinse and flame drying. Cover slips should be handled carefully from this point and the surface protected from contact by storing on lint-free tissues (Kim wipe) in dust free-chambers until used.
2. Fibronectin (or various fibronectin fusion proteins) can be adsorbed to cleaned cover glasses at 2.0 to 50 µg/mL overnight at 4°C (*see* **Note 1**). Care should be taken to reduce evaporation and prevent uneven adsorption of fibronectin by coating substrates in a humidified chamber. Coated cover slips are then washed with PBS and blocked with 1 mg/mL BSA in DFA.
3. Whole embryos and explants at these stages can develop normally in the absence of oxygen allowing the construction of sealed imaging chambers. Simple chambers made from glass bottomed Petri dishes (Matek) allow the best viewing on inverted compound microscopes. Other chambers can be made with nylon washers (Small Parts Inc.) glued to large cover slips (containing prepared substrate) with silicone grease (high vacuum grease, Dow Chemical). These chambers are sealed after adding explants and filling with DFA (*see* **Subheading 3.1.2., step 3**)

with small square cover slips and additional silicone grease; explants can then be viewed through either cover slip allowing use on both upright and inverted compound microscopes.

3.1.2. Microsurgery and Preparation of the Mesendoderm Explant

This has previously been described in detail *(5)* and is discussed here briefly.

1. Vitelline membranes are removed manually with forceps at stage 10–10.25 *(12)* and two incisions are made 90° from the midline using eyebrow knives and hair loops. Most of the vegetal endoderm is then removed with the exception of several rows of sub-blastoporal endoderm. The anterior margin of the tissue is incised just beyond the prospective cement gland. Care is taken not to damage the leading edge of the mesendoderm.

2. Explants are moved around with hair loops and transferred carefully using a disposable plastic pipet or mouth pipet to the nylon ring culture chamber, which contains DFA. Contact with the air–water interface will result in the immediate destruction of the explant (*see* **Note 2**).

3. After arranging the explant on the substrate using hair loops, a small fragment of cover slip with silicon grease located at opposite edges is used to compress the explant gently with forceps (*see* **Note 3**). The volume of DFA in the chamber is then adjusted and another cover glass is used to seal the top of the chamber for microscopy and image analysis. Multiple explants can be situated on a single substrate depending on the size of the chamber and one's ability to work quickly. This facilitates the amount of data that can be collected as discussed in **Subheading 3.1.3**.

4. After imaging, it is often desirable to fix the explant for *in situ* hybridization or immunostaining. The top cover glass is carefully removed from the chamber with forceps and the chamber transferred to a Petri dish containing DFA. Once submerged in DFA the nylon ring can be removed exposing the explant(s) and the cover glass fragment that compresses it. The appropriate fix is then added to the Petri dish and after a suitable incubation time, the cover glass fragment is removed from the explant. The explant is then gently "shaved" off of the underlying substrate cover glass using a piece of mylar film held by forceps. The explant is transferred for further processing and analysis to microfuge tubes, multi-well plates or other suitable small volume containers that allow the explants to be visualized easily while changing solutions.

3.1.3. Leading-Edge Migration Assays and Image Analysis

1. Movements of the leading edge mesendoderm can be followed at 10X magnification on an inverted compound microscope stage. An XYZ motorized stage (Ludl) used in conjunction with a computer controlled image acquisition system (e.g., Metamorph, Universal Imaging or OpenLab, Improvision) makes it possible to follow up to 20 explants simultaneously. Time intervals as long as 10 min can be used to assess the bulk rate movement of the leading edge of the mesoderm.

Movements of individual cells at the leading edge can be tracked with image processing software (NIH-Image or ImageJ, Wayne Rasband, developed at the U.S. National Institutes of Health and available on the Internet at Website: http://rsb.info.nih.gov/nih-image/) running on either an Apple Macintosh or a PC compatible computer. Rates of migration can be directly extracted from the XY positions of the leading edge from time lapses. Statistical comparison of different treatments are carried out using the non-parametric Mann–Whitney *U* Test *(13)*.

2. Cell behavior assays: movements of individual cells can be tracked by outlining the shapes of cells in subsequent frames of time-lapse sequences. Explants expressing membrane-localized GFP or fluorophores-conjugated dextran lineage tracers (*see* **Subheading 2.3., steps 1,2**) can be visualized with a confocal laser scanning confocal microscope at intervals as brief as 5 s. Cell positions, areas, aspect ratios, and protrusions can be tracked from start to finish. Protrusions can be characterized as either discrete events *(14)* or as pixel-areas into which a cell has extended *(15)*. Angle distributions of protrusions or directions of migra-tion are analyzed with the appropriate circular statistics *(14,16)*.

3. Morphogenesis assays: the rearrangements of cells in whole tissues can be observed in time-lapse sequences of cells in explants. Explants expressing membrane-tagged GFP or lineage tracers can be followed with a conventional compound microscope equipped with epifluorescence, a sensitive low light camera (e.g., Orca, Hamamatsu), and a computer controlled shutter (Ludl). Cell shapes and positions in time-lapse sequences collected from computer controlled acquisition software (Metamorph, Universal Imaging Corp.) can be assessed for rearrangement and tissue shape change *(5)*. Assessments can be made for individual events such as the number of neighbor changes per cell in the field or bulk tissue properties such as the mediolateral intercalation index *(17)*.

3.2. Cranial Neural Crest (CNC) Cell Migration

CNC cells are induced at the lateral border of the anterior neural plate. While molecular markers are already visible as early as gastrulation, there is no visible border between the neural plate and the neural crest tissue at this stage. A detailed description of *Xenopus laevis* CNC cells can be found in Sadagiany and Thiebaud *(4)*. CNC cells can be dissected from neurula stage 14 through stage 17 *(12)*. In stage 14 embryos the explant appears as a flat triangle with no distinct segmentation evident. At stage 17 the segmental organization is already visible in the explant (**Fig. 1**).

3.2.1. Microsurgical Removal of CNC Tissue

The following method was adapted from Borchers et al. *(18)*.

1. Neurula stage embryos are placed in an agarose coated Petri dish in 1X MBS containing 50 µg/mL of gentamicin sulfate. The vitelline envelope is removed manually using forceps by pulling on the flank of the embryo to avoid damaging the dorsal neural structures. Devitellinized embryos can be left in this dish until they reach the selected stage.

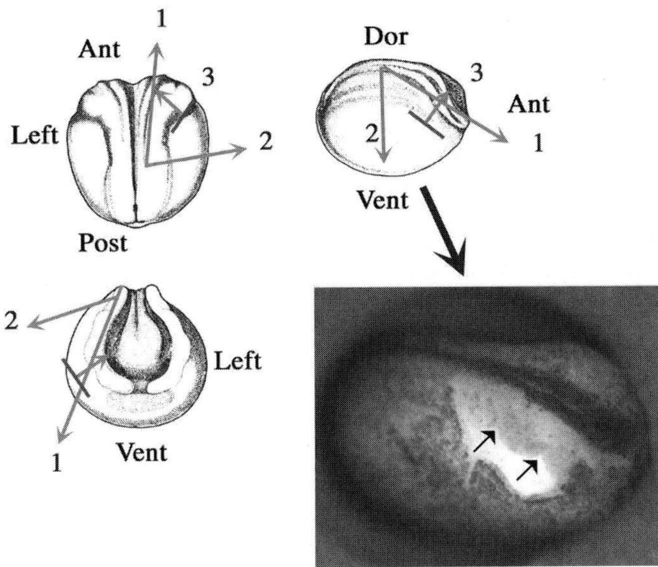

Fig. 1. Microsurgery and identification of CNC tissue used for explants. Illustrations provide dorsal, side, and anterior views of a neurula stage embryo (after Nieuwkoop and Faber *[12]*). Long arrows mark locations of the initial incisions (1 and 2). Base of short arrow (3) shows approximate location of ventral margin of the underlying CNC, which can be lifted away dorsally (in direction of arrow 3). The micrograph shows a typical stage 17 embryo (dorso–lateral view) with the overlying ectoderm removed from the right side. The CNC, which has not been removed, is easy to distinguish in this particular embryo because the CNC are lightly pigmented. Note that segmentation of the CNC has begun by this stage (arrows indicate borders forming between the branchial–hyoid and hyoid–mandibular segments).

2. For dissection, embryos are transferred to plasticine-coated Petri dishes in 1X MBS with 50 µg/mL of gentamicin sulfate. Depressions are made in the plasticine with a glass bead made from the tip of a heat-pulled Pasteur pipet. The size of the depression should allow the embryo to settle in, leaving the dorsal half exposed for microsurgery using an eyebrow knife and hair loop. With the embryo secured in the plasticine, the first incision of the epidermis is made from the trunk toward the front along one of the neural folds (**Fig. 1**). The second incision starts from the same point but goes laterally toward the flank. The epidermis can then be lifted away from the position where the two initial incisions meet.

3. Using the eyebrow knife, the ventral-most edge of the cranial neural crest tissue can be lifted upward (dorsally; *see* **Note 4**). This should leave the underlying mesoderm intact. At no point should the pharyngeal cavity be exposed as

this indicates that the explant contains mesoderm. Once the cranial neural crest explant is lifted, the eyebrow knife is used again to separate the neural crest from the neural fold. The limit between the neural crest and the neural fold is not clear at this stage, but subsequent *in situ* hybridization with neural crest markers can be performed to test whether the correct territory has been dissected. The presence of some neural tissue does not seem to interfere with the experiments as these cells do not migrate out. Explants can be transferred from the dissection dish into a new agarose coated Petri dish containing DFA and 50 µg/mL gentamicin sulfate, until they can be used for cell migration assays (10 to 60 min; *see* **Note 5**). Explant transfer is best achieved using a P20 Pipetman and BSA-coated tips.

3.2.2. Substrate Preparation

Substrates can be prepared either on plastic or glass using standard techniques. On plastic (non tissue culture treated), fibronectin will promote cell migration in a wide range of coating concentrations starting at 5 µg/mL up to 50 µg/mL (*see* **Note 1**). Coating is usually performed overnight at 4°C in PBS as above but also can be achieved in as short a time as 1 h at room temperature. Coating on glass is usually less efficient and thus requires higher concentrations of substrate. Glass cover slips should be prepared as described for the mesendoderm explants (*see* **Subheading 3.1.1., steps 1,2.**).

3.2.3. Migration Assays

CNC migration assays can be performed on intact explants as well as on single cells. Explants can be dissociated to single cells in calcium- and magnesium-free 1X MBS at 15°C in less than 1 h with occasional mixing. Cells or explants are transferred onto fibronectin or other extracellular matrix (ECM)-coated substrates in DFA. Intact explants will attach and cells migrate out on the substrate at temperatures ranging from 17–20°C. Lower temperatures dramatically reduce cell motility whereas higher temperatures decrease explant viability. Cell migration is also observed in 1X MBS but with higher variability in the results. Various measures of CNC cell migration and segmentation can be monitored as described in detail in Alfandari et al. (*6*) and above for the mesendoderm explant. During migration, CNC will deposit their own ECM including fibronectin, thus observations made using various artificial substrates after 5 h should take this factor into account. When intact explants are assayed on fibro-nectin cells initially migrate as a cohesive mass for the first 4 to 5 h. During this time, CNC cells also segregate into segments similar to those observed in vivo (i.e., mandibular, hyoid and branchial). After 5 h, cell cohesion is lost and individual cell migration becomes pronounced. This migration can continue for 15 h before cell death occurs.

3.2.4. Processing and Imaging of CNC

Xenopus embryonic cells, including the CNC, are large, yolk platelet-filled cells and thus more fragile than conventional tissue culture cells. Because of this, special care is required for fixing. We have used two methods of fixation compatible with immunofluorescence, although other techniques may also work.

1. Formaldehyde-based fixes can be used and should be added directly to the dish containing the cells or explants. For example, 2X MEMFA can be added and gently mixed with the DFA culture medium and then replaced with 1X MEMFA. This fresh fixing reagent is left for 30 min at room temperature. It is critical that neural crest cells and explants are not exposed to the air/liquid interface prior to fixation. If permeabilization is required, fixed explants should be submerged in 100% methanol for 15 min. Use of detergent such as Triton X-100 may produce a strong auto-fluorescence, which may be incompatible with subsequent immunostaining. Explants fixed in this manner can be processed for *in situ* hybridization using standard *Xenopus* protocols (*9*) as well as for immunostaining.
2. For some antibodies, acetone is the reagent of choice for fixation. In this case, explants must be prepared on substrated-coated glass cover slips. Cell migration can be observed by placing the cover slips in a 35-mm Petri dish containing DFA. After image analysis the cover slip can be removed carefully using forceps, making sure that the cover slip remains horizontal (with DFA pooled at surface to protect explant from air interface) at all times, and placed into a glass Petri dish containing acetone at –20°C. Fixation should be performed for 10 to 15 min before storage in PBS and further processing.

4. Notes

1. Optimal substrate concentrations of ECM substrates or other adhesion molecules may need to be determined empirically to establish optimal conditions. For example, a range of fibronectin concentrations (5–50 µg/mL) will likely promote excellent attachment and migration of explants, however, higher concentrations may inhibit migration or even attachment. It is always a good idea to titrate a range of substrate molecule concentrations when setting up these assays to identify optimal conditions.
2. Agarose (1–2%) coated dishes are very useful for short-term storage of explants, typically while further microdissections are undertaken. Extirpated *Xenopus* cells and tissues are very "sticky" and will readily attach to one another and to plastic and glass. Explants and single cells will not attach to agarose, however. In the presence of divalent cations explants will initiate a rapid healing response and individual cells will aggregate once in contact, thus, normal levels of Ca^{2+} and Mg^{2+} should only be present once the explant is transferred to substrate unless transfer can be accomplished in under 15 min.

3. The most qualitatively difficult aspect of this procedure is cover slip compression because the force required to compress the explant is not easy to describe. If the explant is not adequately compressed it may tend to round-up and "heal," blocking any attempt to visualize cell movements in the absence of an optically flat preparation. If compression is too great the movements of the explants are repressed or inhibited altogether. Significant trial and error may be necessary to develop a "feel" for the suitable amount of cover slip compression required in these preparations.

4. In more darkly pigmented embryos, the cranial neural crest appears slightly brown and stands out against the lighter mesoderm (for an example, *see* **Fig. 1**). These "optimal" batches of embryos are excellent for training purposes by helping the investigator to distinguish the ventral edge of the tissue. In lighter pigmented embryos this margin is difficult to see but adjustment of the incident light sources, typically by fiber optic placement, can improve contrast of the tissues.

5. Upon long periods of culture in the agarose-coated dish, the CNC explant tends to heal with a loss of neural crest specific markers observed in some cases. This may also slow the initial phase of attachment to the substrate. For this reason it is best to prepare four to eight explants at one time and transfer them before dissecting additional explants. Migration can be prevented by keeping the substrate at 15°C until all explants are transferred. At this temperature, cells attach but do not migrate out and do not appear to lose expression of CNC marker genes.

Acknowledgments

We would like to acknowledge the contributions of many colleagues in helping to develop the methods described, including Helene Cousin, Na Jie, and Joe Ramos. We especially thank Ray Keller for continued help and inspiration in the development of new explant preparations. This work was supported by grants from the USPHS (HD 26402 and DE 14365).

References

1. Keller, R., Davidson, L., Edlund, A., Elul, T., Ezin, M., Shook, D., et al. (2000) Mechanisms of convergence and extension by cell intercalation. *Philos. Trans. R Soc. Lond. B Biol. Sci.* **355,** 897–922.
2. Keller, R. (2002) Shaping the vertebrate body plan by polarized embryonic cell movements. *Science* **298,** 1950–1954.
3. Winklbauer, R., Nagel, M., Selchow, A., and Wacker, S. (1996) Mesoderm migration in the *Xenopus* gastrula. *Int. J. Dev. Biol.* **40,** 305–311.
4. Sadaghiani, B. and Thiebaud, C. H. (1987) Neural crest development in the *Xenopus laevis* embryo, studied by interspecific transplantation and scanning electron microscopy. *Dev. Biol.* **124,** 91–110.
5. Davidson, L. A., Hoffstrom, B. G., Keller, R., and DeSimone, D. W. (2002) Mesendoderm extension and mantle closure in *Xenopus laevis* gastrulation: combined roles for integrin $\alpha_5 \beta_1$, fibronectin, and tissue geometry. *Dev. Biol.* **242,** 109–129.

6. Alfandari, D., Cousin, H., Gaultier, A., Hoffstrom, B. G., and DeSimone, D. W. (2003) Integrin $\alpha_5 \beta_1$ supports the migration of *Xenopus* cranial neural crest on fibronectin. *Dev. Biol.* **260,** 449–464.

7. Sater, A. K., Steinhardt, R. A., and Keller, R. (1993) Induction of neuronal differentiation by planar signals in *Xenopus* embryos. *Dev. Dyn.* **197,** 268–280.

8. Kay, B. K. and Peng, H. B. (1991) *Xenopus laevis: Practical Uses in Cell and Molecular Biology*, Academic Press, New York, NY.

9. Sive, H. L., Grainger, R. M., and Harland, R. M. (2000) *Early Development of Xenopus Laevis: A Laboratory Manual*, Cold Spring Harbor Laboratory Press, Cold Spring Harbor, NY.

10. Wallingford, J. B., Rowning, B. A., Vogeli, K. M., Rothbacher, U., Fraser, S. E., and Harland, R. M. (2000) Disheveled controls cell polarity during *Xenopus* gastrulation. *Nature* **405,** 81–85.

11. Na, J., Marsden, M., and DeSimone, D. W. (2003) Differential regulation of cell adhesive functions by integrin α subunit cytoplasmic tails in vivo. *J. Cell Sci.* **116,** 2333–2343.

12. Nieuwkoop, P. D. and Faber, J. (1994) Normal Table of *Xenopus laevis* (Daudin), Garland Publishing, Inc., New York, NY.

13. Sokal, R. R. (1981) *Biometry*, W. H. Freeman and Company, New York, NY.

14. Ezin, A. M., Skoglund, P., and Keller, R. (2003) The midline (notochord and notoplate) patterns the cell motility underlying convergence and extension of the *Xenopus* neural plate. *Dev. Biol.* **256,** 100–114.

15. Davidson, L. A. and Keller, R. E. (1999) Neural tube closure in *Xenopus laevis* involves medial migration, directed protrusive activity, cell intercalation and convergent extension. *Development* **126,** 4547–4556.

16. Batschelet, E. (1981) *Circular Statistics in Biology*, Academic Press, New York, NY.

17. Elul, T., Koehl, M. A., and Keller, R. (1997) Cellular mechanism underlying neural convergent extension in *Xenopus laevis* embryos. *Dev. Biol.* *191, 243–258.*

18. Borchers, A., Epperlein, H. H., and Wedlich, D. (2000) An assay system to study migratory behavior of cranial neural crest cells in *Xenopus*. *Dev. Genes Evol.* **210,** 217–222.

18

Neural Crest Migration Methods in the Chicken Embryo

Maria Elena de Bellard and Marianne Bronner-Fraser

Summary

Neural crest cells emerge from the neural tube early in development. They migrate extensively throughout the embryo and form most of the head and peripheral nervous system, giving rise to sensory and sympathetic ganglia, heart regions, adrenal cells, head bones, teeth, muscle cells, sensory organs, melanocytes, and other cell types. The neural crest is interesting because of its unique origin, development and differentiation. These cells are initially part of the dorsal neural tube, with a clear epithelial character; later, they transform into actively motile mesenchymal cells. Little is known about the underlying mechanism directing this process. It remains unknown why neural crest cells target particular derivatives (neurons, heart muscle and glia) and body regions (peripheral nerves, heart, skin, head and gut). Neural crest migration can be divided into three stages: 1) emigration from the neural tube; 2) migration along defined pathways; and 3) cessation of migration. At the onset of migration, neural crest cells lose their epithelial nature within the neural tube and transform into a migratory, mesenchymal cell type. Neural crest development has been best studied in avian embryos, which are amenable to surgical manipulation, cell marking techniques, cell culture and transgenesis by electroporation and retrovirally mediate gene transfer. The methods outlined below are those typically used to study and understand the different factors and signals necessary for the neural crest development before and during their migration.

Key Words: Neural crest cells; electroporation; live cell labeling; confocal imaging; whole mount; immunohistochemistry; immunolabeling, migration; chicken embryo.

1. Introduction

The neural crest is a transient population of cells named because it arises on the "crest" of the closing neural tube. Neural crest cells emigrate from the neural tube, migrate along precise pathways and finally localize in characteristic sites in the embryo to give rise to diverse cell types such as melanocytes, craniofacial skeleton, and much of the sensory and autonomic ganglia of the peripheral

From: *Methods in Molecular Biology, vol. 294: Cell Migration: Developmental Methods and Protocols*
Edited by: J-L. Guan © Humana Press Inc., Totowa, NJ

nervous system. These neurons of neural crest-derived sympathetic and parasympathetic ganglia innervate numerous organs such as the gut, kidneys and pancreas. The proper migration of neural crest cells to these targets and subsequent innervations is essential for proper body function and homeostasis.

Neural crest migration can be divided into three stages: 1) emigration from the neural tube; 2) migration along defined pathways; and 3) cessation of migration. At the onset of migration, neural crest cells lose their epithelial nature within the neural tube and transform into a migratory, mesenchymal cell type. Little is known about the mechanism underlying this process, except that alterations occur in adhesive properties and there is a breakdown of the neural tube basal lamina.

Neural crest development has been best studied in avian embryos, which are amenable to surgical manipulation, cell marking techniques, cell culture, and transgenesis by electroporation and retrovirally mediate gene transfer. The methods outlined are those typically used to study and understand the different factors and signals necessary for the neural crest development before and during their migration.

2. Materials

Materials used in in vivo and in vitro experiments are described in **Subheadings 2.1.–2.7.** and **2.8.–2.15.**, respectively.

2.1. In Vivo Labeling of the Whole Neural Tube and Neural Crest With DiI

1. Prepare 10% Sucrose fresh. (1 mg/9 mL H_2O), put at 37°C for a few minutes (*see* **Notes 1** and **2**).
2. Take the CM-DiI tubes (Molecular Probes D-282, 50 µL stored at –20°C) and add 10 µL of 100% ethanol. Add 90 µL of 10% sucrose. Pass to an Eppendorf and spin 2 min to precipitate the undiluted crystals. Take supernatant to fresh tube; this mix is ready to inject into embryo.
3. White leghorn and quail fertilized eggs from local farmers are incubated at 38°C to the desired developmental age in an egg incubator, Lyon Electric Co., using the stages of Hamburger and Hamilton because it is the most accurate *(1)*.
4. Eggs are laid horizontally for at least 1 h to allow the embryo to position on the top. Remove approx 4 mL of albumen with a 5-mL syringe. Add approx 0.1 mL of diluted ink under the embryo.
5. Small curved scissors to cut a hole on the top of the egg (Fine Science Tools; cat. no. 14061-09) and a fine tungsten needle to peel embryo's membrane.
6. Using a dissecting Microscope, Zeiss Stemi SV6, embryos were visualized, under 2–5X magnification.
7. India ink 1:10 (Pelikan Fount; PLK 51822A143) solution in Ringers solution (*see* **step 10**) and a bent needle in a 1-mL syringe.

8. A quartz micropipet (Capillary tube KIMAX-51; cat. no. 34502) is pulled with a needle puller (Sutter Instrument Co.). A very small amount (0.5% or final 10 mg/mL) of Fast Green FCF (Fisher 42053) can be added to the dye solution to make it easier to visualize during injection.
9. Scotch tape.
10. Ringer solution: 7.20 g of NaCl, 0.17 g of $CaCl_2$, 0.37 g of KCl in 1000 mL of dH_2O.
11. 4% Paraformaldehyde (PFA) in Dulbecco's phosphate-buffered saline (PBS; *see* **Note 3**).

2.2. Whole-Mount Labeling of Chick Embryos With HNK-1 Antibody to Visualize Neural Crest Cells

1. PBST = PBS containing 1% Triton X-100 and 10% FBS at 4°C (*see* **Note 4**).
2. HNK-1 supernatant is prepared from HNK-1 cells obtained from ATCC (cat. no. TIB-200; **Table 1**). Secondary antibody anti-mouse IgM-specific Alexa 488 conjugated antibody (**Table 2**; *see* **Note 5**).
3. Confocal microscope: 410 LSM or an upright microscope, Zeiss Axioskop 2 plus, Fluar 10X lens regular fluorescent microscope or with a 5X objective.

2.3. In Vivo Imaging of Neural Crest Cells With DiI

1. Olympus SZX12 DFPLAPO dissecting microscope with a 1X PF long working distance objective.
2. Prepare oval ring of filter paper (1.5 cm along the major axis) around the circumference of the embryo, cutting around the ring and then removing the ring with the embryo attached into Ringer's solution.
3. Use a P-200 pipetman to cleanse the yolk platelets and India ink the embryo.
4. Millicell culture inserts (Millipore PICM 030-50) and six-well culture plates (Falcon 3046). The culture insert membrane was pre-coated with 10 µg/mL of fibronectin (FN, Sigma; cat. no. F4759).
5. The culture insert membrane was underlain with a defined culture medium composed of Dulbecco's modified Eagle's medium (DMEM; Gibco, cat. no. 11995-065) and 0.5 m*M* L-glutamine (Sigma G-3126). Use parafilm to seal the entire plate to prevent dehydration.
6. A Zeiss Axiovert; 410 LSM confocal microscope was set with a 103 Neofluar (NA 5 0.30 lens).
7. We modified the six-well culture plate by making a hole in the bottom of one of the wells into which a round 25-mm glass cover slip (Fisher; cat. no. 48380-080) was sealed using silicone grease (Dow Corning, cat. no. 79810-99). The membrane of the Millipore culture insert is approx 200-mm thick and becomes transparent when moist.
8. The microscope (Zeiss Axiovert) was surrounded by a box fashioned from cardboard (4-mm thick) and covered with thermal insulation (Reflectix Co., 5/16-in. thick) which enclosed a warming heater (Lyon Electric Co. 115-20) that

Table 1
Primary Antibodies

Antibody	Species	Type	Working conc.	Specimen type	Fix/embedding	Notes
3A10	Mouse monoclonal	IgG1	1:1, neat	Sections	Memfa/parafin	DSHB
A2B5	Mouse monoclonal	IgM	1:1000	Sections cells	PFA	Epitope is a sialoganglioside on membranes of neurons and glia. Chemicon MAB312R.
β-Actin	Mouse monoclonal	IgG2a	1:1000 1:5000	Sections cells Western blot	PFA	Good for Westerns; Clone AC-74; Sigma A5316.
Acetylated β-tubulin	Mouse monoclonal	IgG2b	1 in 5000	Sections cells	PFA	Stains neurites, ciliated epidermis. Sigma tT6793; Clone 6-11B-1.
β-tubulin	Mouse monoclonal	IgG1	1 in 1000	Western blot	ECL	Cell extracts from *Xenopus*, From sox9 paper (Saint-Jeannet 2001).
BrdU	Mouse monoclonal	IgG1	1:500	Sections cells	Metoh then 0.1 M HCl to expose epitope.	Works well. Boehringer, cat. no. 1170-376.
CNPase	Mouse monoclonal	IgG1	1:300	Cells	PFA	Not strong signal. Labels glia cells. Clone 11-5B; Sigma C5922.
DBH	Rabbit	IgG	1:500	Sections cells	PFA	Sympathetic neurons, peripheral glia. Chemicon AB1585.
GFAP (G-A-5)	Mouse monoclonal	IgG1	1 in 400	Sections cells	Mempfa/gelatin	Didn't work yet; from Sigma.
GFAP	Rabbit	IgG	1:1,000	Sections cells	PFA	Better than the monoclonal. Dako, cat. no. Z0334.
HA.11	Rabbit	IgG	1:500	Sections cells Whole mount	PFA	Recognizes the well known epitope. Covance, cat. no. PRB-101P.
HistoneH3 (phosphor)	Rabbit polyclonal	IgG	1:500	Sections cells	PFA	Works well; mitosis marker from Upstate Biotech, cat. no. 06-570.

HNK1	Mouse monoclonal	IgM	1:500	Sections cells	PFA; Methanol; Buin's	Used as supernatant, the best NCC marker in avian tissue and stains some mature neurons. Not in neural crest in frogs.
			1:100	Whole mount	PFA	Excellent results.
Hu	Mouse monoclonal	IgG2b	1 in 1,000	Sections	PFA	Molecular Probes anti HuC/D. MolecularProbes cat. no. A21271; did not work in wax; don't freeze-thaw.
NCAM	Rabbit polyclonal	IgG	1 in 200–1000	Sections	PFA	From Rutishauser; doesn't work in wax or EtOH so far.
NF-M	Mouse monoclonal	IgG2a	1:500	Sections cells	PFA	From Virginia Lee; doesn't work in wax on frogs.
Nkx2.2	Mouse monoclonal	IgG2b	1:200	Sections cells	PFA	Stains pmotor neurons. Schwann cells. Clone 74.5A5 from DSHB.
O1	Mouse monoclonal	IgM	1:500	Sections cells	PFA; Metoh	Glial marker (ol's and Schw); Chemicon MAB344.
O4	Mouse monoclonal	IgM	1:500	Sections cells	PFA	Glial marker. Chemicon MAB345.
Rip	Mouse monoclonal	IgG1	1:200	Sections cells	PFA	Stains glial cells. Clone Rip from DSHB.
S-100	Rabbit	IgG	1:1,000	Sections cells	PFA, must use Metoh to expose epitope.	Very good for Schwann cells and other glia. Swant cat. no. 37.
Spectrin	Rabbit	IgG	1:500	Cells	PFA	Is a cytoskeletal protein. Sigma S1390.
Thy 1.1	Mouse monoclonal	IgM	1:10,000	Sections cells	PFA; Metoh	Expressed in neurites and fibroblasts. Used to pan against fibroblasts. Clone TN-26. Sigma M7898.
β-tubulin	Mouse monoclonal	IgG2a	1:1,000 1:500	Sections cells Whole mount	PFA	Excellent stain for neurons. Recognizes class III-β-tubulin. Covance, cat. no. MMS-435P.
Tyr-tubulin	Mouse monoclonal	IgG2a	1:1,000	Sections cells	PFA; Metoh	Works great on any cell. Chemicon MAB1864.

Conc., concentration; DBH, Dopamine β-hydroxylase; NF-M, neurofilament; PFA, paraformaldehyde.

Table 2
Secondaries Antibodies

Conjugate	Species	Against	Working conc.	Specimen type	Notes
Oregon Green	Goat anti-mouse	IgG	1 in 1000	Sections	Molecular Probes.
	Goat anti-rabbit	IgG	1 in 1000	Sections	Molecular Probes.
HRP	Goat anti-rabbit	IgG	1 in 400	Whole mount	Zymed.
	Goat anti-mouse	IgG	1 in 400	Whole mount	Zymed.
Alexa green (488)	Goat anti-rabbit	IgG	1 in 1000	Sections; Whole mount	Molecular Probes; best for longevity and doesn't bleach fast.
	Goat anti-mouse	IgG	1 in 1000	Sections	Better than Oregon green.
Alexa red (594)	Goat anti-rabbit	IgG	1 in 1000	Sections; Whole mount	Molecular Probes; best for longevity and doesn't bleach fast.
	Goat anti-mouse	IgG	1 in 1000	Sections	Better than Oregon green.
Alexa 594 Phalloidin			1:1000	Sections	Molecular Probes (for actin cytoskeleton).
Biotin	Goat anti-mouse	IgG + IgM	1 in 20,000	Western	Jackson Immunoresearch.
	Mouse anti-goat	IgG (H+L)	1 in 20,000	Western	Jackson Immunoresearch.
	Goat anti-mouse	IgG2b	1 in 2000	Western	Molecular Probes.
	Rabbit anti-goat	IgG	1 in 2000	Western	Molecular Probes.
AP	Goat anti-rabbit	IgG	1 in 1000	Whole mount	Zymed.
	Goat anti-mouse	IgG	1 in 1000	Whole mount	Zymed.
	Goat anti-human; Fc-specific	IgG	1 in 1000	Whole mount	Promega, cat no. S382B.

Conc., concentration.

maintained the cultures at 38°C for the duration of time-lapse filming, with only mild temperature fluctuations.

9. The captured images are converted into a QuickTime movie with NIH Image 1.60 and analyzed with DIAS® (Dynamic Image Analysis System; Solltech; Oakdale, IA) program for cell tracking and measurements.

2.4. Electroporation of Neural Crest

1. A solution of 3–5 μg/μL of plasmid deoxyribonucleic acid (DNA) prepared using the Quiagen® Maxi-Prep kit (cat. no. 12163) and mixed with a 0.05% of Fast Green in Ringers.
2. Pico Spitzer (Parker Instruments).
3. DNA electroporator (Pulse Power Supply made at Caltech's shop).
4. Tools for embryo dissection (*see* **Note 6**).

2.5. Injection of DiI Labeled Cells Along the Paths of Neural Crest: In Vivo Confrontation Assay

1. Prepare complete culture media ahead: we use DMEM (Gibco, cat. no. 11995-065), 0.5 m*M* L-glutamine (Sigma G-3126) and fetal bovine serum (FBS; Gibco; cat. no. 16000-036) for culture of HEK293 and COS cells. You will need also one DMEM without additives for washing and keeping cells after DiI injection to prevent them from aggregating into clumps.
2. Prepare a confluent 10-cm tissue culture dish (Falcon) of control and transfected cells, lift cells using a 0.5% Trypsin/2 m*M* ethylenediamine tetraacetic acid (EDTA; Gibco, cat. no. 25200-056). When cells are rounding up, rap the dish and add 5 mL of DMEM/10% FBS to stop trypsin.
3. Spin down the cells in a bench-top centrifuge (i.e., Beckman Model TJ-6).
4. Using a 1-mL syringe (Insulin type) with a 25-gage 1-1/2 needle, take the cells and insert them into a pulled glass pipet (0.5–1.5 capillary tube, KIMAX; *see* **Notes 7** and **8**).

2.6. Whole Embryo Culture Assay

Quail eggs between stages 11 and 16 are opened with small scissors on a Petri dish. The embryos are cut out from the yolk and transferred to another Petri dish with Ringer's solution. Here, the embryos are trimmed with tungsten needles, leaving only a small portion of the surrounding membranes and are kept in Leibovitz-15 media (L-15; Gibco, cat. no. 11415-064) until plating (*see* **Note 9**).

2.7. In Vitro Culture Assay

1. Dispase 1.5 mg/mL (we prepare stocks of 1 and 3 mL and freeze them).
2. Forceps, Dumont no. 5 Biologie, Dumoxel (Fine Science Tools, 11251-30) for dissecting neural tubes and tungsten needle to cut out the membranes.
3. A 5-mL Petri dish with L-15 and a pulled pipet with a diameter good for the neural tube, but not narrower, or else the neural tube will shear when entering it.

4. Glass Nunc chamber slides of two wells (cat. no. 177380) that have been previously coated with FN (10 µg/mL) in DMEM for at least 1 h at 37°C are used for cultures.
5. Complete media: DMEM, L-glutamine and 10% FBS and 100 mg/mL and 100U of penicillin and streptomycin (Gibco, cat. no. 1 5140-122; *see* **Note 10**).
6. Slides are incubated with DAPI in PBS to visualize cell nuclei and mounted in PermaFlour (Immunon, cat. no. 434990; Shandon, PA).

2.8. In Vitro Confrontation Assay

1. Coat a two-well glass chamber slide with 10 µg/mL FN.
2. Add approx 240,000 of cell suspension per well in DMEM/10% FBS.
3. Pasteur pipet and vacuum for removal of some cells from the center. Another modification uses dry cells by dehydrating the monolayer after a gentle wash with DMEM for about 5 h under vacuum with drierite around (Drierite, anhydrous calcium sulfate desiccant; *see* **Notes 11** and **12**).

2.9. In Vitro Migration Assay

1. FN-coated 24-wells with 8-µm pore transwell (Corning, cat. no. 3422) can have either in the lower wells: the secreting cells of interest, the control cells, or the conditioned medium from the same cells.
2. Cultures are kept in complete media.
3. Ice-cold methanol for fixing cells.
4. Toluidine blue (0.1% in aqueous solution) for staining cells.

2.10. In Vitro Migration Assay
Using Neuroprobe Chemotaxis Chamber

1. Complete media for neural tube cultures, FN-coated 10-cm TC dish. Plain DMEM for washing cells (*see* **Note 13**).
2. Tungsten needle for neural tube removal.
3. PBS with 2 m*M* of EDTA for lifting cells and centrifuge.
4. Hemocytometer.
5. A 96-well Neuroprobe chemotaxis chamber (no. 106-2, Neuroprobe; Gaithersburg, MD), 8-µm pore size that has been previously coated with FN.
6. A rubber policeman and a cotton bud. The membrane is fixed in ice-cold methanol and stained with Toluidine blue and cells counted in the underside of the membrane, which corresponds to the migrated cells.

2.11. Electroporation and In Vitro Assay

1. Quail embryos at stages ranging from HH13-16 are windowed as described and are electroporated with pCIG vector carrying the GFP marker (*2*) or with pCS2-GFP (*3*) with two pulses of 100-mVolts each (*see* **Note 14**).
2. Two-well chamber glass slides coated with FN with complete media.
3. Tools for micro dissection of neural tubes.

2.12. In Vitro Wound Assay

1. Quail neural tubes from HH13-16 are cultured on a FN-coated two-well chamber slide in complete media. The next day, the medium was changed to one conditioned for 52 h by experimental or control cells.
2. Pulled glass needle for wounding cell monolayers.

2.13. Standard In Vitro Collagen Gel Assay

1. Trunk neural explants are dissociated and cut as mentioned earlier.
2. Prepare the collagen gel by mixing 900 µL of collagen (Rat tail collagen type I, VWR cat. no. 47743-656, or BD Biosciences cat. no. 354236) with 100 µL of 10X DMEM (cat. no. 12800-017). Add 20 µL of 7.5% sodium bicarbonate, then go 2 µL at a time until the color changes to light pink to fuchsia.
3. Mouth pipet (Sigma aspirator tube assembly; cat. no. A-5177). Some people use a 1:1 collagen:Matrigel (Matrigel, Growth Factor Reduced, Beckton-Dickinson) combination.
4. Complete media.

2.14. Modified In Vitro Collagen Gel Assay

1. Trunk neural explants are dissociated and cut as mentioned above and kept in L-15.
2. Prepare the collagen gel by mixing 900 µL of collagen (Collaborative research) with 100 µL of 10X DMEM (cat. no. 12800-017).
3. HEK or COS cells (one 10-cm dish is more than enough for a good number of drops of cells).
4. Coat the four-well dish with the drops of cells with FN as usual.
5. Prepare complete media mixed with 3% carboxy-methylcellulose in the early morning or better the night before, since CMC takes longer to dissolve.
6. 4% PFA and methanol for fixing and Toluidine blue for staining cells.

2.15. Time-Lapse Video Microscopy

1. Quail embryos at stages 13–15 are electroporated with pCIG vector carrying the GFP marker with two pulses of 25 mVolts each *(4)*.
2. Nunc cover slip chambers (cat. no. 155380) coated with FN with complete media.
3. Zeiss 410 LSM (*see* **Note 15**).
4. The captured images are converted into a QuickTime movie with NIH Image 1.6 and analyzed with Dynamic Image Analysis System® (DIAS®; Solltech Incorporated; Oakdale, IA) program for cell tracking and measurements.

3. Methods

Methods for in vivo and in vitro experiments are described in **Subheadings 3.1.–3.6.** and **3.7.–3.15.**, respectively.

3.1. In Vivo Labeling of Whole Neural Tube and Neural Crest With DiI

1. Chicken or quail fertilized eggs are incubated at 38°C to the desired developmental age, using the stages of Hamburger and Hamilton since it is the most accurate *(1)*.

2. With the tungsten needle, carefully remove the membrane covering the embryo and inject DiI solution inside the lumen of the neural tube (*see* **Notes 1** and **2**).

3. Incubate for the desired time of development, minimum of 2 h for seeing migrating neural crest cells.

4. Embryos are removed from the egg by cutting around them and placing them with forceps in a dish with Ringers solution.

5. Fix embryos in 4% PFA for a minimum of 2 h maximum of 18 h (*see* **Note 3**). Embryos are now ready for whole-mount visualization (*see* protocol in **Subheading 3.2.**). It is best to store the embryos in 2% PFA in PBS at 4°C.

3.2. Whole-Mount Labeling of Chick Embryos With HNK-1 Antibody to Visualize Neural Crest Cells

1. Embryos are thoroughly washed in PBS and then blocked overnight with PBS containing 1% Triton X-100 and 10% FBS at 4°C (*see* **Note 4**).

2. The following day, embryos are incubated with 1:100 HNK-1 supernatant in PBS overnight at 4°C (*see* **Note 5**).

3. Next day, embryos are extensively washed and incubated with 1:500 of an anti-mouse IgM-specific Alexa 488 conjugated antibody (Molecular Probes).

4. The following day, embryos are extensively washed in PBS and Z-scanned with a 410 LSM confocal microscope or looked at under a regular fluorescent microscope with a 5X objective (**Fig. 1**). Embryos can be stored for up to 3 mo if fixed and kept under sterile conditions.

3.3. In Vivo Imaging of Neural Crest Cells With DiI

1. Embryos at desired stages are windowed; a hole was cut in the vitelline membrane, using a fine tungsten needle.

2. An isotonic solution of DiI was injected into the lumen of the neural tube, filling the hindbrain region, using a quartz micropipet positioned just rostral to the first somite pair.

3. Injected eggs are resealed with adhesive tape and incubated at 38°C for 2 h. After this incubation period, we evaluated each embryo in terms of the brightness and uniformity of DiI labeling using an Olympus SZX12 DFPLAPO dissecting microscope with a 1X PF long working distance objective and selected those embryos that are well-labeled but showed no signs of necrosis.

4. Whole embryo explant cultures are prepared according to the method described in Krull and Kulesa (*5*). Younger embryos (stages 9–12) are removed from the egg for explant culture by placing an oval ring of filter paper around the circumference of the embryo, cutting around the ring and then removing the ring with the embryo attached into Ringer's solution. The paper ring was approx 1.5 cm along the major axis with a hole wide enough to provide ample space between the inner side of the ring and the embryo. This leaves the whole embryo intact and includes a portion of the surrounding blastoderm to a diameter equal to the outside diameter of the paper ring. Older embryos are just cut around the membranes and placed in Ringer's solution.

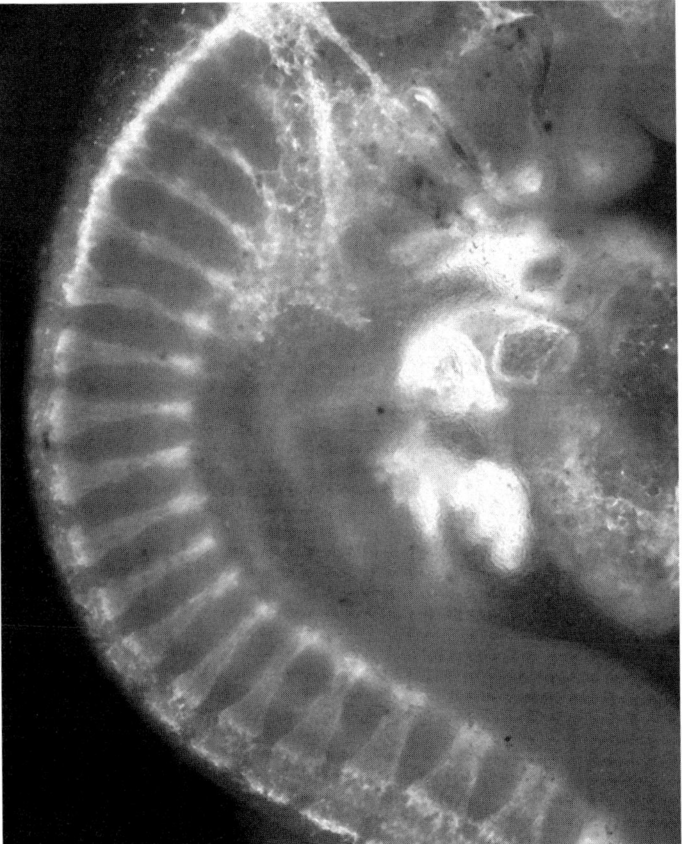

Fig. 1. Whole mount of chicken embryo with HNK-1 antibody. Figure shows flattened confocal Z-series of whole mounts of embryos stained with the HNK-1 antibody to recognize neural crest cells.

5. Explant cultures are created using Millicell culture inserts in six-well culture plates, similar to the protocol developed in Krull et al. *(6)*.

6. The dorsal surface of the embryo was placed on the coated culture insert membrane, leaving the ventral surface exposed to the atmosphere. Excess Ringer's solution was pipetted off the membrane surface at the rostral and caudal ends of the explant so that the flow of the draining solution straightened the embryo's rostrocaudal axis. This naturally spread the explant without flattening the embryo and mimicked the tension of the blastoderm normally created by the stretching of the yolk sac.

7. The culture insert was then placed in one well of the six-well plate. The culture insert membrane was underlain with a defined culture medium composed of

DMEM and L-glutamine. Sterile water was added to the unfilled wells of the culture plate to minimize dehydration during the time-lapse sequences. The entire 6-well plate was then sealed along its sides with parafilm.

8. DiI-labeled explants are visualized using an inverted confocal microscope. We set the aperture to be fully open; although this translated into less confocal effect, it increased the sensitivity of the microscope and allowed for a maximum optical section thickness (about 30–40 mm in z-height with a 103 Neofluar (NA 5 0.30) lens). This let us observe more complete cell trajectories for longer periods of time because the cells did not move out of the optical section during the early events of neural crest cell migration as cells emerged at the dorsal midline and moved into the mesoderm subjacent to the ectoderm.

9. The microscope cardboard box and the thermal maintained the cultures at 38°C for the duration of time-lapse filming with only mild temperature fluctuations.

10. The fluorescent dye, DiI, was excited with the 568 nm laser line using the YHS filter set intended for rhodamine.

11. Images are analyzed and played back in movie form after conversion using the image processing and analysis program, NIH Image 1.60.

3.4. Electroporation of Neural Crest

1. Chicken embryos at stages ranging from HH8-16 are windowed as described above and a solution of 3–5 µg/µL of plasmid DNA was mixed with a 0.05% of Fast Green in Ringers and injected into the neural tube with a Pico Spitzer with 10–20 ms and 30 psi settings, after which DNA was electroporated with two pulses of 25 mVolts each with a pulse power supply electroporator.

2. Embryos are closed with tape and incubated at 38°C for the desired time (*see* **Note 6**).

3. The next day, the embryos are dissected and fixed in 4% PFA and whole mounted for HNK-1 and the plasmid marker or protein tag (usually GFP, but myc and HA-tags have also shown to be good ones; *see* **Fig. 2**).

3.5. Injection of DiI-Labeled Cells Along the Paths of Neural Crest: In Vivo Confrontation Assay

1. Fertilized chicken eggs are incubated at 38°C for approx 28–56 h until embryos reached HH10-12 (for vagal neural crest) and HH13-16 (for trunk neural crest). The eggs are windowed and visualized as usual with India ink.

2. Take the pelleted cells, remove media and add 50 µL of DiI to the cell pellet resuspended in 0.5 mL of DMEM.

3. Incubate for 30 min at room temperature, add 5 mL of DMEM, mix gently, and spin down again. Repeat this step one more time. The cell pellet should be clearly reddish. Pass it on to an Eppendorf tube spin down and withdraw as much media as possible, caring not to leave so little that the cell suspension might dry. Using a 1-mL syringe with a 25-gage 1-1/2 needle, take the cells and insert them into a pulled glass pipet (0.5–1.5 capillary tube, KIMAX). Be careful not to add air bubbles in the mix, otherwise the cells will not come out as a globus.

HNK1

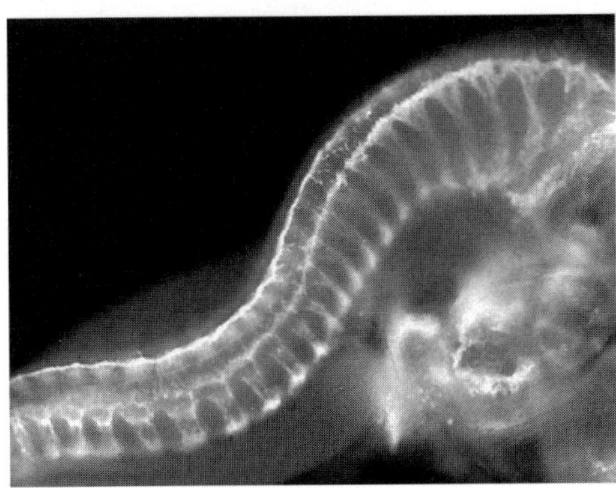

anti-GFP

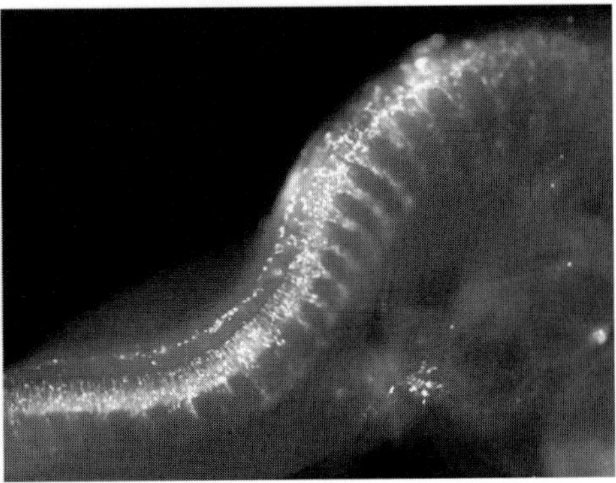

Fig. 2. Neural crest electroporation. Chicken embryo st.14 was electroporated with pCS2-GFP (gift of Dr. Yi Rao, Washington University) and 24 h later the embryo was collected, fixed, and whole mounted for HNK-1 and GFP antibody. Note that not all the cells HNK-1-positive are also positive for GFP because not all the neural crest will be labeled with the plasmid, depending on the electroporation efficiency.

4. Using a Pico Spitzer at 10–20 ms and maximum 30 psi settings, inject the cells into the desired area after removing the end of the pipet with a no. 5 forceps. When looking for disrupting effects on neural crest migration, injections are best done just beneath the somites.
5. Close the egg window with scotch tape and put back into the incubator for the desired period of time.
6. Embryos are removed from the egg as usual and fixed in 4% PFA.
7. Embryos are now ready for whole mounting (*see* **Subheading 3.2.; Notes 7** and **8**).

3.6. Whole Embryo Culture Assay

1. Quail eggs between stages 14 and 16 are opened by cutting the shell with small scissors on a Petri dish because these eggs cannot be cracked open as chicken's (*see* **Note 9**). The embryos are cut out from the yolk and transferred to another Petri dish with Ringer's solution. Here the embryos are trimmed with tungsten needles, leaving only a small portion of the surrounding membranes and are kept in L-15 until plating.
2. Millicell inserts are coated with FN (10 µg/mL) for at least 1 h at 37°C after which the excess liquid is drained.
3. Embryos are placed upsidedown on the membranes and all media and liquid is withdrawn completely. The inserts are then carefully placed on media in a 6-well plate (Falcon) so not to make bubbles underneath. This procedure (placing embryos and draining media) has to be done quickly; otherwise embryos will

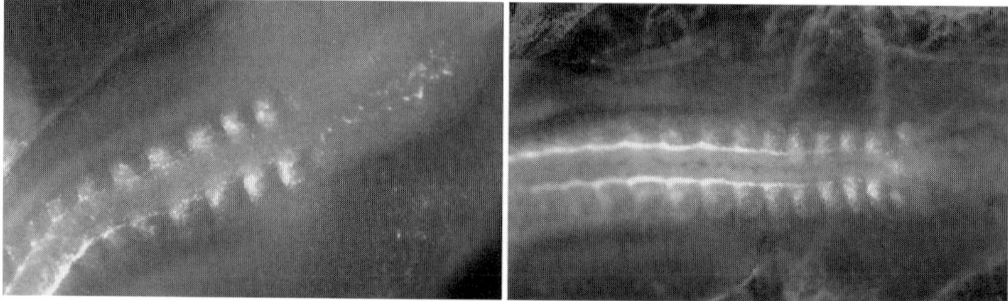

Fig. 3. Effects of Slit2-expressing cells on neural crest migration in vivo. Quail embryos are cultured on Millicel inserts and whole mounted after fixing for HNK-1. Note the green cells migrating out of the neural tube on the rostral side of the somites but on the right picture the cells are scattered over the somites, indicating disruption of their usual rostral migratory path.

    dry and die. The media can be conditioned or with different factors to test their effect on neural crest migration.

4. The plate is put in a 37°C incubator with 5% CO_2 and plenty of water in the incubator to prevent drying. They can be cultured for up to 3 d; the survival is around 50% under these conditions.

5. After culture the embryos are fixed in 4% PFA and whole mounted as described in **Subheading 3.2., Notes 7** and **8**. Embryos stained with HNK-1 will typically look as shown in **Fig. 3**.

3.7. In Vitro Culture Assay

One complication of in vivo experiments is that cells are exposed to a wide range of factors as they migrate along their paths, making conditions variable and sometime difficult to control. Thus, understanding cell migratory behavior under defined conditions is more easily tackled in vitro, during which the conditions can be better defined. The experiments listed in this and following sections describe a variety of assays used in our lab that have proved helpful and simple while providing concrete answers to different questions.

1. Quail neural crest cells often grow better in tissue culture than chick neural crest cells and therefore are recommended for in vitro studies. Quail eggs are opened with small scissors on a Petri dish. The embryos are cut out from the yolk and transferred to another Petri dish with Ringer's solution. Then, the embryos are dissected with tungsten needles, leaving only the trunk region encompassed by the neural tube and the somites and a bit of ectoderm, and transferred to another dish with 1.5 mg/mL of dispase.

2. Embryos are incubated in the dispase solution for about 30–60 min and then washed in L-15 medium. You can tell if it is time when the neural tubes start to coil/curve in the explanted embryo pieces, a sign of detachment from mesoderm.

3. Neural tubes from HH10-12 (for vagal crest, using only the neural tubes from the first until the seventh somite) and HH12-16 (for trunk, the neural tube from eighth somite to the end of the tube) are dissociated using fine forceps. After removal of somites, the neural tubes are transferred to a new L-15 Petri dish using a pulled pipet with a diameter good for the neural tube, but not narrower, or else the neural tube will shear when entering it. Here the tubes are cut in small pieces (size of two to three somites).

4. The neural tube fragments can now be transferred with a pipet to the center of glass Nunc chamber slides of two wells that have been previously coated with FN (*see* **Note 10**).

5. Neural tubes are cultured in complete media: DMEM and 10% FBS, L-glutamine and 100 mg/ml and 100 U of penicillin and streptomycin respectively for 18 h.

6. Cultures are fixed in 4% PFA for 30 min and subsequently blocked for 30 min with PBS, 1% Triton X-100, 10% FBS.

7. The HNK-1 antibody (1:500) is good for visualizing neural crest cells, followed by an anti-mouse IgM-Alexa 488 secondary. At the end, slides are incubated with DAPI in PBS to visualize cell nuclei and mounted in PermaFlour.

3.8. In Vitro Confrontation Assay

1. Coat a two-well glass chamber slide with 10 µg/mL of FN for at least 1 h at 37°C.

2. Add approx 240,000 of cell suspension per well in DMEM/10% FBS. The next day, the cells have formed a tight monolayer.

3. On the center of this using a Pasteur pipet and vacuum, remove some cells from the center. Be careful not to rub the Pasteur on the slide because it will also remove the FN coat, necessary for the neural crest cells.

4. Gently add complete media again because the cell monolayer could be easily dislodged from the edges. Then add the cut neural tube pieces in the center of the well, and position them closer to the cell border, so when the neural crest comes out of the neural tube it will encounter the cells expressing the desired protein (*see* **Note 11**).

5. The next day (approx 18 h) cultures are scored for cell response and fixed and stained with HNK-1 antibody (*see* **Note 12**).

3.9. In Vitro Migration Assay

1. Quail neural tubes from HH10-12 (for vagal crest) and HH13-16 (for trunk and the posterior parts of the tube) are obtained as mentioned in **Subheading 3.7.**

2. Cultured in DMEM/10% FBS on a fibronectin coated 24-wells with 8-µm pore transwell (Corning) that can either have in the lower wells: the secreting cells of interest, the control cells, or the conditioned medium from same cells.

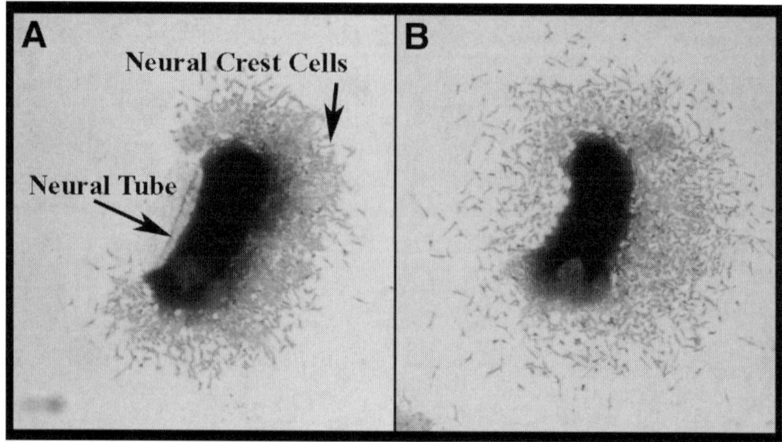

Fig. 4. Neural crest cells migrate further in the presence of Slit2 conditioned medium. (**A**), A neural tube explanted in the presence of conditioned medium from control cells shows that migrating neural crest cells have moved several cell diameters away from the neural tube after 18 h in culture. (**B**), A similar neural tube cultured in medium conditioned by Slit2-expressing cells had neural crest cells that had migrated significantly further away in 18 h.

3. Cultures are grown overnight, next day the neural crest could be seen as a halo around the neural tubes (**Fig. 4**). The cultures are fixed with ice-cold methanol for 20 min, then cells are stained with Toluidine blue and the farthest distance traveled by the neural crest was measured per tube and compared between both kinds of treatments, as well as with unconditioned medium.

3.10. In Vitro Migration Assay Using Neuroprobe Chemotaxis Chamber

1. Quail or chicken neural tubes from HH10-12 (for vagal crest) and HH13-16 (for trunk and the posterior parts of the tube) are obtained and cut into small pieces (*see* **Note 13**).
2. These neural tube cuts are cultured in DMEM/10% FBS on a FN-coated 10-cm TC dish for 18 h. The next day, a halo of migrated neural crest cells could be seen around the neural tubes. The neural tubes are then removed carefully with fine forceps or tungsten needles and the dish is washed slowly to remove any floating cells or neural tube parts.
3. The neural crest cells are then lifted with PBS and 2 m*M* of EDTA for about 20 min. Cells are gently rapped and dislodged with a pipet and spin down on a Falcon conical tube at 800 rpm in DMEM/FBS to concentrate them.

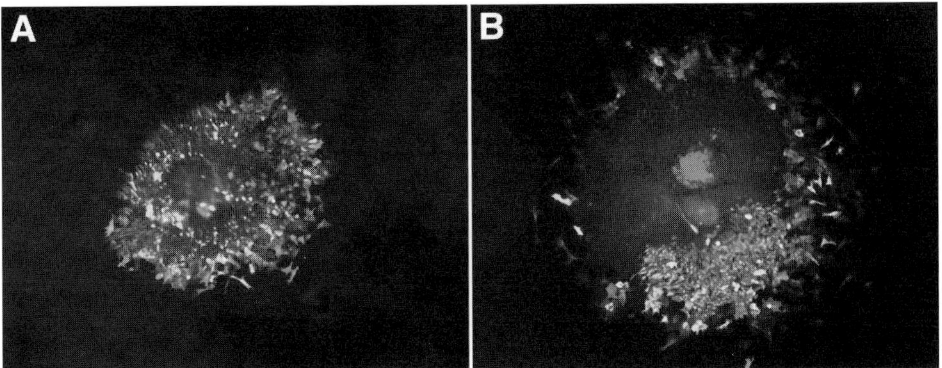

Fig. 5. Neural crest migration in vitro after electroporation in ovo. Quail embryos are electroporated and the neural tubes dissected and cultured. After 24 h, cultures are fixed and stained for GFP. Panel (**A**) corresponds to pCIG-GFP and panel (**B**) to pCS2-GFP.

4. The cell pellet is resuspended in approx 250 mL of DMEM/FBS and cells are counted. Plate 10–30,000 cells per well in the top wells of a 96-well Neuroprobe chemotaxis chamber, 8-μm pore size that has been previously coated with fibronectin. The bottom wells should have been already filled with the media plus the factors to test on the chemotaxis assay. The wells have a 30-μL capacity, so be careful to avoid drying during this time, better not work under a hood whose air flow will quickly dry the drop.

5. After 18 h, the upper portion of the chemotaxis chamber is removed from revealing the membrane. The upper side is stripped of cells with a rubber policeman and a cotton bud. Then the membrane can be fixed in ice-cold methanol and stained with Toluidine blue and cells counted in the under-side of the membrane, which corresponds to the migrated cells.

3.11. Electroporation and In Vitro Assay

1. Quail embryos at stages ranging from HH13-16 are windowed as described and are electroporated with pCIG vector carrying the GFP marker *(2)* or with pCS2-GFP *(3)* with 2 pulses of 100 mVolts each (*see* **Note 14**).

2. After 1–2 h in a 38°C incubator, neural tubes are isolated and cultured on two-well chamber glass slides coated with fibronectin in DMEM/10% FBS for 18–24 h.

3. The next day the neural crest could be seen as a halo around the neural tubes. Cultures are then fixed and stained for GFP and HNK-1 epitopes to visualize all the migrated neural crest versus the electroporated cells (**Fig. 5**).

3.12. In Vitro Wound Assay

1. Quail neural tubes from HH13-16 (for trunk, the posterior parts of the tube are obtained as mentioned in **Subheading 3.7.**).

Fig. 6. Trunk neural crest motility in a wound assay. Trunk neural tubes are cultured overnight on fibronectin. After 1 d, a wound of one to two cells' width was made with a fine pipet; after 2 h, the percent of wounds with cells crossing and sealing the gap was determined.

2. Cultured in DMEM/10% FBS in a FN-coated two-well chamber slide overnight, next day the neural crest could be seen as a halo around the neural tubes, medium was changed to one conditioned for 52 h by experimental or control cells, neural crest was thus pre-incubated with conditioned medium for 2 h before making the wounds.
3. The wounds are created by scraping cells (two to three cell-width lanes) with a pulled glass needle (**Fig. 6**), and lanes scored over 1- to 1.5-h periods to count for cells crossing the wound space.
4. The criterion used was no cells crossing, some cells crossing (regardless of the number) and sealed wound. Data was scored as percentage of total neural tubes that had sealed the wound on a period of 2 h.

3.13. Standard In Vitro Collagen Gel Assay

1. Trunk neural explants are dissociated and cut as mentioned in **Subheading 3.7.** (*see* **Note 13**).
2. Prepare the collagen gel by mixing 900 μL of collagen with 100 μL of 10X DMEM. Keep these solutions on ice! Add 20 μL of 7.5% sodium bicarbonate, then go 2 μL at a time until the color changes to light pink to fuchsia. This means it is ready for adding the tissue.
3. Take the tissue pieces and put them together in a drop of 50–100 μL of collagen placed previously in a four-well dish using a mouth pipet. Try to have them very

close to each other (approx 100 μm), otherwise no visible effect. Some people use a 1:1 collagen:Matrigel combination.

4. Leave at 37°C for 15–30 min in a humidified incubator until the collagen has become a gel. Then add media and culture.

3.14. Modified In Vitro Collagen Gel Assay

We had found that neural crest do not migrate vigorously in collagen gels, probably because it is not a good substrate for them *(7)*. We had modified the above assay to increase the neural tubes that have a good number of migrated neural crest cells.

1. Trunk neural explants are dissociated and cut as mentioned in **Subheading 3.7.** (*see* **Note 13**) and kept in L-15.
2. Prepare the collagen gel as described in **Subheading 3.13., steps 2–4.**
3. Isolate the secreting HEK or COS cells from tissue culture dishes (one 10-cm dish is more than enough for a good number of drops of cells). Spin cell suspension in a conical tube and withdraw the media as much as possible. Resuspend the cells in ice-cold collagen at desired density. We recommend not too high, otherwise the cells will dislodge more easily from the collagen gel drop.
4. Place a 2- to 4-μL drop of cells in collagen and allow to gel for less than 5 min in a humidified warm incubator, otherwise the drop will dry and cells will not survive.
5. Coat the four-well dish with the drops of cells with fibronectin as usual for 1 h at 37°C in plain DMEM.
6. Gently remove the DMEM with fibronectin and add media (DMEM/10% FBS) with prewarmed 3% carboxy-methylcellulose. Take the tissue pieces and put them close to the drop of cells using a mouth pipet. Try to have them close to the secreting cells.
7. Next day remove part of the thick media and add 4% PFA for 2 h. Remove PFA, add PBS, and incubate at 65°C to start washing away the CMC. Repeat this step until the culture wells look free of any sugar residues.
8. Add ice-cold methanol for 2 min.
9. Add Toluidine blue for 30 min. Wash in tap water and let dry.
10. Cultures can be photographed now and the distances of the neural crest cells towards or away from the secreting cells can be measured with an Adobe Photoshop measuring tool.

3.15. Time-Lapse Video Microscopy

1. Quail embryos at stages 13–15 are electroporated with pCIG vector carrying the GFP marker with two pulses of 100 mVolts each *(4)*.
2. After 1–2 h in a 38°C incubator, neural tubes are isolated as before and cultured on glass Nunc cover slip chambers coated with FN in complete media.
3. The next day cells are taken for imaging or primed for 1 h with control or experimental-conditioned medium as before and labeled cells are imaged in a Zeiss 410 LSM every 90 s for approx 3 h (*see* **Note 15**).

4. The captured images are converted into a QuickTime movie with NIH Image 3 and analyzed with DIAS® program for cell tracking and measurements.

4. Notes

1. The DiI dries quickly in the pipet we use to inject. Solubilize by dipping it in ethanol and change needles quickly.
2. This same procedure can be used for focal labeling of small groups of cells with DiI by leaving out the step of diluting the DiI in sucrose. Use the concentrated DiI solution to touch the top of the dorsal neural tube and make very small injections of concentrated DiI. This labels a group of approx 50 cells in a focal manner.
3. Chicken embryos tend to bend when cutout of their membranes, which makes it more difficult to visualize the neural crest in whole mounts. We have found that it is best to fix the embryos surrounded by their membranes in 4% PFA overnight, which keeps them flat. Afterwards, the membranes can be carefully removed with fine no. 5 forceps and tungsten needles.
4. You can also use old tissue culture media containing 10% FBS and 1% Triton X-100.
5. We use higher concentrations of primary antibodies for whole mounting since the staining has to penetrate deeper and be more extensive than for immunohistochemistry. We have found that the staining remains more stable when we use higher concentrations as well.
6. We had observed that different plasmids have different times at which the levels of the protein will be detectable in whole mount. We thus recommend a minimum of 24 h in order to observe any disturbance in the migration of the neural crest owing to the expressed gene.
7. Embryos can be stored for up to 3 mo if fixed and kept under sterile conditions.
8. Neural crest response is determined as follows: if both types of cells (neural crest and HEK) either mixed and/or overlap, it was considered that there was no repulsion; if however, neural crest cells kept a marked space in between both cells, did not overlap, or stopped migrating altogether, it was considered that there was repulsion. The number of embryos with such responses was counted and the percentage determined based on total number of embryos with neural crest in close proximity to the 293 cells.
9. Quail are the choice for long-term whole embryo cultures. We found that chicken embryos will not do as well and the survival rate is very low.
10. The incubating media should be already in the chamber, since the tubes should be added to this media, or else they will not sink down if it is done the other way around (tubes first, then change the media).
11. This assay can be performed as well with cell monolayer that had been prepared as above but the media was removed and cells are dried overnight using a sealed chamber with Drierite and vacuum.
12. Neural crest response is determined as follows: if both types of cells (neural crest and HEK) either mixed and/or overlap, it was considered that there was no repul-

sion; if however, neural crest cells kept a marked space in between both cells, did not overlap, or even changed their shape from a mesenchymal type to a slender, more collapsed shape, it was considered that there was repulsion. The number of neural tubes with such responses was counted and the percentage determined based on the total of neural tubes with neural crest in close proximity to the 293 cells. Neural tubes with neural crest cells, which are not in close proximity to the HEK cells, are not counted.

13. At least 80 embryos are needed to collect sufficient neural crest cells.

14. We electroporate neural tubes for culture so high to increase the amount of neural tube cells that take the plasmid. These levels are lethal for keeping the live embryos afterwards. We found that the two pulses at 100 mVolts was more efficient in increasing the amount of GFP-positive cells and facilitated scoring the effects of diverse neural crest inducers or molecules involved in migration.

15. To compare the average speed traveled by neural crest cells in control vs experimental containing medium, individual cells are picked from same day experiments and compared based on similar numbers of frames moved, with each frame corresponding to 90 s.

References

1. Hamburger, V. and Hamilton, H. L. (1992) A series of normal stages in the development of the chick embryo. *Dev. Dyn.* **195,** 231–272.

2. Megason, S. G. and McMahon, A. P. (2002) A mitogen gradient of dorsal midline Wnts organizes growth in the CNS. *Development* **129,** 2087–2098.

3. Ward, M., McCann, C., DeWulf, M., Wu, J. Y., and Rao, Y. (2003) Distinguishing between directional guidance and motility regulation in neuronal migration. *J. Neurosci.* **23,** 5170–5177.

4. De Bellard, M. E., Rao, Y., and Bronner-Fraser, M. (2003) Dual function of Slit2 in repulsion and enhanced migration of trunk, but not vagal, neural crest cells. *J. Cell Biol.* **162,** 269–279.

5. Kil, S. H., Krull, C. E., Cann, G., Clegg, D., and Bronner-Fraser, M. (1998) The alpha4 subunit of integrin is important for neural crest cell migration. *Dev. Biol.* **202,** 29–42.

6. Krull, C. E., Collazo, A., Fraser, S. E., and Bronner-Fraser, M. (1995) Segmental migration of trunk neural crest: time-lapse analysis reveals a role for PNA-binding molecules. *Development* **121,** 3733–3743.

7. Newgreen, D. F., Gibbins, I. L., Sauter, J., Wallenfels, B., and Wutz, R. (1982) Ultrastructural and tissue-culture studies on the role of fibronectin, collagen and glycosaminoglycans in the migration of neural crest cells in the fowl embryo. *Cell Tissue Res.* **221,** 521–549.

19

In Vivo and In Vitro Models of Mammalian Angiogenesis

Mien V. Hoang and Donald R. Senger

Summary

Angiogenesis is a complex process involving the organization of proliferating endothelial cells into new blood vessels. Both in vivo models and in vitro models are important for investigating angiogenesis and for defining the involvement of specific molecules. This chapter describes a basic mouse model of vascular endothelial growth factor-driven angiogenesis in mouse skin together with a modified version of this model in which retrovirus-packaging cells are included as a means to efficiently achieve retroviral transduction in vivo. With this approach, the contributions of specific proteins to angiogenesis can be defined. In addition, we describe a model of capillary morphogenesis in vitro that uses microvascular endothelial cells transduced with retrovirus in culture. This in vitro model provides a complementary strategy for investigating the importance of specific molecules for angiogenesis.

Key Words: Angiogenesis; neovascularization; VEGF; endothelial cells; retrovirus; mouse model; collagen I; morphogenesis.

1. Introduction

Angiogenesis, also known as neovascularization, is the formation of new blood vessels through a complex process involving endothelial cell proliferation and morphogenesis. Definition of the key molecules and mechanisms involved in angiogenesis are important because such understanding may lead to the development of new therapies. Thus for disorders such as cancer and retinopathies in which growth of new blood vessels contributes to pathology, there is much interest in suppressing angiogenesis *(1,2)*. Conversely, for disorders in which ischemia and/or insufficient blood supply is problematic, the therapeutic goal is to stimulate angiogenesis and thereby alleviate vascular insufficiency *(3)*.

Experimental models are critical for elucidating the basic biology of angiogenesis and for defining the contributions of specific molecules. In particular,

From: *Methods in Molecular Biology, vol. 294: Cell Migration: Developmental Methods and Protocols*
Edited by: J-L. Guan © Humana Press Inc., Totowa, NJ

"

we believe that the utilization of complementary in vivo and in vitro models of angiogenesis in parallel provides the most effective experimental strategy. Thus, this chapter describes both a versatile mouse model of angiogenesis in vivo and a complementary in vitro angiogenesis model utilizing isolated microvascular endothelial cells in culture. In particular, we describe retroviral transduction both in vivo and in vitro as useful strategies for probing the functions of specific proteins in angiogenesis.

2. Materials
2.1. Basic Model of Angiogenesis in Mouse Skin

1. Athymic nude mice (*see* **Note 1**).
2. Matrigel (BD Biosciences, Bedford, MA; *see* **Note 2**).
3. Stable SK-MEL-2 vascular endothelial cell growth factor (VEGF) transfectants, or equivalent (*see* **Note 3**).
4. Avertin (a formulation of tribromoethanol) *(4)* or other appropriate drug for mouse sedation.
5. Syringes (1 mL) with 25-gage needles.
6. 0.5% (w/v) Evan's blue dye in sterile saline to be used as tracer for monitoring blood vessel perfusion and syringes with 27-gage needles for tail vein injection (*see* **Note 4**).
7. Dissection tools (fine surgical scissors and forceps, common pins, sheets of dental wax).
8. Dissecting microscope with camera.
9. 10% Buffered formalin for tissue fixation, routine automated tissue processor for embedding tissue with paraffin, and microtome for preparing paraffin sections.
10. Reagents for immunohistochemical staining of endothelial cells in tissue sections, as follows: xylene, ethanol, phosphate-buffered saline (PBS), TPCK-trypsin (Sigma, St. Louis, MO), 0.3% hydrogen peroxide in PBS, 5% goat serum in PBS, rat anti-mouse CD31 monoclonal antibody (primary antibody, clone MEC 13.3; Pharmingen, La Jolla, CA), anti-rat IgG (mouse adsorbed secondary antibody, no. BA-4001; Vector Laboratories, Burlingame, CA), Vectastain Elite ABC (horseradish peroxidase) Kit (Vector Laboratories), DAB substrate (Zymed; San Francisco, CA), hematoxylin (Vector Laboratories), and Cytoseal 60 Mounting Medium (Stephens Scientific, Kalamazoo, MI).

2.2. Model of Angiogenesis in Mouse Skin
Employing Retrovirus-Packaging Cells

1. All materials required for the basic model of angiogenesis in mouse skin (listed in **Subheading 2.1.**).
2. Retroviral vector (e.g., pLCNX2; Clontech Laboratories, Palo Alto, CA). However, when retrovirus is desired for both in vitro and in vivo experiments, we favor a modified version of this traditional vector (*see* **Note 5**).
3. Packaging cell line (e.g., RetroPack PT67; Clontech) for stable production of replication-incompetent retrovirus.

4. Investigator-chosen complimentary deoxyribonucleic acid (cDNAs) encoding proteins/mutants to be inserted into the expression cassette of the retroviral vector.
5. G418 sulfate (Mediatech, Herndon, VA) for drug selection of cells.

2.3. In Vitro Model of Capillary Morphogenesis

1. Human dermal microvascular endothelial cells and culture medium (*see* **Note 6**).
2. Rat tail collagen I (BD Biosciences).
3. Phorbol myristate acetate (PMA; Sigma), basic fibroblast growth factor (bFGF; R+D Systems, Minneapolis, MN), VEGF (R+D Systems).
4. Packaging cells stably expressing retrovirus of interest (same as those described in **Subheading 3.2.1.**).
5. Polybrene (Sigma) and chondroitin sulfate C (from shark cartilage; Sigma).
6. Inverted tissue culture phase microscope with camera.

3. Methods

The methods described here outline: 1) a basic model of angiogenesis in adult mouse skin, 2) the use of retroviruses to probe the functions of specific molecules in the mouse skin angiogenesis model, and 3) an in vitro model of capillary morphogenesis involving isolated microvascular endothelial cells, transduced with retroviruses and embedded in three dimensional collagen I. In this in vitro model of angiogenesis, endothelial cells organize into multi-cellular structures that closely imitate endothelial cell organization during the early stages of angiogenesis in vivo.

3.1. Basic Model of Angiogenesis in Mouse Skin

Passaniti and colleagues *(6)* originally described a mouse model involving subdermal implantation of reconstituted basement membrane (Matrigel) mixed with heparin and bFGF to drive angiogenesis. We describe here a modified version of that model that substitutes heparin and bFGF with cells stably transfected for expression of VEGF (also known as VEGF-A). We favor this modified model for several reasons. First, VEGF, not bFGF, is widely recognized as the critical angiogenic cytokine responsible for the neovascularization associated with important pathologies including cancers *(7–14)*, psoriasis *(15)*, and retinopathies *(16–19)*. Moreover, Matrigel containing cells stably transfected for VEGF expression provokes angiogenesis in mouse skin within 6 d, and we find such angiogenesis to be more robust and reproducible than that provoked by purified bFGF and heparin. Finally, Matrigel containing purified VEGF does not provoke angiogenesis significantly, probably because of the rapid diffusion of VEGF from the Matrigel implants. Thus, the use of stable VEGF-transfectants overcomes this problem and provides an effective strategy for delivering a constant source of VEGF during the experimental interval (6–10 d).

3.1.1. Preparation and Implantation
of Matrigel-Containing VEGF Transfectants

1. Thaw frozen Matrigel carefully by placing vials on top of ice packs overnight in a refrigerator. Because it rapidly forms a solid gel at room temperature it is important that Matrigel be thawed carefully. The above method works best, however more rapid thawing can be accomplished if care is taken to monitor temperature closely. Subsequently, all pipettes, tubes, dilution media, and syringes should be precooled to prevent solidification of thawed Matrigel.
2. Dilute Matrigel to a final concentration of 9–10 mg/mL with serum-free cell culture medium (e.g., Dulbecco's modified Eagle's medium [DMEM], or Hank's balanced salt solution). Significantly higher concentrations of Matrigel are difficult to inject because of high viscosity, and significantly lower concentrations may not form a suitably rigid implant.
3. Resuspend stable VEGF-transfectants in Matrigel (per injection site: 1.5×10^6 cells in 0.3 mL of Matrigel). 1.5×10^6 stable VEGF-transfectants per injection site is only an estimate; the actual number should be defined in preliminary experiments. Thus, it should be determined by injecting a range of doses just how many transfected cells are required to provoke the desired intensity of neovascularization (*see* **Subheading 3.1.2.**).
4. Sedate athymic nude mice (*see* **Note 1**) appropriately (e.g., Avertin, 0.3 mL/20 g mouse, i.p.).
5. Inject the mice subdermally (25-gage needle) in each flank with 0.3 mL of Matrigel (*see* **Note 2**) containing approx 1.5×10^6 stable VEGF-transfectants (*see* **Note 3**). Upon injection the Matrigel rapidly solidifies, thus encapsulating the VEGF transfectants. To obtain good reproducibility of responses, we find it best to inject the Matrigel suspensions just below the skeletal muscle cell layer of the dermis, rather than within the dermis. Intradermal injections of Matrigel in mice are difficult and typically result in implantation both above and below the skeletal muscle cell layer. Thus injection of the Matrigel just below the skeletal muscle cell layer provides greater uniformity of responses.

3.1.2. Gross and Histochemical Analyses of Angiogenesis

We typically need 6 d for this assay, by which time the significant development of new blood vessels is evident in the skin immediately above and neighboring the Matrigel implants (*see* **Fig. 1**). A 10-d interval allows for even more extensive neovascularization, if desired.

1. At harvest, animals are euthanized and rapidly dissected to expose the Matrigel implants together with associated skin, as described in **steps 2–4**. The function of new vessels in supporting blood flow can be tested by injecting animals intravenously 5 min before euthanasia with a tracer such as Evan's blue dye (0.2 mL of 0.5% w/v solution made in sterile saline). Perfused vessels appear dark blue-purple (*see* **step 5** [*opposite page*], *see* **Note 4**).
2. Make a midline incision along the back of the animal with sharp, fine-pointed surgical scissors.

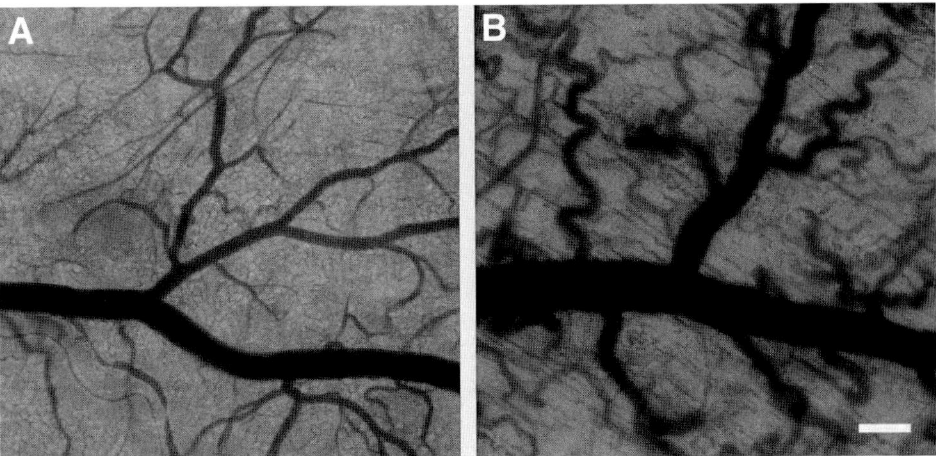

Fig. 1. Basic model of VEGF-driven angiogenesis in mouse skin observed at d 6. Dissected skin immediately adjacent to the Matrigel implants was photographed: **(A)** skin adjacent to Matrigel alone, **(B)** skin adjacent to Matrigel containing SK-MEL-2 cells transfected for expression of VEGF. Bar, 300 microns. New blood vessels **(B)** typically are tortuous ("cork screw-like") in appearance.

3. Make perpendicular incisions to the right and left of the mid-line incision at both the shoulders and hindquarters.
4. Carefully separate the Matrigel implant from the underlying fascia, leaving the Matrigel associated with the skin, which is spread out on a piece of dental wax and secured with common pins.
5. Capture images of dissected skin specimens (*see* **Fig. 1**) immediately with a dissecting photomicroscope because the visibility of new blood vessels can deteriorate with time.
6. After photography, trim skin specimens with the Matrigel implants still attached and fix the samples with 10% buffered formalin for 60–90 min. Longer fixation times (particularly longer than 2 h) greatly reduce antigen-recognition by the CD31 antibody used to stain endothelial cells.
7. After fixation, tissues are incubated with PBS for 1–2 h and then again with fresh PBS overnight at 4°C.
8. Embed specimens with paraffin in a tissue processor the next day (*see* **step 9**) or store them in 70% ethanol at 4°C for subsequent embedding at later times. It is critical that specimens not be placed directly in 70% ethanol without the preceding PBS incubations. Otherwise, overfixation will occur, and the CD31 antigen will be irretrievably lost. Also, specimens should be bisected before embedding, because segmenting of specimens is difficult after tissue processing.

9. For embedding, we typically use the following automated immersion protocol: 70% ethanol (10 min, twice), 80% ethanol (20 min, once), 95% ethanol (20 min, twice), 100% ethanol (10 min, twice; 20 min, once), xylene (20 min, twice), paraffin (60 min, twice). All steps prior to paraffin embedding are performed at 40°C; paraffin-embedding is performed at 60°C. Finally, embedded specimens are cast in paraffin blocks with the orientation of specimens such that cut sections will contain cross-sections of Matrigel implants and overlying skin.

10. For immunohistochemistry, standard 5-micron tissue sections are prepared with a microtome, and sections are stained with antibody to CD31 (PECAM-1) to detect endothelial cells, as described in **steps 11–33**. CD31 is also expressed by platelets, but this poses no complication because CD31-positive endothelial cells with nuclei are readily distinguished from platelets which are much smaller and lack nuclei.

11. Incubate sections with xylene for 15 min and then with fresh xylene for an additional 10 min at room temperature to remove paraffin.

12. Next, proceed with the following incubations at room temperature unless indicated otherwise:

13. 100% Ethanol, 3 min, twice.

14. 95% Ethanol, 3 min, twice.

15. 70% Ethanol, 3 min, twice.

16. Rinse with running water, 20 min.

17. PBS, 3 min, twice.

18. 0.3% Hydrogen peroxide in PBS, 30 min (to block endogenous peroxidase activity).

19. PBS, 5 min, twice.

20. 0.1% TPCK-trypsin in PBS, 37°C for 20 min (for antigen retrieval; suitability of trypsin concentration and incubation times should be confirmed with representative slides and adjusted if necessary before proceeding with large numbers of slides).

21. Running water, 10 min.

22. PBS, 5 min, twice.

23. Block with 5% goat serum for 1 h at room temperature.

24. Primary antibody: rat monoclonal MEC.13.3 anti-mouse CD31 diluted to 2.5 µg/mL in PBS containing 5% goat serum; incubate overnight at 4°C.

25. Rinse with PBS for 5 min twice.

26. Secondary antibody: biotinylated goat anti-rat (mouse adsorbed; cat. no. BA-4001; Vector Laboratories) 1 µg/mL diluted in PBS containing 5% goat serum; incubate 1 h.

27. PBS, 5 min, twice.

28. Premix Vectastain Elite AB (horseradish peroxidase) reagent 30 min; incubate with sections 30 min.

29. Wash with PBS twice, 5 min each.

30. Incubate with fresh DAB solution (Zymed) until desired staining intensity develops (2–10 min, monitor with microscope). **Note:** For effective staining and low

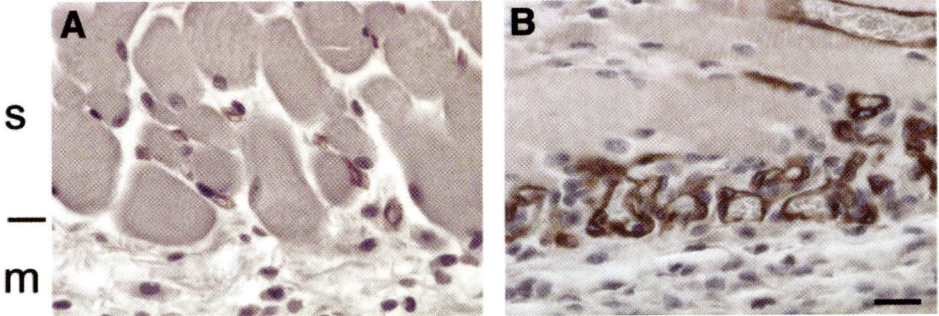

Fig. 2. Immunohistochemical staining for CD31. (**A**) Matrigel control, (**B**) Matrigel containing VEGF-transfectants. Samples were harvested at d 6 as in **Fig. 1**. Note abundant staining of endothelial cells (brown color) in (**B**) at the interface between the Matrigel implant containing VEGF-transfectants (m) and the skeletal muscle cell layer (s). Bar, 20 microns.

background, DAB solution must be yellow, not brown. Brown color indicates reagent has expired.

31. Rinse with tap water 10–20 min.
32. Counterstain with fresh hematoxylin for 2 min.
33. Rinse with hot water (10–20 min), then twice with 70% ethanol (2 min each), twice with 95% ethanol (2 min each), twice with 100% ethanol (2 min each), twice with 100% xylene (2 min each). Apply cover slips with mounting medium (Cytoseal 60; Stephens Scientific) while xylene is still wet.
34. After this staining procedure, endothelial cells are readily recognized as brown cells (DAB staining) with blue nuclei (hematoxylin staining; *see* **Fig. 2**, and for more examples *[20]*). From these stained tissue specimens, various parameters relevant to angiogenesis can be quantified, including: total numbers of endothelial cells per unit tissue area, numbers of endothelial cells integrated into mature blood vessels with recognizable lumens, total area occupied by blood vessels in cross section (i.e., vessel lumen area) as a percent total of total tissue cross-sectional area, etc. Finally, we believe that analyses of endothelial cells and blood vessels should be focused on the host skin immediately above the Matrigel implants as opposed to those within the Matrigel implants for two reasons. First, angiogenesis in host tissue is more relevant than angiogenesis in the artificial Matrigel implants. Second, and even more importantly, we find that angiogenesis within the Matrigel is variable from implant to implant because it is dependent on how much host tissue was trapped within the implant during injection. In sharp contrast to angiogenesis within the Matrigel implants themselves, angiogenesis in the overlying host skin is highly consistent.

3.2. Model of Angiogenesis in Mouse Skin Employing Retrovirus-Packaging Cells

This model system is fundamentally the same as the basic angiogenesis model described in **Subheading 3.1.** with the important addition of retrovirus-packaging cell lines stably expressing high titers of retrovirus encoding proteins or mutant proteins of interest. Transduction with purified retrovirus in vivo is widely recognized as inefficient, and therefore we chose to implant packaging cells as a means to provide a continuous source of retrovirus and thereby improve transduction efficiency. Our findings have established the effectiveness of this strategy (*see* **Subheading 3.2.2.**). The methods for analyses of angiogenesis in this model are the same as those described for the basic model (*see* **Subheading 3.1.2.**).

3.2.1. Preparation of Retrovirus Packaging Cell Lines

Packaging cells expressing replication-incompetent retrovirus are prepared by stable transfection of RetroPack PT67 cells (Clontech) with retroviral vector (e.g., pLCNX2 or pLNCX2/IRES-EGFP; *see* **Note 5**) into which has been inserted the cDNA to be expressed. This system is designed to produce retrovirus capable of transducing a broad range of mammalian cells including mouse and human; packaging cells producing retrovirus with host range restricted to mouse/rat can be substituted and may be preferred in some circumstances (*see* **Note 7**). Stable high titer clones ($\geq 1 \times 10^5$ c.f.u./mL) are selected according to manufacturer's instructions. For experiments involving direct comparisons of retroviruses encoding different proteins, it is important to select packaging cell clones expressing closely similar titers of retrovirus.

3.2.2. Use of Retroviral Packaging Cell Lines to Define Functional Contributions by Specific Proteins to Angiogenesis

The cloned retroviral packaging cells stably expressing high titers of retrovirus ($\geq 1 \times 10^5$ c.f.u./mL) are suspended together with the VEGF-transfectants in Matrigel and injected subdermally as described in **Subheading 3.1.1.** These Matrigel implants provide both a continuous source of VEGF and retrovirus throughout the experimental interval. As a general guide, we employ equal numbers of VEGF transfectants and retroviral packaging cells (per injection site: 1.5×10^6 of each cell population suspended in 0.3 mL of Matrigel). We have validated the efficacy of this strategy by demonstrating that packaging cells expressing retrovirus encoding a variety of mutant proteins regulate angiogenesis in accordance with expectations. For example, our experiments with models of angiogenesis in vitro, such as those described in **Subheading 3.3.1.**, had suggested that active RhoA promotes angiogenesis and that dominant-negative RhoA inhibits angiogenesis *(21)*. As illustrated in **Fig. 3**, our findings with the retrovirus-based model described here have confirmed this prediction in vivo *(21)*.

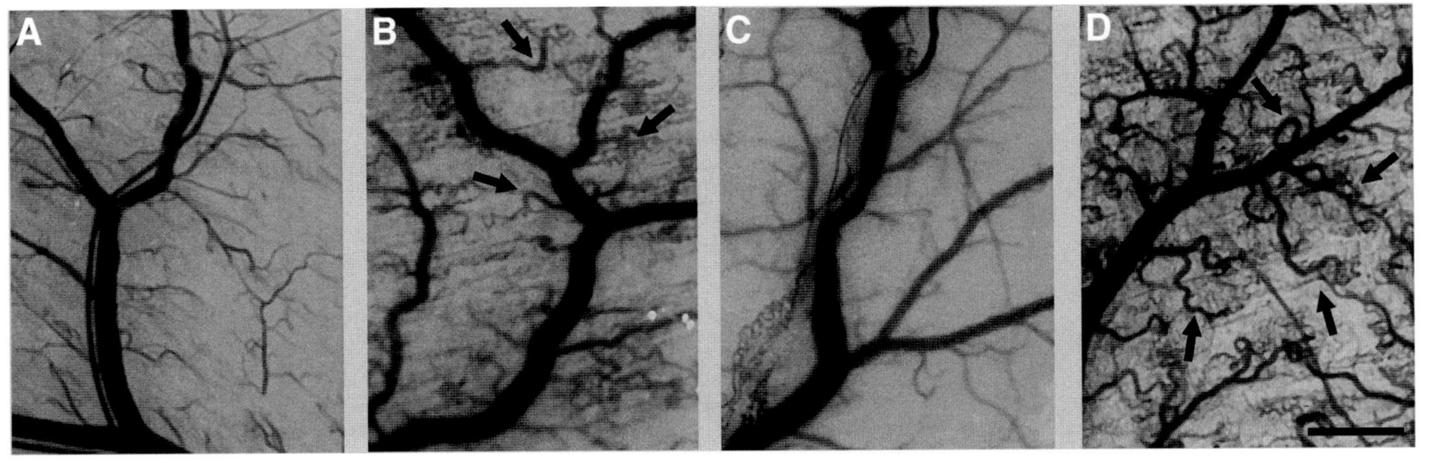

Fig. 3. Model of angiogenesis in mouse skin employing retrovirus-packaging cells. The dissected skin immediately adjacent to Matrigel implants was photographed; note these are low power views in comparison with **Fig. 1** (bar at far lower right, 680 microns). (**A**) Matrigel control, (**B**) Matrigel containing VEGF transfectants together with packaging cells expressing control retrovirus, (**C**) Matrigel containing VEGF transfectants together with packaging cells expressing retrovirus encoding dominant negative RhoA (N19RhoA), (**D**) Matrigel containing VEGF transfectants together with packaging cells expressing retrovirus encoding an active RhoA mutant (V14RhoA). Animals were harvested at d 6. Note that the appearance of new blood vessel sprouts, which are present in the positive control (panel b, arrows) were suppressed markedly by N19RhoA (**C**) and increased by active V14RhoA (**D**, arrows).

3.3. In Vitro Models of Angiogenesis

In vitro assays with isolated endothelial cells in culture can provide important complementary information that is useful for interpreting findings made with angiogenesis assays in vivo. There are several in vitro assays that are highly relevant to angiogenesis, including those that measure endothelial cell proliferation, survival, and motility. However, because cell proliferation, survival, and motility assays have been described extensively in the literature and because there are many different and similarly valid variations of these assays, we will not describe them here. Rather, we have chosen to focus on a simple but important in vitro model of capillary morphogenesis. In this model, endothelial cells organize into cord-like structures that closely imitate the solid pre-capillary cords that form during embryonic angiogenesis in vivo *(22–26)*. These solid pre-capillary cords are the precursors to mature capillaries with lumens *(27)*, and formation of these structures represents an important intermediate step in blood vessel formation. Thus, this in vitro model of pre-capillary cord formation is particularly interesting because it offers the opportunity to gain information on the signaling pathways and underlying mechanisms that regulate endothelial cell morphogenesis and organization. Moreover, the findings made with this model of capillary morphogenesis can be tested for relevance in vivo with the retroviral-based angiogenesis model described in **Subheading 3.2.2.**

3.3.1. Stimulation of Endothelial Cell Capillary Morphogenesis by Collagen I

Basically, there are two experimental formats for inducing capillary morphogenesis by collagen I. Each has its own distinct advantage; therefore, both are described here. The simplest format involves growing microvascular endothelial cells to confluence and then overlaying them with soluble collagen I (the "overlay" format). The procedures are as follows.

1. Acid-solubilized rat tail collagen (e.g., BD Biosciences, Bedford, MA) is neutralized and rendered isotonic in accordance with manufacturer's instructions and diluted in ice cold serum-free culture medium (*see* **Notes 6** and **8**).
2. Medium is removed from confluent cultures and replaced with the medium containing collagen I (final concentrations in the range 200–500 µg/mL are best).
3. The cultures are returned to the tissue culture incubator and must not be disturbed until the collagen I has polymerized completely (30–45 min). Otherwise, the collagen will not polymerize properly, which may result in endothelial cell death. However, cell death is not observed when complete polymerization occurs *(27)*.
4. Typically, human dermal microvascular endothelial cells will undergo detectable morphogenesis in this collagen I-overlay assay within 1 h and will form cords by 4–6 h, at which time the assay is terminated. Longer time intervals require the addition of survival-promoting factors.

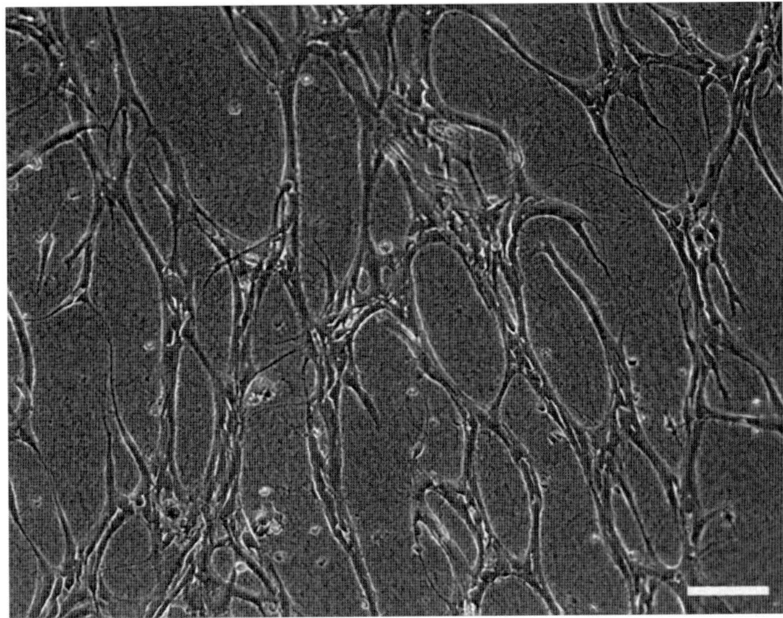

Fig. 4. In vitro cord formation by human microvascular endothelial cells sandwiched between two layers of three-dimensional collagen I; photographed 48 h after addition of the upper collagen layer. Bar, 150 microns.

Another format for studying capillary morphogenesis in vitro involves the sandwiching of a monolayer of endothelial cells between two layers of three-dimensional collagen I (the "sandwich" format; *see* **Fig. 4**). Because it takes approx 24–48 h for formation of cords in this assay, addition of factors that promote endothelial cell survival in three-dimensional collagen I is required. For human dermal microvascular endothelial cells, addition of 16 nm of PMA, 20 ng/mL VEGF, and 2.5 ng/mL bFGF to the collagen I-containing medium (EBM-2, Clonetics) works well. This cocktail and similar versions have been described previously for preservation of human umbilical vein endothelial cells in collagen I *(28,29)*. The steps for formulating this "sandwich" version of the capillary morphogenesis assay are as follows:

1. Prepare "cushion" of three-dimensional collagen I (1–2 mg/m: dissolved in medium) by adding liquid collagen solution to preferred tissue culture vessel (e.g., 24- or 12-well plate).
2. Allow the collagen cushion to polymerize and then add a suspension of microvascular endothelial cells in collagen-free medium (cell number should be chosen such that a suitably confluent monolayer will form).

3. After cells have attached and formed a confluent monolayer (overnight) carefully and completely remove medium, and immediately overlay the monolayer with the same collagen I-solution used to formulate the underlying collagen cushion. Thus, these steps result in the formation of an endothelial cell monolayer sandwiched between two layers of three-dimensional collagen I.

4. The initiation of capillary morphogenesis is evident within several hours and cords are evident within 24 h; however, 48 h may be required before cord formation is complete. Thus, potential disadvantages of this "sandwich" format in comparison with the "overlay" format are that cord formation takes longer and a cocktail of survival factors must be included. Nevertheless, this assay offers the advantage that cells are completely embedded in three-dimensional matrix and the cords that form are more symmetrical (cylindrical) in shape. Moreover, under these conditions, the cords will eventually form hollow lumens *(29)*. Methods for analyses of lumen formation by endothelial cells embedded in collagen I have been described in detail previously *(28,29,31)*.

The collagen I-based in vitro assays described previously are particularly useful for investigating the significance of specific proteins and signaling pathways for capillary morphogenesis. Thus, they can provide important complementary information for the in vivo angiogenesis assays described beforehand in this section. For example, and as discussed in **Subheading 3.2.2.**, the importance of the GTPase RhoA for angiogenesis was predicted by these and similar assays *(21)*. We have confirmed this prediction with the in vivo angiogenesis model employing packaging cells producing retrovirus that encode RhoA mutants *(21)* (*see* **Subheading 3.2.2.** and **Fig. 3**). Thus, for collagen I-based in vitro assays of capillary morphogenesis, it is often desirable to employ endothelial cells transduced with retrovirus encoding dominant-negative or constitutively active mutant proteins. For this purpose, a method for efficient retroviral transduction of microvascular endothelial cells is described next.

3.3.2. Efficient Retroviral Transduction of Microvascular Endothelial Cells

The method for stable retroviral transduction of microvascular endothelial cells described here is based on a previously described method for retroviral gene transfer in primary fibroblasts *(30)*. This method involves the formation of a high molecular weight retrovirus–polymer complex involving Polybrene and chondroitin sulfate C and when applied to proliferating microvascular endothelial cells, provides for 75–95% transduction. Complete (i.e.100%) transduction can be achieved with drug selection (*see* **steps 1–8**) to kill the remaining untransduced cells. Specific steps are as follows.

1. PT67 clones producing high titer retrovirus (>10^5 c.f.u./mL; from **Subheading 3.2.1.**) are grown in roller bottles to achieve a cell density of 4×10^8 cells/200 mL of culture medium.

2. Culture medium from a 24-h incubation is collected and passed through a 45-µ filter to assure complete removal of PT67 cells.
3. Incubate filtered culture medium containing retrovirus with 80 µg/mL chondroitin sulfate C for 10 min at 37°C and then add an equal amount of polybrene and incubate for an additional 10 min to allow formation of the polymer complex.
4. Collect retrovirus complexed with chondroitin sulfate/polybrene by centrifugation (10,000g, 15 min).
5. Resuspend complex in 10 mL of endothelial cell culture medium containing 20 ng/mL VEGF and incubate directly with a subconfluent and proliferating culture of microvascular endothelial cells (10-cm dish).
6. After 4 h, rinse culture to remove polymer and replace fresh medium with VEGF.
7. Repeat **steps 3** and **4** for two more consecutive days.
8. Allow culture to recover for 2 to 3 d before subjecting to drug selection with 250 µg/mL G418 sulfate (*see* **Note 9**). Change G418-containing medium every 2 or 3 d for 1 wk. Cell population should consist of 100% transductants by this time. Transduction can be confirmed by fluorescence when retrovirus encodes green fluorescent protein (GFP).

4. Notes

1. Female athymic nude mice are best. Males inflict each other with skin wounds that can complicate the assay. We routinely use athymic NCr nude mice at 7 to 8 wk of age, but other athymic nude strains and older mice are also suitable. However, because the robustness of angiogenesis may vary with strain and age, it is important to use appropriately matched animals in experiments.
2. Basement membrane Matrigel is available in standard and growth factor-reduced formulations. We usually observe no significant differences among them for purposes described here, although we have observed occasionally that the standard formulation of Matrigel can, by itself, provoke angiogenesis slightly. However, we never have observed any angiogenesis-provoking activity associated with growth factor-reduced Matrigel.
3. The form of VEGF specified here is VEGF-A$_{165}$, which is the most common isoform of VEGF-A. The designation VEGF-A is now often applied to what was once referred to solely as VEGF, to distinguish it from other more recently discovered VEGF family members. There are potentially numerous options for choosing cells to be employed in this assay. SK-MEL-2 human melanoma cells (American Type Culture Collection; Rockville, MD) offer the advantage that they normally express little or no detectable VEGF. Consequently, untransfected SK-MEL-2 cells do not provoke angiogenesis in this model and serve as negative controls for the SK-MEL-2 VEGF-transfectants. Moreover, these cells are easy to transfect and propagate in vitro. However, in addition to SK-MEL-2 cells, many other immortalized cell lines should suffice; and depending on experimental design, others may be required. For example, when testing the consequences of administering a specific blocking antibody for angiogenesis, one should employ a VEGF-transfected cell line that is not recognized by the

antibody in order to avoid possible complications owing to antibody-mediated cytotoxicity toward the transfected cells. Thus, a preferred cell line for preparing VEGF transfectants would be one that did not bind the antibody to be tested. Finally, we use the mammalian expression vector, pcDNA 3.1 (Stratagene, La Jolla, CA) for preparing VEGF transfectants; however, others with a suitable mammalian promoter and drug resistance gene could be substituted.

4. Evan's blue dye rapidly complexes with serum albumin in blood and serves as a tracer for blood flow. Because new blood vessels are hyperpermeable, dye complexed with albumin will leak from the vasculature with time. However, we find that this is not a problem for short time intervals (10 min or less) and that 5 min is generally sufficient time for blood vessel perfusion. Other tracers to consider include high-molecular-weight fluorescent dextrans, however, visualization of such tracers requires a dissecting microscope with fluorescence-imaging capabilities.

5. In many circumstances, we find it desirable to use a different retroviral vector that encodes a marker for monitoring retroviral transduction. Therefore, we designed a modified version of pLCNX2 that co-expresses GFP and the protein encoded by the inserted cDNA as separate elements. This retroviral vector was engineered by subcloning unidirectionally the 1360 base-pair *Xho*I/*Not*I restriction fragment from the mammalian expression vector pIRES2-EGFP (Clontech) into the multicloning site of pLCNX2. The 1360 base-pair *Xho*I/*Not*I restriction fragment from pIRES2-EGFP includes multiple cloning sites, an internal ribosomal entry site, and the coding sequence for enhanced GFP. Because this modified vector (pLNCX2/IRES-EGFP) expresses GFP separately from the protein encoded by the inserted cDNA, it avoids possible complications associated with GFP-fusion proteins.

6. Microvascular endothelial cells are preferable to endothelial cells isolated from umbilical vein or aorta because the microvasculature is most relevant to angiogenesis. Human dermal microvascular endothelial cells are available commercially (e.g., Cascade Biologics, Portland, OR; Clonetics/BioWhittaker/Cambrex, Walkersville, MD). Commercial suppliers provide their own media formulations that are generally required for optimal cell growth. In addition, microvascular endothelial cell populations isolated from other vascular beds are available from these and other vendors. For isolation of dermal microvascular endothelial cells, an excellent method has been described in detail *(5)*. The culture medium for cells isolated according to this protocol is EBM (Clonetics/BioWhittaker/ Cambrex) supplemented with 1 µg/mL hydrocortisone acetate, 5×10^{-5} *M* *N*-6,2' *O*-dibutryl-adensosine 3',5'-cyclic AMP (Sigma), 20% heat-inactivated fetal calf serum, 100 U/mL penicillin, and 100 µg/mL streptomycin.

7. Because retroviruses produced by this system are capable of infecting both murine and human cells, they can be used both in the mouse angiogenesis model and also for transduction of human endothelial cells in culture, as described in **Subheading 3.3.2.** This is a distinct advantage. However, in some cases, retroviruses with a restricted host-range may be preferred as, for example, when a retrovirally encoded

mutant protein interferes with VEGF expression by the human SK-MEL-2 transfectants used to drive angiogenesis. In such cases, an ecotropic virus capable of infecting murine cells but not human cells would be required. To control for the possibility that retrovirus alters VEGF expression, we routinely quantify VEGF expression by the SK-MEL-2 transfectants after retroviral transduction with the same retroviruses employed in vivo. However, thus far we have not observed any effects of transduction by these retroviruses on VEGF expression.

8. Dermal microvascular endothelial cells will undergo collagen I-stimulated capillary morphogenesis similarly in the presence or absence of serum. Therefore, culture medium may contain serum, if desired. However, serum-free medium offers the advantage that it allows for analyses of collagen I-signaling in endothelial cells in the absence of the many agonists present in serum.

9. In some cases, retrovirus-mediated expression of specific mutant proteins may reduce proliferation or even induce cell death. Cells may be rescued from mutant protein toxicity with appropriate counteracting drugs, if available.

Acknowledgments

This work was supported by a grant from the National Cancer Institute (CA77357 to D. R. S.) and by the V. Kann Rasmussen Foundation.

References

1. Folkman, J. (1995) Clinical applications of research on angiogenesis. *N. Engl. J. Med.* **333,** 1757–1763.
2. Smith, L. E., Shen, W., Perruzzi, C., Soker, S., Kinose, F., Xu, X., et al. (1999) Regulation of vascular endothelial growth factor-dependent retinal neovascularization by insulin-like growth factor-1 receptor. *Nat. Med.* **5,** 1390–1395.
3. Isner, J. M. and Asahara, T. (1999) Angiogenesis and vasculogenesis as therapeutic strategies for postnatal neovascularization. *J. Clin. Invest.* **103,** 1231–1236.
4. Papaioannou, V. E. and Fox, J. G. (1993) Efficacy of tribromoethanol anesthesia in mice. *Lab. An. Sci.* **43,** 189–192.
5. Richard, L., Velasco, P., and Detmar, M. (1998) A simple immunomagnetic protocol for the selective isolation and long-term culture of human dermal microvascular endothelial cells. *Exp. Cell Res.* **240,** 1–6.
6. Passaniti, A., Taylor, R. M., Pili, R., Guo, Y., Long, P. V., Haney, J. A., et al. (1992) A simple, quantitative method for assessing angiogenesis and antiangiogenic agents using reconstituted basement membrane, heparin, and fibroblast growth factor. *Lab. Invest.* **67,** 519–528.
7. Kim, K. J., Li, B., Winer, J., Armanini, M., Gillett, N., Phillips, H. S., et al. (1993) Inhibition of vascular endothelial growth factor-induced angiogenesis suppresses tumour growth in vivo. *Nature* **362,** 841–844.
8. Brown, L. F., Berse, B., Jackman, R. W., Tognazzi, K., Manseau, E. J., Dvorak, H. G., et al. (1993) Increased expression of vascular permeability factor (vascular endothelial growth factor) and its receptors in kidney and bladder carcinomas. *Am. J. Pathol.* **143,** 1255–1262.

9. Brown, L. F., Berse, B., Jackman, R. W., Tognazzi, K., Manseau, E. J., Senger, D. R., et al. (1993) Expression of vascular permeability factor (vascular endothelial growth factor) and its receptors in adenocarcinomas of the gastrointestinal tract. *Cancer Res.* **53,** 4727–4735.

10. Berkman, R. A., Merrill, M. J., Reinhold, W. C., Monacci, W. T., Saxena, A., Clark, W. C., et al. (1993) Expression of the vascular permeability factor/vascular endothelial growth factor gene in central nervous system neoplasms. *J. Clin. Invest.* **91,** 153–159.

11. Warren, R. S., Yuan, H., Matli, M. R., Gillett, N. A., and Ferrara, N. (1995) Regulation by vascular endothelial growth factor of human colon cancer tumorigenesis in a mouse model of experimental liver metastasis. *J. Clin. Invest.* **95,** 1789–1797.

12. Millauer, B., Longhi, M. P., Plate, K. H., Shawver, L. K., Risau, W., Ullrich, A., et al. (1996) Dominant-negative inhibition of Flk-1 suppresses the growth of many tumor types in vivo. *Cancer Res.* **56,** 1615–1620.

13. Yuan, F., Chen, Y., Dellian, M., Safabakhsh, N., Ferrara, N., and Jain, R. K. (1996) Time-dependent vascular regression and permeability changes in established human tumor xenografts induced by an anti-vascular endothelial growth factor/vascular permeability factor antibody. *Proc. Natl. Acad. Sci. USA* **93,** 14,765–14,770.

14. Claffey, K. P., Brown, L. F., del Aguila, L. F., Tognazzi, K., Yeo, K.-T., Manseau, E. J., et al. (1996) Expression of vascular permeability factor/vascular endothelial growth factor by melanoma cells increases tumor growth, angiogenesis, and experimental metastasis. *Cancer Res.* **56,** 172–181.

15. Detmar, M., Brown, L. F., Claffey, K. P., Yeo, K-T., Kocher, O., Jackman, R. W., et al. (1994) Overexpression of vascular permeability factor/vascular endothelial growth factor and its receptors in psoriasis. *J. Exp. Med.* **180,** 1141–1146.

16. Aiello, L. P., Pierce, E. A., Foley, E. D., Takagi, H., Chen, H., Riddle, L., et al. (1995) Suppression of retinal neovascularization in vivo by inhibition of vascular endothelial growth factor (VEGF) using soluble VEGF-receptor chimeric proteins. *Proc. Natl. Acad. Sci. USA* **92,** 10,457–10,461.

17. Pierce, E. A., Avery, R. L., Foley, E. D., Aiello, L. P., and Smith, L. E. H. (1995) Vascular endothelial growth factor/vascular permeability factor expression in a mouse model of retinal neovascularization. *Proc. Natl. Acad. Sci. USA* **92,** 905–909.

18. Robinson, G. S., Pierce, E. A., Rook, S. L., Foley, E., Webb, R., and Smith, L. E. H. (1996) Oligodeoxynucleotides inhibit retinal neovascularization in a murine model of proliferative retinopathy. *Proc. Natl. Acad. Sci. USA* **93,** 4851–4856.

19. Miller, J. W. (1997) Vascular endothelial growth factor and ocular neovascularization. *Am. J. Pathol.* **151,** 13–23.

20. Senger, D. R., Perruzzi, C. A., Streit, M., Koteliansky, V. E., de Fougerolles, A. R., and Detmar, M. (2002) The alpha1beta1 and alpha2beta1 integrins provide critical support for vascular endothelial growth factor signaling, endothelial cell migration, and tumor angiogenesis. *Am. J. Pathol.* **160,** 195–204.

21. Hoang, M.V., Whelan, M.C., and Senger, D.R. (2004). Rho activity critically and selectively regulates endothelial cell organization during angiogenesis. *Proc. Natl. Acad. Sci. USA* **101,** 1874–1879.

22. Vernon, R. B., Lara, S. L., Drake, C. J., Iruela-Arispe, M. L., Angello, J. C., Little, C. D., et al. (1995) Organized type I collagen influences endothelial patterns during "spontaneous angiogenesis in vitro:" planar cultures as models of vascular development. *In Vitro Cell. Dev. Biol.* **31,** 120–131.

23. Vernon, R. B. and Sage, E. H. (1995) Between molecules and morphology: extracellular matrix and creation of vascular form. *Am. J. Pathol.* **147,** 873–883.

24. Drake, C. J., Brandt, S. J., Trusk, T. C., and Little, C. D. (1997) TAL1SCL is expressed in endothelial progenitor cells/angioblasts and defines a dorsal-to-ventral gradient of vasculogenesis. *Dev. Biol.* **192,** 17–30.

25. Drake, C. J. and Little, C. D. (1999) VEGF and vascular fusion: implications for normal and pathological vessels. *J. Histochem. Cytochem.* **47,** 1351–1355.

26. Drake, C. J. and Fleming, P. A. (2000) Vasculogenesis in the day 6.5 to 9.5 mouse embryo. *Blood* **95,** 1671–1679.

27. Whelan, M. C. and Senger, D. R. (2003) Collagen I initiates endothelial cell morphogenesis by inducing actin polymerization through suppression of cyclic AMP and protein kinase A. *J. Biol. Chem.* **278,** 327–334.

28. Ilan, N., Mahooti, S., and Madri, J. A. (1998) Distinct signal transduction pathways are utilized during the tube formation and survival phases of in vitro angiogenesis. *J. Cell Sci.* **111,** 3621–3631.

29. Davis, G. E. and Camarillo, C. W. (1996) An $\alpha 2\beta 1$ integrin-dependent pinocytic mechanism involving intracellular vacuole formation and coalescence regulates capillary lumen and tube formation in three-dimensional collagen matrix. *Exp. Cell Res.* **224,** 39–51.

30. Le Doux, J. M., Landazuri, N., Yarmush, M., and Morgan, J. R. (2001) Complexation of retrovirus with cationic and anionic polymers increases the efficiency of gene transfer. *Human Gene Ther.* **12,** 1611–1621.

31. Montesano, R., Orci, L., and Vassalli, P. (1983) In vitro rapid organization of endothelial cells into capillary-like networks is promoted by collagen matrices. *J. Cell Biol.* **97,** 1648–1652.

V

Biochemical and Novel Approaches in Cell Migration

20

Measurement of Protein Tyrosine Phosphorylation in Cell Adhesion

Vidhya V. Iyer, Christoph Ballestrem, Joachim Kirchner, Benjamin Geiger, and Michael D. Schaller

Summary

This chapter describes biochemical, immunochemical, and microscopic approaches to measure protein tyrosine phosphorylation after cell adhesion. We have outlined detailed procedures to biochemically examine the phosphotyrosine content of cellular proteins by Western blotting, which in some cases can be performed using phospho-specific antibodies. Furthermore, we have described in detail the examination of subcellular localization of phosphotyrosine-containing proteins in focal adhesions using immunofluorescence. Finally, a quantitative fluorescence microscopic technique using an SH2-containing phosphotyrosine reporter to monitor tyrosine phosphorylation in focal adhesions in live cells is described.

Key Words: Phosphotyrosine; focal adhesions; extracellular matrix; integrins; immunofluorescence; phosphotyrosine reporter; live cell imaging.

1. Introduction

Focal adhesions (FAs) are regions of the cell, which serve as attachments to the extracellular matrix (ECM) and as sources of traction to move the cell body forward during migration *(1,2)*. In most cases, these attachments are mediated by integrins, which are transmembrane proteins that link the ECM proteins to intracellular cytoskeletal components. The binding of integrins to ECM proteins initiates cytoplasmic events that regulate cellular functions *(3)*. These cytoplasmic events include changes in intracellular pH *(4,5)*, modulation of the activity of Rho family GTPases *(6,7)* as well as elevation of intracellular calcium *(8,9)*. In addition, protein kinases are activated resulting in phosphorylation of tyrosine and serine residues of proteins such as focal adhe-

From: *Methods in Molecular Biology, vol. 294: Cell Migration: Developmental Methods and Protocols*
Edited by: J-L. Guan © Humana Press Inc., Totowa, NJ

sion kinase (FAK) *(10,11)*, p130(Cas) *(12,13)*, and paxillin *(14–16)*, which play important roles in regulating biological processes, for example, motility *(17–21)*. Tyrosine phosphorylation of proteins is a key mode of signaling and a frequently studied integrin-mediated cellular event. Further, tyrosine phosphorylation has been implicated in the control of important integrin-regulated biological functions including motility, cell proliferation and cell survival *(3)*. As aberrant regulation of these processes can produce pathological conditions, these cell adhesion-controlled signaling pathways may play a role in the development and progression of human disease.

As cells attach to the ECM, both structural and signaling proteins localize in FA with integrins *(1,3,22,23)*. These proteins include FAK (focal adhesion kinase) and Pyk2 (proline rich tyrosine kinase 2), which are highly related protein tyrosine kinases associated with cell adhesion *(24–27)*. Paxillin is an important adaptor protein that interacts with FAK and aids in the recruitment of proteins to focal adhesions *(28,29)*. p130(Cas) is a docking protein with multiple protein interaction motifs and plays an important role in migration *(30)*. These focal adhesion-associated proteins are major substrates for tyrosine phosphorylation following cell adhesion. Select examples using several of these substrates will be used to illustrate methods for the analysis of cell adhesion-dependent tyrosine phosphorylation.

2. Materials

1. Tissue culture dishes for propagation of cells.
2. 100-mm or 60-mm Falcon plastic (non-tissue culture) dishes are used for coating of dishes with ECM proteins.
3. Glass cover slips for immunofluorescence.
4. 35-mm MatTek glass-bottom dishes (MatTek Corporation; Ashland, MA; Website: www.mattek.com/) for live cell imaging.
5. Primary chicken embryo fibroblasts (CE), SV80 cells, and NIH 3T3 cells (American Type Culture Collection, Rockville, MD). CE cells were grown in Dulbecco's modified Eagle's medium (DMEM) supplemented with 5% fetal bovine serum and 1% chick serum. SV80 and NIH 3T3 cells were grown in DMEM supplemented with 10% bovine calf serum.
6. Modified RIPA lysis buffer: 50 mM Tris-HCl, pH 7.3, 150 mM NaCl, 1% NP40 or Triton X-100, and 0.5% deoxycholate. This buffer is made by dissolving the Tris and NaCl, then adjusting the pH to 7.3. Add the detergents, adjust the pH if necessary, filter the buffer and store at 4°C. Add protease and phosphatase inhibitors (**steps 8–12**) to the lysis buffer immediately prior to use. These inhibitors block proteolysis and dephosphorylation of proteins upon cell lysis.
7. Triton X-100 lysis buffer: 20 mM Tris-HCl, pH 7.5, 150 mM NaCl, 1% Triton X-100, 10% glycerol. Filter the buffer and store at 4°C. Add protease and phosphatase inhibitors (**steps 8–12**) to the lysis buffer immediately prior to use.

8. Ethylene diamine tetraacetic acid: 0.2 *M* stock; use at a final concentration of 2 m*M*.
9. Aprotinin, stored at 4°C; use at a final concentration of 0.05 TIU/mL.
10. Leupeptin: 10 mg/mL stock dissolved in water and stored in aliquots at –20°C. Use at a final concentration of 50 µg/mL.
11. Phenylmethylsulfonyl fluoride: 200 m*M* stock dissolved in methanol, stored at –20°C. Use at a final concentration of 1 m*M*.
12. Sodium orthovanadate: 200 m*M* stock dissolved in water, stored at 4°C. Use at a final concentration of 1 m*M*.
13. Phosphate-buffered saline (PBS).
14. Trypsin.
15. Soybean trypsin inhibitor: 0.5 mg/mL of soybean trypsin inhibitor dissolved in PBS. Store at 4°C.
16. Fibronectin: reconstitute to 1 mg/mL in sterile water, by incubation for 30 min at room temperature. Do not shake. Aliquot and store frozen.
17. Poly-L-lysine: reconstitute to 1 mg/mL in sterile water.
18. Bicinchoninic acid assay (Pierce) or other reagents for determination of the protein concentration of cell lysates.
19. Bovine serum albumin (BSA): reconstitute to 2 mg/mL in PBS. It is very important to use lipid-free BSA.
20. Equipment for electrophoresis (sodium dodecyl sulfate-polyacrylamide gel electrophoresis) and electrophoretic transfer of proteins from gels to membranes (Western blotting).
21. A phosphotyrosine antibody. We use the RC20 antibody (BD Bioscience), which works well for the proteins of interest in the lab. However, there are many alternative antibodies that have been used.
22. Phospho-specific antibodies. For FAK and paxillin antibodies, we have used the phospho-specific antibodies from Biosource. There are other sources for similar antibodies.
23. TBS-T buffer: 10 m*M* Tris-HCl, pH 7.5, 150 m*M* NaCl, 0.1% Tween-20 (polyoxyethylene-sorbitan monolaurate).
24. Blocking agents: fish gelatin, BSA and non-fat dried milk. Different blocking buffers are made by dissolving these agents in TBS-T.
25. Protein A-Sepharose or protein G-Sepharose beads: swell the beads (3 g) in 40 mL of PBS for 4 h at room temperature. Wash four times with 20 mL of PBS. Resuspend in an equal volume of PBS (i.e., create a 50% slurry). Add sodium azide. Store at 4°C.
26. 2X Sample buffer: 125 m*M* Tris-HCl, pH 6.8, 4% sodium dodecyl sulfate (SDS), 20% glycerol, 1.4 *M* β-mercaptoethanol, Bromophenol blue.
27. Horseradish peroxidase (HRP)-conjugated secondary antibody.
28. Enhanced chemiluminescence (ECL) reagents.
29. Stripping buffer: 62.5 m*M* Tris-HCl (pH 7.5), 2% SDS, 84 mM β-mercaptoethanol.
30. Fluorophore-conjugated secondary antibody for immunofluorescence.
31. Lipofectamine Plus reagent (Invitrogen Corporation, Carlsbad, CA).

32. EYFP and ECFP vectors (Clontech, Palo Alto, CA).
33. The YFP-dSH2- and CFP-paxillin-encoding plasmids were prepared according to **ref. *31***.
34. Elvanol (Mowiol® 4-88; Hoechst, Frankfurt, Germany).
35. Zeiss Axiovert 100 microscope, equipped with a 100X/1.4 NA plan-Neofluar objective (Zeiss, Oberkochen, Germany). To maintain stable temperature, the microscope was set under a Box and Temperature control system from Life Imaging Services (Switzerland; website: www.lis.ch/). Filters to detect CFP and YFP were obtained from Chroma (website: www.chroma.com/). Images were acquired with a DeltaVision system (Applied Precision, Issaqua, WA).

3. Methods

The methods outlined are procedures to examine biochemical changes in phosphotyrosine levels, cellular localization of phosphotyrosine after adhesion, and a strategy to observe dynamic changes in phosphotyrosine in living cells.

3.1. Preparation of ECM Protein-Coated Dishes

1. Different ECM proteins can be used in these studies including fibronectin, laminin, collagen and vitronectin. Typically we have used fibronectin. To prepare fibronectin-coated surfaces, coat 100-mm plastic dishes (*see* **Note 1**) with fibronectin at 1–6 µg/cm^2 in 2 mL of 1X PBS by incubation at 37°C for 1 h with occasional rocking to ensure uniform coating of the plate surface.
2. After incubation for 1 h, aspirate the excess solution, rinse the plates with PBS, and allow it to dry uncovered in a sterile environment before use.
3. Poly-L-lysine-coated plates are used as a control to evaluate phosphotyrosine changes owing to non-integrin-specific binding of suspended cells to a matrix. To prepare poly-L-lysine-treated surfaces, coat 100-mm dishes with 0.05 mg/cm^2 of poly-L-lysine in 5 mL of H$_2$O with constant rocking at room temperature for approx 1 h. Check the plates regularly as surface tension makes it difficult to uniformly coat the surface.
4. Remove the excess solution, rinse the plates with H$_2$O, and allow it to dry uncovered in a sterile environment.
5. Block the fibronectin and Poly-L-lysine-coated plates with 2 mg/mL of lipid-free BSA by incubation at 37°C for 1–2 h (*see* **Note 2**). Aspirate the BSA and rinse the plate with PBS prior to use.

3.2. Plating Cells on ECM Protein-Coated Dishes

Adherent cells growing in culture in serum-containing media are attached to ECM proteins and may exhibit high basal levels of cell adhesion-dependent signaling, including high levels of phosphotyrosine on focal adhesion-associated proteins. To study cell adhesion-dependent tyrosine phosphorylation, basal levels are reduced by detaching cells from the ECM and holding the cells in suspension before replating the cells on ECM protein coated dishes. The procedure is as follows:

1. Trypsinize the cells and then resuspend in 1X PBS containing 0.5 mg/mL soybean trypsin inhibitor to terminate trypsinization.
2. Wash the suspended cells with 1X PBS containing 0.5 mg/mL soybean trypsin inhibitor.
3. Resuspend in serum-free medium (approx 2.5×10^6 cells/mL) and incubate for 45 min at 37°C, keeping the cells in suspension during this interval.
4. Plate the suspended cells onto 100-mm ECM-coated dishes (*see* **Subheading 3.1.**) at a density of 2.5×10^6 cells/mL in a total volume of 2 mL of serum-free medium.
5. At various times following attachment of the cells to the plate, cell extracts can be prepared to examine tyrosine phosphorylation by Western blotting or phosphotyrosine content can be examined by immunofluorescence or live cell imaging.

3.3. Cell Lysis

1. Wash the cells twice with 1X PBS, siphon off the residual buffer and incubate the cells for 10 min on ice in lysis buffer. Use 1 mL of lysis buffer per 100-mm dish of confluent cells.
2. Scrape the lysates off the dish and clarify by centrifugation in a microcentrifuge for 10 min at 4°C. The supernatant is then taken for analysis.
3. For suspended cells, spin the cells down at a low speed to avoid rupturing of cells, wash with 1X PBS once, resuspend in lysis buffer, incubate on ice for 10 min, and then clarify.
4. Determine the protein concentration of the cell lysates using the bicinchoninic acid assay (Pierce). Different lysis buffers have been used to examine cell adhesion-dependent tyrosine phosphorylation of focal adhesion-associated proteins. Typically we use modified RIPA buffer or Triton X-100 lysis buffer. The lysis buffer must be supplemented with protease and phosphatase inhibitors. Sodium orthovanadate is especially important because its activity as a phosphatase inhibitor facilitates the detection of phosphotyrosine.

3.4. Examining Phosphotyrosine Content

Phosphotyrosine content is examined by Western blotting using a phosphotyrosine antibody. Cell lysates can be used directly to examine overall phosphotyrosine content or tyrosine phosphorylation of specific proteins of interest can be determined by immunoprecipitation out of the lysates prior to blotting.

3.4.1. Immunoprecipitating Proteins of Interest

1. Typically immunoprecipitations are performed using 500 µg to 1 mg of lysate. The amount of antibody required to completely immunodeplete an antigen must be determined empirically. Usually 5 µL of antiserum or 2–5 µg of purified monoclonal antibody is used per immunoprecipitation. Add the antibody to the cell lysate and incubate for 1 h on ice.

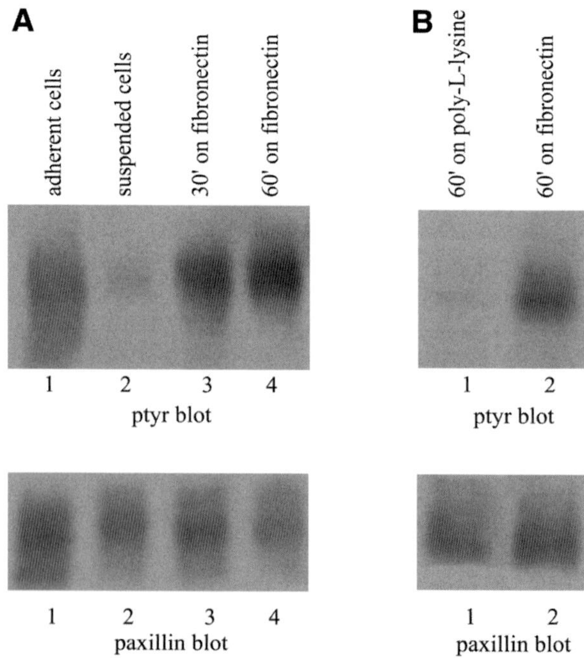

Fig. 1. Cell adhesion-dependent tyrosine phosphorylation of paxillin. Adherent CE cells, CE cells held in suspension (**A**) or plated onto poly-L-lysine or fibronectin-coated dishes (**B**) were lysed. Paxillin was immunoprecipitated and analyzed by Western blotting for phosphotyrosine (top panels) or for paxillin as a loading control (bottom panels).

2. Add approx 50 µL of protein-A or protein-G Sepharose beads (a 50% slurry in 1X PBS) to capture the immune complexes and incubate for 1 h with constant rocking at 4°C (*see* **Note 3**).
3. Spin down the beads and wash twice with lysis buffer, once with 1X PBS, and then resuspend in a small volume of 2X sample buffer.

3.4.2. Overall Phosphotyrosine Content

To determine overall phosphotyrosine levels in lysates or of immunoprecipitated proteins, samples are resolved by sodium dodecyl sulfate-polyacrylamide page electrophoresis and then transferred to nitrocellulose membranes. The proteins on the nitrocellulose membranes are immunoblotted for phosphotyrosine (*see* **Fig. 1**). The procedure is as follows:

1. Block the nitrocellulose membrane by incubating in TBS-T, TBS-T containing 2% fish gelatin, or TBS-T containing 5% BSA for at least 1 h at room temperature on an orbital shaker (*see* **Notes 4** and **5**).

2. Incubate the membrane in blocking buffer containing the phosphotyrosine antibody for 1 h at room temperature. Typically we use RC20, a recombinant phosphotyrosine antibody conjugated to horseradish peroxidase (*see* **Notes 6** and **7**).

3. Wash the membrane in TBS-T five times for 5 min each wash.

4. As RC20 is conjugated to HRP and can be directly detected by ECL, proceed to **step 7**.

5. Unconjugated phosphotyrosine antibodies must be detected with an HRP-conjugated secondary antibody, for example, HRP-conjugated anti-mouse antibody to detect monoclonal antibodies recognizing phosphotyrosine (*see* **Note 8**). Incubate the secondary antibody with the membrane for 1 h at room temperature in blocking buffer.

6. Wash the membrane again five times in TBS-T for 5 min each wash.

7. Treat the membrane with ECL chemicals as per the manufacturer's directions and expose to film.

8. As a loading control, strip the membrane of any adherent antibodies and reprobe with an antibody directed against the protein of interest. To strip the membrane, incubate in stripping buffer at 65°C for 1 h. After rinsing in H_2O, re-probe after **steps 1–7**. Use 5% milk in TBS-T as blocking buffer and an antibody that recognizes the protein of interest.

3.4.3. Immunoblotting for Specific Tyrosines

Recently, phosphorylation site-specific antibodies have been developed for a number of proteins, including tyrosine-phosphorylated isomers of FAK and paxillin. For these proteins, phosphorylation of specific tyrosines can be examined by Western blotting either cell lysates or immune complexes (*see* **Fig. 2**). The protocol is very similar to that outlined in **Subheading 3.4.2.**, with several modifications. First, the blocking buffer used is TBS-T containing 2% fish gelatin (*see* **Notes 4** and **5**). Secondly, HRP-conjugated protein A, or HRP-conjugated anti-rabbit antibodies, are used to detect the phospho-specific antibodies, as we use phospho-specific antibodies that were raised in rabbits (*see* **Note 8**).

3.5. Immunofluorescence

In addition to measuring biochemical changes in tyrosine phosphorylation, it is sometimes useful to examine subcellular localization of tyrosine-phosphorylated substrates following cell stimulation. For many years it has been possible to monitor the localization of phosphotyrosine by immunofluorescence in fixed cells. In addition, recent advances have led to methodologies to examine cellular phosphotyrosine in living cells. To determine subcellular localization of proteins and cellular phosphotyrosine following fixation, the following protocol may be used:

1. Detach the cells and hold in suspension, then plate in 60-mm dishes containing glass cover slips coated with ECM proteins, for example, fibronectin (as described

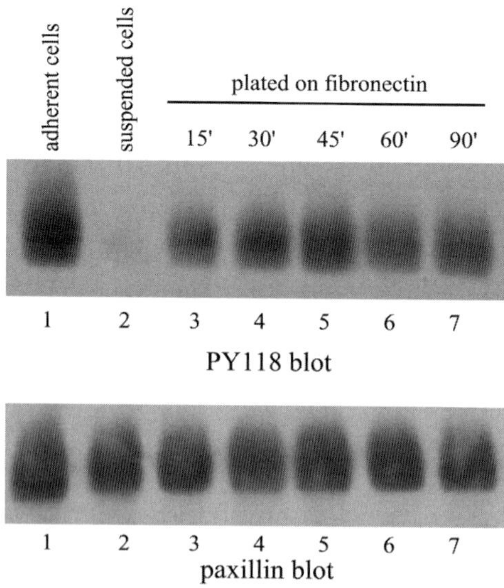

Fig. 2. Cell adhesion-dependent phosphorylation of paxillin at tyrosine 118. Adherent CE cells, CE cells held in suspension or plated onto fibronectin for various times were lysed. Paxillin was immunoprecipitated and analyzed by Western blotting using a phospho-specific antibody recognizing phosphorylated tyrosine 118 (**top panel**) or using a paxillin antibody as a loading control (**bottom panel**).

in **Subheading 3.1.**; *see* **Note 9**). Incubate at 37°C for a short time to allow attachment and spreading before processing for immunofluorescence.

2. Wash the cells thrice with 1X PBS and fix the cells by incubation in 1 mL of 3% paraformaldehyde (or alternatively formaldehyde) in 1X PBS for 20 min at room temperature.

3. Wash thrice with 1X PBS. Permeabilize the fixed cells with 1–2 mL of 0.5% Triton X-100 (in 1X PBS) for 5 min at room temperature.

4. Wash the permeabilized cells thrice with 1X PBS and block the cells if necessary to eliminate high fluorescence background (*see* **Note 10**). Dispense 60–75 μL aliquots of the primary antibody (10 μg/mL) in 1X PBS on a piece of parafilm. Place the cover slips, cells facing downwards, on an aliquot of antibody solution and incubate at 37°C for 40 min. Wash the cover slips twice with 1X PBS.

5. Dispense 100- to 200-μL aliquots of the appropriate fluorophore-conjugated secondary antibody (5 μg/mL), for example, rhodamine-conjugated rabbit anti-mouse antibody to detect a monoclonal phosphotyrosine antibody, on a piece of parafilm. Place the cover slips, cells facing downward, on an aliquot of the secondary antibody solution and incubate for 40 min at room temperature in the dark. Wash the cover slips twice with 1X PBS (*see* **Note 11**).

6. Dry the cover slips and mount in Mowiol® 4-88 and store in the dark at 4°C until the time of microscopic examination.

3.6. Monitoring Tyrosine-Phosphorylation in Focal Adhesions of Live Cells

SH2 domains bind phosphorylated tyrosine residues embedded in specific amino acid sequences *(32)*. This binding constitutes an important mode of regulating and mediating protein–protein interactions including the interactions of a number of proteins with their focal adhesion-associated binding partners *(3,22)*. This observation was exploited to develop a protocol to monitor tyrosine phosphorylation in living cells. We describe here a quantitative fluorescence microscopy-based approach for studying tyrosine phosphorylation in focal adhesions of live cells. For this purpose, a specific "phosphotyrosine reporter" consisting of yellow fluorescent protein (YFP), fused to two tandem phosphotyrosine-binding SH2-domains (dSH2) derived from pp60^{c-Src}, is used. This YFP-dSH2 reporter localizes to cell-matrix adhesions (*see* **Fig. 3**) and its intensity is linearly correlated with that of an anti-phosphotyrosine antibody labeling *(31)*. The following protocol was devised to monitor phosphotyrosine in focal adhesions.

1. Co-transfect cells with complimentary deoxyribonucleic acid constructs encoding YFP-dSH2 (the phosphotyrosine reporter; *see* **Note 12**) and CFP-paxillin (a focal adhesion marker) fusion proteins using Lipofectamine Plus reagent (Invitrogen) according to the manufacturer's instructions. YFP-dSH2- and CFP-paxillin-encoding plasmids were prepared as previously described *(31)*.
2. Plate cells on 35-mm MatTek glass-bottom dishes coated for 1 h with 10 µg/mL of FN.
3. Twelve hours later, exchange the regular DMEM for Ham's F12 medium (*see* **Note 13**) containing 25 mM HEPES. Live-cell imaging was conducted with a back-illuminated frame transfer grade 1 Quantics CCD camera equipped with an EEV 57-10 G1 Chip (Photometrics, Tucson, AZ), generating 12-bit digital data. During the recording, the cells were maintained at 37°C on the microscope stage.
4. Changes in tyrosine phosphorylation with time can be quantified by fluorescence ratio imaging (FRI) *(33)*.

4. Notes

1. To study the rise in tyrosine phosphorylation levels with readhesion of cells to fibronectin, 100-mm bacterial dishes and not tissue culture dishes, should be used. Cells will not attach to the plastic and thus cell adhesion occurs via the extracellular matrix coated on the dish.
2. BSA is used to block the plate prior to use. The BSA binds to plastic surfaces that are not covered with the extracellular matrix applied to the dish. Some cells express and secrete their own extracellular matrix proteins. The BSA

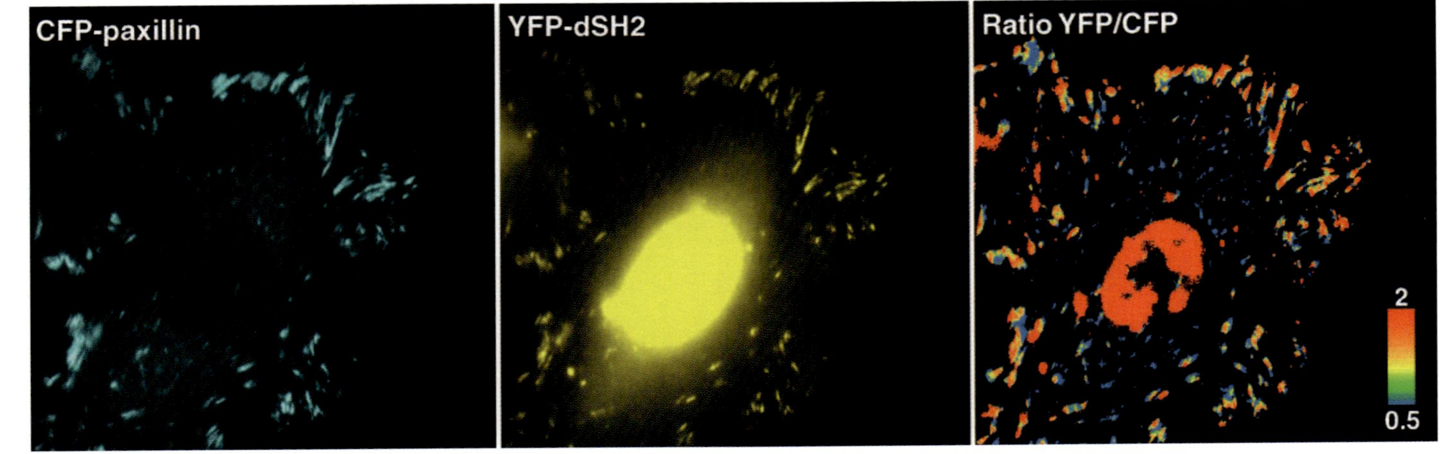

Fig. 3. Paxillin and phosphotyrosine in focal adhesions of NIH 3T3 fibroblasts. Cells were transfected with CFP-paxillin and the marker of tyrosine phosphorylation YFP-dSH2 (reporter of PY sites). Both CFP-paxillin (**left panel**) and YFP-dSH2 (**middle panel**) localize to focal adhesions and their overall distribution is similar. The ratio image outlines areas with higher and lower levels of tyrosine phosphorylation in focal adhesions marked by CFP-paxillin (**right panel**).

prevents binding of secreted matrix proteins to the dish and thus the cells adhere and spread on a homogeneous matrix. It is important to use lipid-free BSA, as lipid contaminants such as lysophosphatilic acid (LPA) stimulate tyrosine phosphorylation of focal adhesion-associated proteins.

3. Some classes of monoclonal antibodies bind poorly to protein A. This problem can be obviated by using protein G Sepharose beads or protein A Sepharose beads pre-coated with rabbit anti-mouse antibodies to capture the monoclonal antibody.

4. As a general rule, milk should not be used with phosphotyrosine antibodies, owing to the high levels of phosphotyrosine in milk.

5. TBS-T can be used to block nitrocellulose membranes, particularly when using RC20, but 2% fish gelatin and 5% BSA have proved to provide better blocking efficiency. If a high level of background signal is detected, blocking nitrocellulose membranes can be improved by increasing the blocking time or temperature (37°C rather than room temperature).

6. Monoclonal phosphotyrosine antibodies can exhibit sequence specificity. We have found that RC20 recognizes focal adhesion-associated proteins of interest. When initially examining tyrosine phosphorylation of new substrates, it is prudent to try several different antibodies.

7. Antibodies used for blotting can be conserved by placing the nitrocellulose membrane protein side-down in contact with blocking buffer containing the antibody on top of a piece of parafilm.

8. When using rabbit polyclonal antibodies as the primary antibody, an HRP-conjugated anti-rabbit secondary antibody can sometimes lead to high background staining, particularly if rabbit polyclonal antibodies were used for immunoprecipitation of the samples. In these cases, using HRP-conjugated protein A to detect the primary polyclonal antibody can reduce the background.

9. To ensure an appropriate density of cells for visualization following processing, it is prudent to plate the cells at several different densities and to prepare multiple cover slips per sample.

10. If background immunofluorescence presents a problem, the permeabilized cells can be blocked to reduce background staining. Inclusion of 1% BSA in the buffer during the incubations with primary and secondary antibodies can reduce background. The cells can also be blocked with 4% horse serum in PBS for 30 min at 37°C before incubation with the primary and secondary antibodies.

11. To eliminate salt deposits on the cover slip after incubation of secondary antibody, dip the cover slips briefly in double-distilled water and dry them by gentle aspiration and blotting the edge on an absorbent surface.

12. SH2 domains exhibit sequence-specific binding to tyrosine-phosphorylated proteins. The autophosphorylation site of FAK conforms to a high affinity-binding site for the SH2 domain of Src. Thus, the YFP-dSH2 reporter likely binds FAK efficiently. Phosphotyrosine reporters utilizing distinct SH2 domains may exhibit differential phosphotyrosine-dependent subcellular localization.

13. Ham's F12 contains a reduced amount of riboflavin. Riboflavin has high autofluorescence in the wavelength used for detection of CFP and YFP.

Acknowledgments

Thanks to Jill Dunty and Veronica Gabarra-Niecko for their comments during the preparation of this chapter.

References

1. Jockusch, B. M., Bubeck, P., Giehl, K., Kroemker, M., Moschner, J., Rothkegel, M., et al. (1995) The molecular architecture of focal adhesions. *Annu. Rev. Cell Dev. Biol.* **11,** 379–416.
2. Webb, D. J., Parsons, J. T., and Horwitz, A. F. (2002) Adhesion assembly, disassembly and turnover in migrating cells—over and over and over again. *Nat. Cell Biol.* **4,** E97–100.
3. Miranti, C. K. and Brugge, J. S. (2002) Sensing the environment: a historical perspective on integrin signal transduction. *Nat. Cell Biol.* **4,** E83–E90.
4. Schwartz, M. A., Lechene, C., and Ingber, D. E. (1991) Insoluble fibronectin activates the Na/H antiporter by clustering and immobilizing integrin alpha 5 beta 1, independent of cell shape. *Proc. Natl. Acad. Sci. USA* **88,** 7849–7853.
5. Schwartz, M. A., Both, G., and Lechene, C. (1989) Effect of cell spreading on cytoplasmic pH in normal and transformed fibroblasts. *Proc. Natl. Acad. Sci. USA* **86,** 4525–4529.
6. del Pozo, M. A., Price, L. S., Alderson, N. B., Ren, X. D., and Schwartz, M. A. (2000) Adhesion to the extracellular matrix regulates the coupling of the small GTPase Rac to its effector PAK. *EMBO J.* **19,** 2008–2014.
7. Ren, X. D., Kiosses, W. B., and Schwartz, M. A. (1999) Regulation of the small GTP-binding protein Rho by cell adhesion and the cytoskeleton. *EMBO J.* **18,** 578–585.
8. Schwartz, M. A. (1993) Spreading of human endothelial cells on fibronectin or vitronectin triggers elevation of intracellular free calcium. *J. Cell Biol.* **120,** 1003–1010.
9. Pelletier, A. J., Bodary, S. C., and Levinson, A. D. (1992) Signal transduction by the platelet integrin alpha IIb beta 3: induction of calcium oscillations required for protein-tyrosine phosphorylation and ligand-induced spreading of stably transfected cells. *Mol. Biol. Cell* **3,** 989–998.
10. Guan, J. L. and Shalloway, D. (1992) Regulation of focal adhesion-associated protein tyrosine kinase by both cellular adhesion and oncogenic transformation. *Nature* **358,** 690–692.
11. Hanks, S. K., Calalb, M. B., Harper, M. C., and Patel, S. K. (1992) Focal adhesion protein-tyrosine kinase phosphorylated in response to cell attachment to fibronectin. *Proc. Natl. Acad. Sci. USA* **89,** 8487–8491.
12. Vuori, K. and Ruoslahti, E. (1995) Tyrosine phosphorylation of p130Cas and cortactin accompanies integrin-mediated cell adhesion to extracellular matrix. *J. Biol. Chem.* **270,** 22,259–22,262.
13. Petch, L. A., Bockholt, S. M., Bouton, A., Parsons, J. T., and Burridge, K. (1995) Adhesion-induced tyrosine phosphorylation of the p130 src substrate. *J. Cell Sci.* **108,** 1371–1379.

14. Burridge, K., Turner, C. E., and Romer, L. H. (1992) Tyrosine phosphorylation of paxillin and pp125FAK accompanies cell adhesion to extracellular matrix: a role in cytoskeletal assembly. *J. Cell Biol.* **119,** 893–903.
15. De Nichilo, M. O. and Yamada, K. M. (1996) Integrin alpha v beta 5-dependent serine phosphorylation of paxillin in cultured human macrophages adherent to vitronectin. *J. Biol. Chem.* **271,** 11,016–11,022.
16. Bellis, S. L., Perrotta, J. A., Curtis, M. S., and Turner, C. E. (1997) Adhesion of fibroblasts to fibronectin stimulates both serine and tyrosine phosphorylation of paxillin. *Biochem. J.* **325,** 375–381.
17. Cary, L. A., Chang, J. F., and Guan, J. L. (1996) Stimulation of cell migration by overexpression of focal adhesion kinase and its association with Src and Fyn. *J. Cell Sci.* **109,** 1787–1794.
18. Owen, J. D., Ruest, P. J., Fry, D. W., and Hanks, S. K. (1999) Induced focal adhesion kinase (FAK) expression in FAK-null cells enhances cell spreading and migration requiring both auto- and activation loop phosphorylation sites and inhibits adhesion-dependent tyrosine phosphorylation of Pyk2. *Mol. Cell Biol.* **19,** 4806–4818.
19. Sieg, D. J., Hauck, C. R., and Schlaepfer, D. D. (1999) Required role of focal adhesion kinase (FAK) for integrin-stimulated cell migration. *J. Cell Sci.* **112,** 2677–2691.
20. Petit, V., Boyer, B., Lentz, D., Turner, C. E., Thiery, J. P., and Valles, A. M. (2000) Phosphorylation of tyrosine residues 31 and 118 on paxillin regulates cell migration through an association with CRK in NBT-II cells. *J. Cell Biol.* **148,** 957–970.
21. Yano, H., Uchida, H., Iwasaki, T., Mukai, M., Akedo, H., Nakamura, K., et al. (2000) Paxillin alpha and crk-associated substrate exert opposing effects on cell migration and contact inhibition of growth through tyrosine phosphorylation. *Proc. Natl. Acad. Sci. USA* **97,** 9076–9081.
22. Zamir, E. and Geiger, B. (2001) Molecular complexity and dynamics of cell-matrix adhesions. *J. Cell Sci.* **114,** 3583–3590.
23. Geiger, B., Bershadsky, A., Pankov, R., and Yamada, K. M. (2001) Transmembrane crosstalk between the extracellular matrix—cytoskeleton crosstalk. *Nat. Rev. Mol. Cell Biol.* **2,** 793–805.
24. Schaller, M. D. (2001) Biochemical signals and biological responses elicited by the focal adhesion kinase. *Biochim. Biophys. Acta* **1540,** 1–21.
25. Schlaepfer, D. D., Hauck, C. R., and Sieg, D. J. (1999) Signaling through focal adhesion kinase. *Prog. Biophys. Mol. Biol.* **71,** 435–478.
26. Avraham, H., Park, S. Y., Schinkmann, K., and Avraham, S. (2000) RAFTK/Pyk2-mediated cellular signalling. *Cell Signal.* **12,** 123–133.
27. Parsons, J. T. (2003) Focal adhesion kinase: the first ten years. *J. Cell Sci.* **116,** 1409–1416.
28. Schaller, M. D. (2001) Paxillin: a focal adhesion-associated adaptor protein. *Oncogene* **20,** 6459–6472.
29. Turner, C. E. (2000) Paxillin and focal adhesion signalling. *Nat. Cell Biol.* **2,** E231–E236.

30. O'Neill, G. M., Fashena, S. J., and Golemis, E. A. (2000) Integrin signalling: a new Cas(t) of characters enters the stage. *Trends Cell Biol.* **10,** 111–119.

31. Kirchner, J., Kam, Z., Tzur, G., Bershadsky, A. D., and Geiger, B. (2003) Live-cell monitoring of tyrosine phosphorylation in focal adhesions following microtubule disruption. *J. Cell Sci.* **116,** 975–986.

32. Pawson, T., Gish, G. D., and Nash, P. (2001) SH2 domains, interaction modules and cellular wiring. *Trends Cell Biol.* **11,** 504–511.

33. Zamir, E., Katz, B. Z., Aota, S., Yamada, K. M., Geiger, B., and Kam, Z. (1999) Molecular diversity of cell-matrix adhesions. *J. Cell Sci.* **112,** 1655–1669.

21

Cell-Adhesion Assays

*Fabrication of an E-Cadherin Substratum and Isolation
of Lateral and Basal Membrane Patches*

Frauke Drees, Amy Reilein, and W. James Nelson

Summary

Cell adhesion between cells and with the extracellular matrix (ECM) results in dramatic changes in cell organization and, in particular, the cytoskeleton and plasma membrane domains involved in adhesion. However, current methods to analyze these changes are limited because of the small areas of membrane involved in adhesion, compared to the areas of membrane not adhering (a signal to noise problem), and the difficulty in accessing native protein complexes directly for imaging or reconstitution with purified proteins. The methods described here overcome these problems. Using a mammalian expression system, a chimeric protein comprising the extracellular domain of E-cadherin fused at its C-terminus to the Fc domain of human IgG1 (E-cadherin:Fc) is expressed and purified. A chemical bridge of biotin-NeutrAvidin-biotinylated Protein G bound to a silanized glass cover slip is fabricated to which the E-cadherin:Fc chimera binds in the correct orientation for adhesion by cells. After cell attachment, the basal membrane (a contact formed between cellular E-cadherin and the E-cadherin:Fc substratum) is isolated by sonication; a similar method is described to isolate basal membranes of cells attached to ECM. These membrane patches provide direct access to protein complexes formed on the membrane following cell–cell or cell–ECM adhesion.

Key Words: Epithelial cells; polarity; plasma membrane; membrane domains; cell–cell adhesion; cell–extracellular matrix adhesion; cadherin; integrin; collagen; substrate; plasma membrane; cytoskeleton; actin; microtubules; membrane patches.

1. Introduction

Polarized epithelial cells have a higher-order organization involving cell–cell and cell–extracellular matrix contacts that orient cells into a monolayer that separates different biological compartments in the body. The cell surface

From: *Methods in Molecular Biology, vol. 294: Cell Migration: Developmental Methods and Protocols*
Edited by: J-L. Guan © Humana Press Inc., Totowa, NJ

bounded by these contacts (basal–lateral domain) is structurally and function-
ally distinct from the unbounded surface (apical domain; **refs.** *1*,*2*). Understanding how different membrane domains are organized in polarized epithelial
cells requires knowledge of how cells adhere to one another and the extracellular matrix, and how the resulting cell surfaces are converted into specific membrane domains by localized assembly and targeted delivery of specific proteins.

Epithelial cell–cell adhesion is mediated by a variety of membrane proteins,
including classical cadherins, claudins/occludin, nectin, and desmosomal
cadherins *(3–5)*. Classical cadherins are single membrane spanning proteins
with a divergent extracellular domain of five repeats and a conserved cytoplasmic domain. Binding between extracellular domains, which requires Ca^{2+} for
protein conformation, is thought to involve multiple *cis*-dimers of cadherin,
forming *trans*-oligomers between cadherins on opposing cell surfaces *(4)*.
Binding between cadherin extracellular domains is weak, but strong cell–cell
adhesion develops during lateral clustering of cadherins by proteins that link
cadherin to the actin cytoskeleton *(5)*. However, little is known about how
these protein complexes assemble in cells, how the cadherin complex binds
and organizes the actin cytoskeleton, or how other proteins identified at cell-
cell contacts modify cadherin function and actin organization.

The analysis of mechanisms involved in cell adhesion has many limitations.
Current methods that are available to investigate mechanisms of assembly of
cadherin complexes include analysis of protein distributions in fixed or live
cells, isolation of protein complexes with antibodies following extraction from
cells, and in vitro dissection of protein-protein binding with bacterial expressed
proteins or yeast two-hybrid analysis. None of these methods, however, allow
direct access to native protein complexes on the membrane of cells. Another
significant problem is that the initial interactions between cells, when protein
complexes of interest are assembled and modified, occur on a very small area
of plasma membrane relative to area of the membrane involved in cell attachment to extracellular matrix and the free cell surface. The "signal" from cell–
cell adhesion complexes is correspondingly small compared to that from
cell–ECM protein complexes, and hence difficult to isolate and analyze.

A step to overcome problems of the relatively small "signal" of initial cell–
cell adhesion is to induce cells to adhere and spread on a substratum that promotes adhesion through cadherins rather than extracellular matrix. This
requires purification of native E-cadherin and correct orientation on the substratum for cell adhesion. Recombinant deoxyribonucleic acid (DNA) technology offers a simple means for production of specific protein domains.
Many eukaryotic and prokaryotic heterologous expression systems have been
used to produce proteins in quantities suitable for biochemical analysis. However, a mammalian system is the best choice for production of the extracellular

domain of mammalian transmembrane membrane proteins, especially when proper posttransitional modification of the protein is essential for function. To use E-cadherin for cell attachment, large amounts of protein that have been posttranslationally modified are required: cleavage of presequence, and complex glycosylation are essential for E-cadherin function *(3,4)*. To present the protein in the correct orientation for cell adhesion (N-terminus to N-terminus), E-cadherin is tagged at the C-terminus, thereby providing not only a way to correctly orientate the protein on artificial substratum for adhesion assays, but also an easy one-step method for protein purification *(6)*.

To overcome the limitation of accessibility to native protein complexes on membranes, isolated substrate-bound membranes can be prepared simply and reproducibly from cells plated on cover slips or filters to provide a cell-free system for analyzing and reconstituting cytoskeleton–membrane interactions *(7)*. Isolated membrane patches retain plasma membrane and the associated cytoskeleton, including actin filaments and in many cases microtubules. Nuclei are not retained and most internal organelles are removed. A good preparation yields thousands of patches per cover slip or filter. Isolated membrane patches provide an improved visualization of the cytoskeleton at the membrane, since the cytoskeleton not associated with the basal membrane is removed. Moreover, proteins on the basal membrane are more accessible to fixative, which is particularly advantageous in the case of tall cells such as Madin–Darby Canine Kidney (MDCK) cells on filters.

2. Materials

2.1. Expression and Purification of E-Cadherin:Fc

1. pCEP4 Plasmid (Invitrogen, San Diego, CA).
2. CDM8 Plasmid (Invitrogen).
3. 293 Cells expressing EBV nuclear antigen-1 (Invitrogen).
4. Lipofectamine (Gibco-BRL, Gaithersburg, MD).
5. Hygromycin B (Calbiochem, La Jolla, CA).
6. High glucose Dulbecco's modified Eagle's medium (DMEM; Gibco-BRL).
7. Fetal bovine serum (FBS), cell culture grade (Sigma, St. Louis, MO).
8. Dimethyl sulfoxide (DMSO), cell culture grade (Sigma).
9. HiTrap protein G Sepharose 4B column (capacity 24 mg human IgG/mL drained gel; Pierce, Rockford, IL).
10. Centricon YM-10 filter (Amicon).
11. BCA Protein Assay Kit (Pierce).

2.2. Fabrication of E-Cadherin:Fc Substratum

1. 12-mm Glass cover slips, number 1.5, ceramic cover slip holders (Electron Microscope Sciences, Fort Washington, PA).
2. Nonidet P-40 (Sigma).

3. Nochromix (Pierce).
4. Branson Sonifier 250 (Pierce).
5. *N*-(2-aminethyl)-3-amino-propyltrimethoxysilane (Pierce).
6. Anhydrous methanol (Sigma).
7. "Atmosbag" (Aldrich).
8. Tris-saline: 20 m*M* Tris-HCl, pH 7.4, 137 m*M* NaCl.
9. Ringer's buffer: 10 m*M* HEPES, pH 7.4, 154 m*M* NaCl, 7.2 mM KCl.
10. Sulfo-NHS-biotin (Pierce).
11. Biotin (Pierce).
12. NeutrAvidin (Pierce).
13. Biotinylated protein G (Pierce).

2.3. Plating Cells and Cross-Linking Procedures for E-Cadherin:Fc Substratum

1. MDCK cells (American Type Culture Collection).
2. Twenty-four-well tissue culture plates.
3. 12-mm 0.45-μm Transwell polycarbonate filter membranes (Corning).
4. Type I collagen solution (*see* **Note 1**).
5. BS^3, DTSSP (Pierce).

2.4. Preparation of Basal Membranes From Cells Bound to Substrata by Sonication

1. Hypotonic buffer: 15 m*M* HEPES, 15 mM KCl, pH 7.2.
2. Branson Sonifier 250 with 1/8-inch diameter microprobe.
3. Adjustable platform (support jack).
4. Ringers buffer: 10 m*M* HEPES, pH 7.4, 154 m*M* NaCl, 7.2 m*M* KCl.

2.5. Fixing Basal Membranes for Microscopy

1. Methanol.
2. Formaldehyde.
3. Glutaraldehyde, NaBH$_4$.

3. Methods

3.1. Expression and Purification of E-Cadherin:Fc

The following methods described outline: 1) construction of the E-cadherin:Fc expression plasmid, 2) transfection and expression of E-cadherin:Fc in 293 cells, and 3) the purification of E-cadherin:Fc from tissue culture cells (**Fig. 1**).

3.1.1. E-Cadherin:Fc Expression Plasmid

The expression vector CDM8FT *(8)* is used to express a chimeric protein comprising the complete extracellular domain of E-cadherin fused at the C-terminus to human immunoglobulin Fc domain (**Fig. 1A; ref. 6**). CDM8FT is

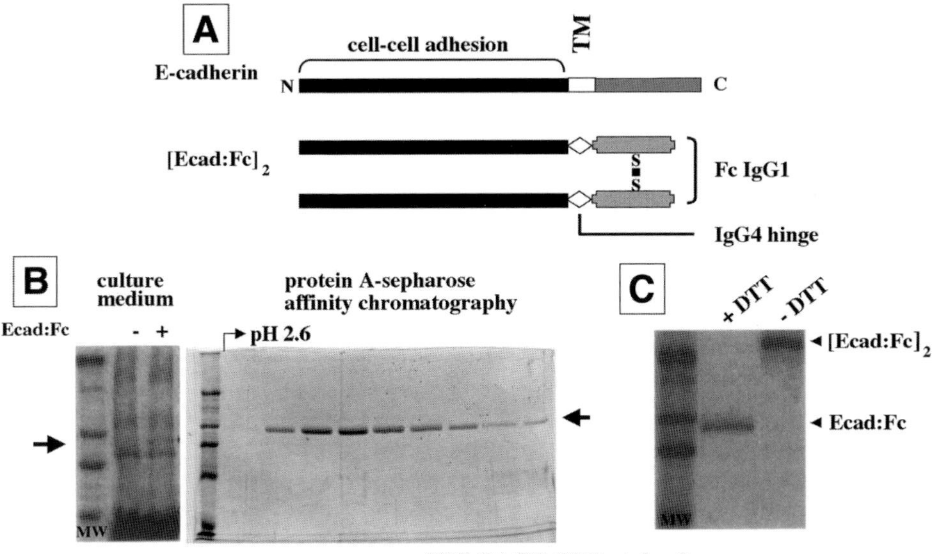

Fig. 1. Construction, expression and purification of E-cadherin:Fc. (**A**), E-cadherin is a single transmembrane protein, and the extracellular domain is required for cell-cell adhesion. E-cadherin extracellular domain is fused to the hinge region of IgG4 and the Fc domain of human IgG1 to generate a dimeric E-cadherin:Fc chimeric protein. (**B**), Culture medium from control (–) cells and cells expressing E-cadherin:Fc (+) separated by SDS-PAGE and stained with Coomassie-Brilliant blue (CBB) shows secretion of E-cadherin:Fc. E-cadherin:Fc is purified from culture medium in one-step by affinity chromatography on Protein A Sepaharose 4B and elution with buffer at pH 2.6, as shown by SDS-PAGE and CBB staining (arrow). (**C**), E-cadherin:Fc migrates as a dimer ([Ecad-Fc]₂) or monomer (Ecad-Fc) in SDS-PAGE in the absence or presence of reducing agent DTT, respectively (MW, molecular weight standard proteins from top: myosin [205 kDa], β-galactosidase [115 kDa], phosphorylase b [105 kDa], and bovine serum albumin [BSA; 68 kDa]).

derived from CDM8, an expression vector with the human cytomegalovirus early promoter and SV40 origin of replication.

1. CH2 and CH3 domains of human immunoglobulin G1, with or without the hinge region, are generated by polymerase chain reaction (PCR) using as templates either plasmid SIgpoly7 or a DNA fragment containing the Fc domain.

2. A *Xho*1 site (reading phase = CTC[Leu]-GAG[Glu]) is added to the 5' end, and a *Not*1 site is added to the 3' end downstream to a stop codon (TAA-A-GCG GCCGC).

3. The resulting PCR product (CH2-CH3) is subcloned into the *Xho*1/*Not*1 sites of CDM8FT, which results in an in-frame fusion with the hemagglutinin epitope tag on CDM8FT through the *Xho*1 restriction site.
4. The extracellular domain of canine E-cadherin is generated by PCR using cloned cDNA of canine E-cadherin as a template, with a Hind3 site and an *Xho*1 site (phase = Leu-Glu) at the 3' and 5' ends, respectively.
5. The PCR product, containing the start codon and the signal peptide of E-cadherin, is subcloned into the *Hind*3/*Xho*1 sites of the vector containing the CH2 and CH3 domains to replace the hemagglutinin tag; this results in an in-frame fusion of the E-cadherin domain with the CH2 and CH3 domain through the *Xho*1 site.
6. The complete coding region of the chimeric protein is excised from the vector plasmid with *Hind*3 and *Not*1 and subcloned into the *Hind*3/*Not*1 sites of the EBV Ori P-based expression vector, pCEP4.

The scheme can be modified to accommodate other types of affinity or epitope tags, and the E-cadherin extracellular domain can be replaced with the extracellular domain of different type I transmembrane proteins *(6)*.

3.1.2. Transfection and Expression of E-Cadherin:Fc in 293 Cells

The E-cadherin:Fc expression plasmid is transfected into 293 EBNA cells (293 cells expressing EBV nuclear antigen-1) maintained in high glucose DMEM/10% FBS. Transcription of the chimeric construct is driven by human cytomegalovirus early promoter in mammalian cells, and secretion into the growth medium is driven by the endogenous signal peptide of canine E-cadherin (**Fig. 1**).

1. Transfect cells using lipofectamine with the following modifications of the protocol supplied with the reagent: 3×10^5 cells in 3-cm dishes are transfected with 0.5 μg of cesium chloride purified E-cadherin:Fc DNA/5 μL of lipofectamine in 1 mL of serum-free DMEM for 6 h. Using this protocol, >70% of cells are routinely transfected.
2. Two days after transfection, 200 μg/mL of hygromycin B is added to the growth medium.
3. After about 10 d, a hygromycin-B-resistant cell population emerges, which is expanded further in the presence of a reduced concentration of hygromycin B (100 μg/mL; *see* **Note 2**).
4. Store replicate cultures in liquid nitrogen at a density of 1×10^6 cells/mL in 1-mL aliquots in a solution of 90% FBS/10% DMSO.
5. Thaw cells rapidly in a 37°C water bath, dilute into 30 mL of high glucose DMEM/10% FBS in a 10-cm diameter plastic Petri dish, and incubate at 37°C in a humidified, 5% CO_2 in air atmosphere. After 2 h, change the medium.
6. Grow cells to confluency, and then divide between 10 15-cm Petri dishes, grow cells to confluency, and then divide once more between 50 15-cm Petri dishes. The cells do not adhere strongly to the plastic and they can be removed simply by pipetting a stream of warm medium. Grow cells to approx 75% confluency.

7. To collect medium containing secreted Fc-E-cadherin all traces of FBS are first removed from the culture: the growth medium is discarded and the cells are washed in three changes of high glucose DMEM (without FBS); cells are then incubated for 1 to 2 h in 20 mL of high-glucose DMEM (without FBS) at 37°C in a humidified, 5% CO_2 in air atmosphere; the growth medium is discarded and 15 to 20 mL of fresh high-glucose DMEM (without FBS) is added, and the cells are incubated for 2 d at 37°C in a humidified, 5% CO_2 in air atmosphere.

8. Collect media and combine (**Fig. 1B**). Floating cells and large debris are removed by centrifugation. The supernatant is passed through a 0.45-µm filter and then stored in sterile containers at 4°C, or frozen and stored at –80°C depending on the time interval between collection and protein purification.

9. To continue collecting conditioned medium, the cells need to be subdivided and grown for 2 d in the presence of FBS: divide approximately half of the cells into 50 15-cm Petri dishes and incubate in high glucose DMEM/10% FBS for 2 d or until they are approx 75% confluent; wash cells and incubate in high glucose DMEM (without FBS), and collect conditioned medium.

10. This procedure can be repeated for three collection cycles, and then the cells are discarded. A new vial of frozen cells is thawed and expanded for more protein collection.

3.1.3. Purification of E-Cadherin:Fc

E-cadherin:Fc is purified from conditioned medium in one step using affinity chromatography on protein G Sepharose which binds the Fc domain of the chimeric protein (**Fig. 1B**). The following methods describe the purification, concentration, and storage of the protein:

1. Equilibrate a 1 mL HiTrap protein A Sepharose 4B column in 0.1 *M* Na phosphate, buffer, pH 7.0.

2. Apply conditioned medium containing E-cadherin:Fc to the column (flow rate, approx 60 mL/h) and discard the flow through.

3. Wash the column in 10 volumes of 0.1 *M* Na phosphate buffer, pH 7.0.

4. Elute E-cadherin:Fc with 0.1 *M* glycine-HCl, pH 2.6, at a flow rate of 60 mL/h and collect 400- to 500-µL fractions; neutralize eluant immediately with 1 *M* Tris-HCl, pH 9.0 (approx 400 µL of 0.1 *M* glycine-HCl, pH 2.6, is neutralized to approx pH 7.0 with 10 µL of 1 *M* Tris-HCl, pH 9.0).

5. Measure OD_{280} and combine fractions containing eluted protein.

6. Concentrate combined protein fractions using a Centricon YM-10 filter.

7. Exchange buffer to Ringer's buffer using the Centricon filter or dialysis.

8. Determine protein concentration with the BCA protein assay using BSA as a protein standard, and assess protein purity by sodium dodecyl sulfate-polyacrylamide page electrophoresis (SDS-PAGE) and Coomassie blue staining.

9. Adjust concentration of purified E-cadherin:Fc to 200 µg/mL with Ringer's buffer, flash freeze aliquots (50 µL) in liquid nitrogen, and store at –80°C (protein is stable for several months at –80°C).

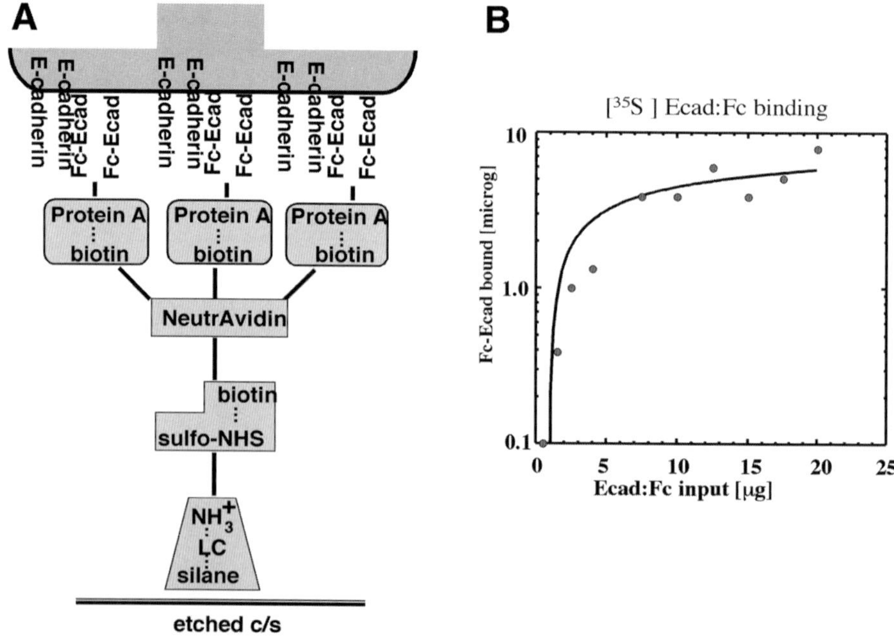

Fig. 2. Fabrication of E-cadherin:Fc substratum. (**A**), Scheme of chemical assembly of E-cadherin:Fc on a cover slip. A glass cover slip is silianized with a long-chain silane containing an amine, to which sulfo-NHS-biotin is linked. NeutrAvidin and then biotinylated protein A are added sequentially, followed by purified E-cadherin:Fc. E-cadherin on the surface of cells binds to the correctly oriented E-cadherin:Fc substratum. (**B**), ^{35}S-methionine/cysteine-labeled E-cadherin:Fc was purified from cells incubated with ^{35}S-methionine/cysteine as described in **Subheading 3.1.3.** and used in a dilution series to determine the saturation density of E-cadherin:Fc on the substratum as approx 50,000 molecules/µm^2.

10. Approximately 0.2–0.5 mg E-cadherin:Fc is purified per liter of conditioned medium (**Fig. 1B**). The purified E-cadherin:Fc migrates as a monomer when treated with reducing agents (e.g., DTT) owing to di-sulfide bonds in the Fc domains (**Fig. 1A**), or a dimer in the absence of reducing agent (**Fig. 1C**).

3.2. Fabrication of E-Cadherin:Fc Substratum

A glass surface is used for attachment of cells to fabricated E-cadherin:Fc substrates so that cells and membrane patches can be analyzed by immunofluorescence microscopy as well as used for biochemical reconstitution of protein complexes. The following methods described outline: 1) the preparation of glass surfaces to form chemical binding sites by silanizing the glass surface, and 2) fabrication of the E-cadherin:Fc substratum (**Fig. 2**).

3.2.1. Etching and Silanizing Glass Cover Slips

1. Load glass cover slips in a ceramic holder (used for all subsequent operations).
2. Submerge cover slips in warm (50–60°C) 2% (v/v) Nonidet P-40 in dH_2O and stir for 15–30 min.
3. Sonicate cover slips in warm 2% NP-40 in dH_2O solution for 5 min using a Branson Sonifer (1/8 in. microtip, setting no. 5).
4. Rinse three times for 5 min in dH_2O and allow to drain.
5. In a fume hood, slowly submerge cover slips in concentrated H_2SO_4 containing 3% Nochromix for 10 min, agitating occasionally by carefully dunking the ceramic holder up and down.
6. Slowly submerge cover slips in three changes of 100 mL of dH_2O for 3–6 min and agitate by dunking the ceramic holder up and down; allow to drain.
7. In a fume hood, slowly submerge cover slips in a solution of one part anhydrous methanol to one part dH_2O-saturated KOH for 10 min, and agitate occasionally by dunking the ceramic holder up and down.
8. Slowly submerge cover slips in three changes of 100 mL of dH_2O for 3–6 min and agitate by dunking the ceramic holder up and down; allow to drain.
9. Briefly submerge cover slips in 100% methanol.
10. Quick dry cover slips with a hot air dryer (a commercial hair dryer works well).
11. Prepare a solution of 95% methanol in dH_2O and adjust to pH 4.5–5.5 with glacial acetic acid (approx 40 mL 95% methanol + 1 µL of CH_3COOH).
12. Under nitrogen atmosphere, add *N*-(2-aminethyl)-3-amino-propyltrimethoxysilane to a final concentration of 4% (v/v) in the pH-adjusted 95% methanol solution, allow 5 min for hydrolysis and silanol formation.
13. Under nitrogen atmosphere, dip cover slips into silanol solution for 1–2 min and agitate occasionally by dunking the ceramic holder up and down.
14. Briefly submerge cover slips in anhydrous methanol.
15. Cure silanized cover slips at 40–50°C for 24 h.
16. Use silanized cover slips immediately.

3.2.2. Fabricating the E-Cadherin:Fc Substratum

Silanizing glass cover slips with a long chain silane containing a free amine provide the base for chemical assembly of the E-cadherin:Fc substratum using a biotin-neutrAvidin-biotinylated Protein G-E-cadherin:Fc sandwich (**Fig. 2**; *see* **Note 3**).

1. Incubate cover slips in a freshly prepared solution of 50 m*M* sulfo-NHS-LC-biotin (a long chain (LC), nonreversible biotinylating reagent that crosslinks to free amines) in DMSO in the dark at 37°C for 2–3 h (*see* **Note 4**).
2. Wash cover slips in three changes of Tris-saline.
3. Wash cover slips in three changes of 150 mL of Ringer's buffer.
4. Incubate cover slips in a fresh solution of 84 µ*M* NeutrAvidin in Ringer's buffer in the dark at room temperature for 2–3 h.
5. Wash cover slips in three changes of 150 mL of Ringer's buffer.

6. Incubate cover slips in a fresh solution of 17 µ*M* biotinylated protein G in Ringer's buffer in the dark at room temperature for 2–3 h.
7. Wash cover slips in three changes of 150 mL of Ringer's buffer.
8. Block excess free sites with a solution of 50 µ*M* biotin, 1 mg/mL BSA in Ringer's solution in the dark at room temperature for 2–3 h.
9. Wash cover slips in three changes of 150 mL of Ringer's buffer, and drain by touching the edge of the cover slip to a piece of filter paper.
10. Incubate cover slips in approx 4 µg of purified E-cadherin:Fc in 50 µL Ringer's buffer at room temperature for 1 h, or overnight at 4°C.
11. Wash cover slips in three changes of 150 mL of Ringer's buffer.
12. Use immediately for attachment of cells.

3.3. Plating Cells and Cross-Linking Procedures for E-cadherin:Fc or Collagen Substrata

Any cell type that expresses E-cadherin on the cell surface can be used for adhesion to the fabricated E-cadherin:Fc substratum (*see* **Note 5**); adhesion is specific as cells do not bind to substratum containing IgG of Fc instead of E-cadherin:Fc, or to E-cadherin:Fc when extracellular Ca^{2+} is left out of the medium. Before plating cells on E-cadherin:Fc, we passage cells each day for two days at very low cell density in order to render cells "contact naïve," during which time the cells have had little or no cell-cell adhesion and hence induction of cell-cell adhesion on an E-cadherin:Fc substratum will elicit formation of de novo–adhesion complexes. Methods are also described for plating cells on type I collagen surfaces, as the methods used to prepare 'lateral' membranes attached to E-cadherin:Fc (**Subheading 3.4.**) can be also used to prepare basal membranes attached to extracellular matrix. Cells plated on E-cadherin:Fc accumulate E-cadherin and associated proteins on the attached "basal" membrane, and proteins characteristic of focal–adhesion complexes involved in cell-ECM adhesion are not present. In contrast, cells attached to a type I collagen substratum accumulate integrin–adhesion complexes on attached basal membranes, and not E-cadherin.

3.3.1. Attachment of Cells to an E-Cadherin:Fc Substratum

1. Trypsinize approx 50% confluent culture of MDCK cells, replate cells on 15-cm Petri dishes at a density of 1.5×10^6 cells/dish in DMEM/10% FBS, and incubate at 37°C in a humidified, 5% CO_2 in air atmosphere (the number of plates and cells will depend on the number of E-cadherin:Fc/cover slips that will be used in an experiment).
2. After 24 h, trypsinize the cells and re-plate them on 15 cm Petri dishes at a density of 1.5×10^6 cells/dish in DMEM/10% FBS and incubate at 37°C in a humidified, 5% CO_2 in air atmosphere for a further 24 h.
3. Trypsinize the cells, combine and re-suspend in DMEM/2% FBS at different cell densities ($1.5–2.0 \times 10^6$ cells will make an instantly confluent cell culture on a 12-mm cover slip).

4. Place 12-mm glass cover slips with an E-cadherin:Fc substratum (*see* **Subheading 3.2.**) in the well of a 24-well cluster dish, and aspirate excess Ringer's buffer.
5. Add 0.5 mL of a single cell suspension (3×10^6 cells/mL) to the well.
6. Cells attach to the E-cadherin:Fc substratum at $1g$ for 1 h, or by centrifugation at $500g$ for 5 min in a swinging bucket rotor at room temperature (do not slow the spin by breaking as this will swirl cells into the center of the cover slip).
7. Incubate cells at 37°C in a humidified, 5% CO_2 in air atmosphere for up to 6 h (this period is sufficient for cells to fully spread on the E-cadherin:Fc substratum, and to assemble protein complexes specific to cadherin-mediated cell-cell contacts; *see* **Note 5**).

3.3.2. Attachment of Cells to Fabricated Collagen Substratum

1. Dilute a stock solution of rat tail type I collagen (*see* **Note 1**) 1/10 in 0.1% (v/v) acetic acid.
2. Add diluted collagen solution to cover slips for 2 min or to 12-mm 0.45-μm Transwell polycarbonate filter membranes, and then remove excess by aspiration.
3. Air-dry cover slips or filters under UV light for no more than 1–2 h (we use the clean bench UV light).
4. Plate cells on 12-mm collagen-coated glass cover slips in 24-well tissue-culture dishes, or 12-mm Transwells at a confluent cell density.
5. Cells are cultured at least 2 d to allow for secretion of endogenous ECM and more stable attachment to the cover slip (*see* **Note 6**).

3.3.3. Cross-Linking Cells to E-Cadherin:Fc Substratum

1. Rinse cells free of DMEM/FBS with Ringer's buffer containing 1.8 mM $CaCl_2$.
2. Incubate cells for 20 min at room temperature in Ringer's buffer containing 1 mM solution of the membrane-impermeable cross-linkers BS^3 or DTSSP to stabilize the interaction between cellular cadherin and E-cadherin:Fc.
3. Quench excess cross-linker by adding glycine to a final concentration of 20 mM and incubate for an additional 5 min at room temperature.

3.4. Preparation of Basal Membranes From Cells Bound to Substrata by Sonication

The subsequent method outlines how to generate cover slips covered in basal membranes of cells attached to either an E-cadherin:Fc (**Fig. 3**) or extracellular matrix (**Figs. 4** and **5**). The method is simple: hypotonic swelling of cells followed by brief sonication, which results in the removal of the dorsal (apical) cell surface, nucleus and intracellular membranes and organelles, leaving behind on the cover slip the basal membrane attached to the substratum (*see* **Figs. 3–5**). The method described below is adapted for E-cadherin:Fc substratum and the epithelial cells used in our laboratory (*see* **ref. 7** for background information and variations on the methodology).

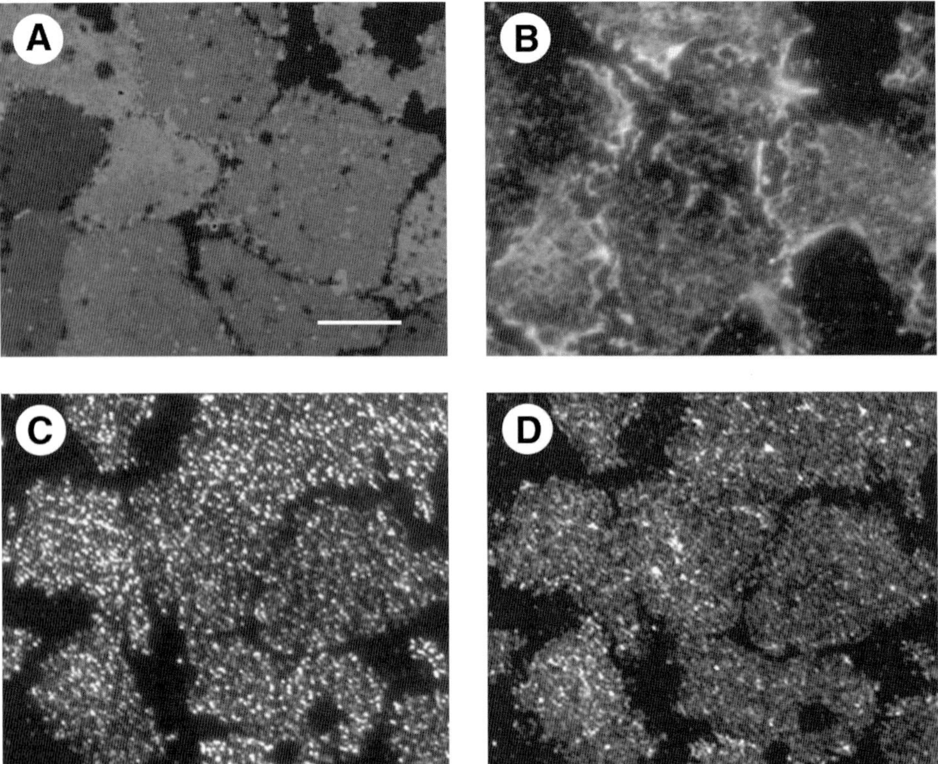

Fig. 3. Characterization of basal membrane patches from MDCK cells on E-cadherin:Fc substratum. (**A**), Staining with membrane dye, DiIC18(3), of isolated basal membrane patches on E-cadherin:Fc substratum. (**B**), Staining with monoclonal actin antibody (Chemicon). (**C**), Basal membranes on E-cadherin:Fc substratum stained with antibody directed against the intracellular domain of E-cadherin. (**D**), Staining with β-catenin antibody. All fixations were with 4% formaldehyde in Ringer's buffer. Bar, 10 µm.

1. Rinse cells three times in hypotonic buffer containing 1.8 mM CaCl$_2$, and incubate in the same buffer for 5 to 10 min at room temperature (*see* **Note 7**).
2. Fill the tissue-culture well nearly to the brim with hypotonic buffer.
3. Place the tissue culture dish on an adjustable platform/support jack.
4. Raise the dish until the tip of the 1/8-in. diameter microprobe of the Branson Sonifier 250 is 4–8 mm above the cells (*see* **Note 8**).
5. Wearing earphones to protect your ears, sonicate cells with a brief (less than 1 s) pulse, duty cycle 20, and output 19–22% (*see* **Note 9**).

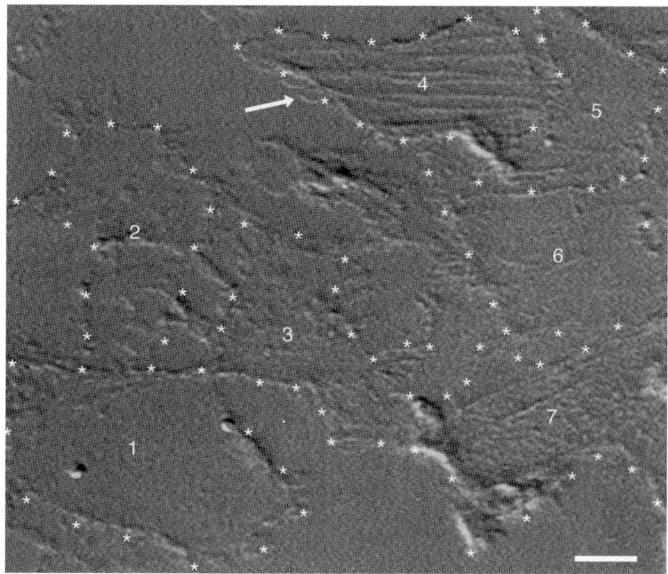

Fig. 4. Visualization of patches by differential interference contrast microscopy. MDCK clone II/G were plated on collagen-coated cover slips for 2 d, sonicated, and isolated membranes were imaged by differential interference contrast microscopy. Patches are outlined by asterisks. Note the actin fibers present on the membrane indicated by the arrow. Bar, 10 μm.

6. Rinse membranes briefly in Ringer's buffer before fixation (*see* **Subheading 3.5.**). When buffer is aspirated from the cover slip or filter, the area that has been rendered into basal patches is visible as a hole in the cell layer. The hole will vary in size from preparation to preparation (*see* **Note 10**).

3.5. Fixing Basal Membranes for Microscopy

Isolated membrane patches expose the cytoplasmic face of the membrane, and thereby the protein complexes that form the cellular attachments to E-cadherin or extracellular matrix. These membrane patches can be used for direct analysis of protein distributions by immunofluorescence microscopy or electron microscopy. However, of greater significance in terms of understanding the regulation of protein–complex interactions and assembly, these membrane patches can be used to selectively remove, or add back proteins to the complexes. The following methods outline fixation of membrane patches in 1) cold methanol, 2) glutaraldehyde, or 3) formaldehyde for immunofluorescence microscopy (*see* **Note 11**).

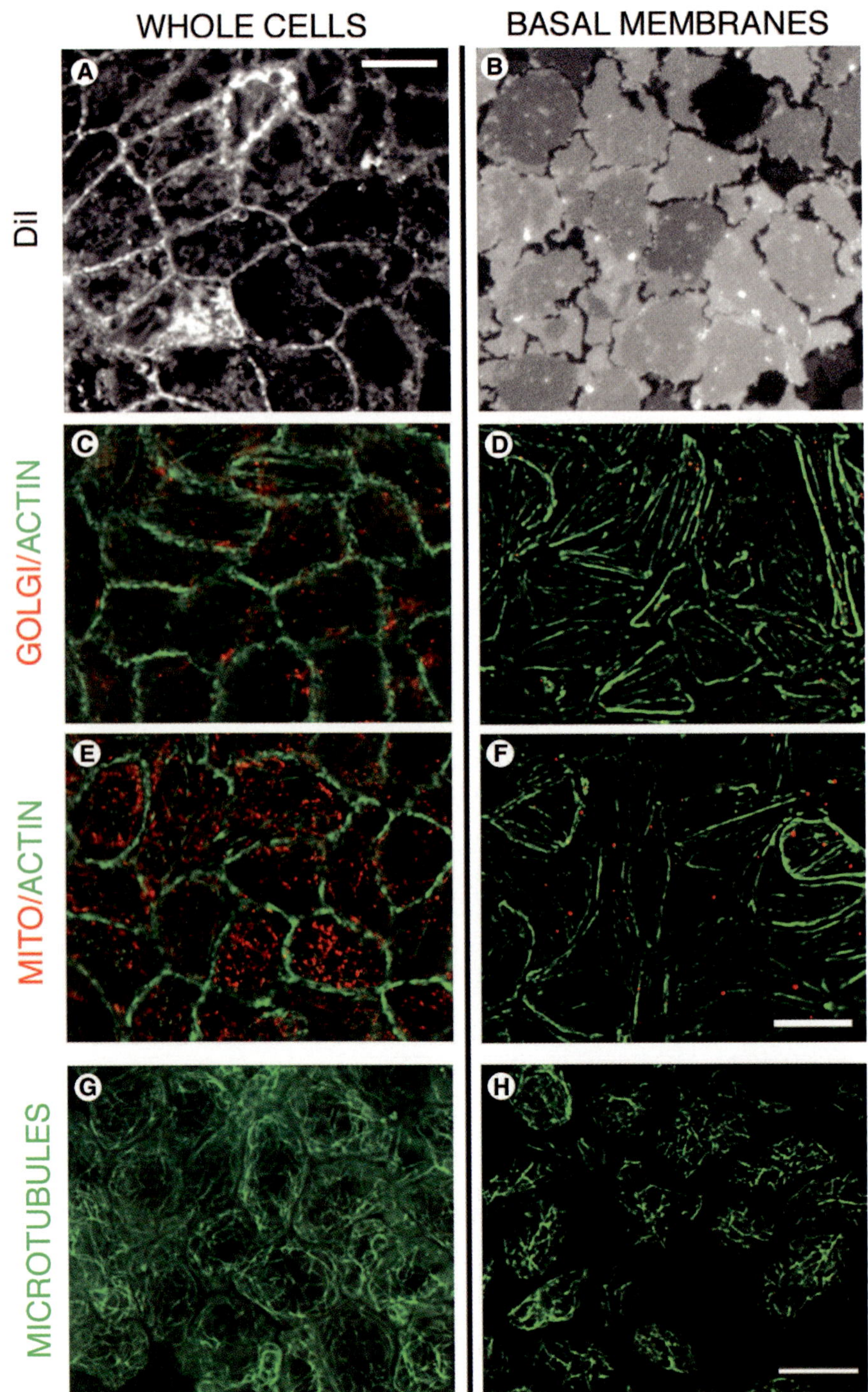
WHOLE CELLS
BASAL MEMBRANES
DiI
GOLGI/ACTIN
MITO/ACTIN
MICROTUBULES
A
B
C
D
E
F
G
H

3.5.1. Methanol Fixation

1. Pre-cool 100% methanol to –20°C.
2. Fix membrane patches in –20°C pre-cooled 100% methanol for 10 min.
3. Rinse briefly in 0.1% Triton in Ringer's buffer to break the surface tension.
4. Rinse briefly in two changes of Ringer's buffer.

3.5.2. Formaldehyde Fixation

1. Dilute 37% (v/v) formaldehyde solution to 4% (v/v) in Ringer's buffer immediately before use.
2. Fix membrane patches in 4% (v/v) formaldehyde solution for 10 min at room temperature.
3. Rinse in three changes of Ringer's buffer.

3.5.3. Glutaraldehyde Fixation (Suitable for Staining Microtubules)

1. Fix membrane patches in freshly prepared 0.3% (v/v) glutaraldehyde in Ringer's buffer.
2. Rinse briefly three times in Ringer's buffer.
3. Incubate three times for 5 min each in 1 mg/mL $NaBH_4$ in Ringer's buffer to quench unreacted aldehyde groups.
4. Rinse in three changes of Ringer's buffer.

Fig. 5. *(opposite page)* Characterization of basal membrane patches from MDCK cells on filters. **(A,B)** Staining with the membrane dye, DiIC18(3), of whole MDCK cells grown on filters **(A)**, and isolated membrane patches **(B)**, which show few internal organelles remaining after sonication. Bar, 10 µm. **(C,D)** Staining for the Golgi protein, p115 (red) and actin (green) on the basal membrane of intact MDCK cells grown on filters **(C)**, and isolated basal membranes which retain actin fibers whereas p115-containing Golgi membranes are largely removed **(D)**. **(E,F)** Staining for mitochondria (red) and actin filaments (green) of the basal membrane of intact MDCK cells **(E)**, and isolated basal patches which show preservation of actin filaments but absence of mitochondria **(F)**. Bar, 10 µm. Fixation of cells for Golgi and mitochondria staining was with 4% formaldehyde diluted in Ringers buffer. A polyclonal antibody to Golgi component p115 was a gift from Dr. Suzanne Pfeffer (Stanford University, Stanford, CA). A monoclonal antibody to mitochondrial Hsp70 was from Affinity Bioreagents. **(G,H)** GFP microtubules on the basal membrane of intact MDCK cells expressing GFP-tubulin **(G)**, and isolated basal membranes, which show improved visibility of the cytoskeleton **(H)**. Bar, 10 µm. Green fluorescent protein-microtubules were fixed with 0.3% glutaraldehyde in BRB80 buffer, for 10 min for isolated membranes or 20 min for intact cells, with the addition of 0.1% Triton X-100 for intact cells.

4. Notes

1. Rat tails are used to prepare type I collagen. Briefly, place 5–10 tails in 95% ethanol. Prepare a 0.1% (v/v) acetic acid solution using sterile water, a sterile beaker and a sterile stir bar. Have the dilute acetic acid solution stirring at room temperature. Keep solution covered. Starting at the cut end of the tail, clamp two hemostats about 2–3 cm apart on a tail. While holding the hemostat in your left hand, twist or rotate the right hemostat 360 degrees and then pull. Keep pulling until it breaks off. You should have white collagen fibers at the end of the broken (2–3 cm) piece of the tail. Cut the white fibers off the broken piece of the tail using a sharp razor blade and place them on a glass gel plate. Continue breaking and pulling 2–3 cm pieces of the tail. You get less material the closer you get to the tip of the tail. Tease the collagen fibers by holding one end of the fibers stationary with one razor blade and then use a scraping motion with a razor blade at a 45° angle. You want to flatten the fibers and open them up. Put the teased fibers in the stirring acetic acid solution. They should turn translucent. When finished, put the beaker of collagen at 4°C and stir overnight. Next day, centrifuge three quarters full 50-mL Nalgene plastic tubes for 2 h at 25,000g. Remove supernatant and save. Discard pellets. (Portion of the pellet will be gelatinous). Store supernatant as the stock solution at 4°C.

2. A major advantage of this approach is that stable producer cells can be generated in less than 2 wk once recombinant cDNA constructs are made. Production of chimeric E-cadherin:Fc from a pooled drug-resistant cell population is stable for at least 10 cell passages, and generally, a newly thawed replicate culture is used for a new round of protein purification. There is, however, some heterogeneity of expression levels in the population, and it may be possible to achieve a higher yield of secreted proteins by clonal isolation of cells and screening for higher expression. However, in light of the quantity of soluble chimeric protein produced, and the ease of generating stable cells, labor intensive cloning and screening of higher expressing cells can be avoided.

3. This chemical assembly serves two purposes: 1) the E-cadherin:Fc chimera is attached at a distance from the glass surface with the purpose to allow some degree of lateral movement during binding to cell surface E-cadherin; and 2) binding of the E-cadherin:Fc chimera through the Fc domain to the protein G correctly orients the N-terminus of the E-cadherin extracellular domain towards attaching cells. This chemical assembly results in a saturated density of approx 50,000 E-Cadherin:Fc molecules/μm_2.

4. In the methods outlined here, cover slips are placed with the silanized surface facing up on a flat surface (the caps of small polystyrene tubes can be used), approx 50 μL of solution is added (sufficient volume to cover a 12-mm cover slip), and incubated at room temperature in a humidified atmosphere. For washing, the cover slips are transferred to a ceramic cover slip holder (make sure the orientation of the silanized surface is noted) and placed in beaker containing the wash buffer.

5. We use MDCK cells, an established cell line derived from canine kidney, and mouse L-cell fibroblasts in which canine E-cadherin expression is controlled by

an inducible dexamethasone promoter (*see* **ref. 9**). To seed MDCK GII cells on E-cadherin:Fc substrate, we trypsinize the cells briefly. Cells can also be lifted from the tissue culture dishes using non-enzymatic methods, but we have found cells to adhere to the cadherin substrate better after light trypsinization. Cells are plated in DMEM with 2% instead of the usual 10% FBS to reduce the amount of ECM proteins introduced through the serum. However, certain growth factors in the serum are required for cell–cell adhesion, which is why we recommend not omitting the serum completely. Cells will start to spread on the E-cadherin:Fc substratum after 1–2 h of incubation, which is slightly slower than the time for cells to spread on collagen-coated cover slips. MDCK G cells will secrete their own extracellular matrix over time, but no focal contact staining is observed until >12 h on the substratum.

6. In the absence of chemical cross-linking to the substratum, cells should be confluent and plated for at least 24 h, although longer is better because cells will have a more stable attachment to the substratum. Nonconfluent cells or cells plated less than 24 h are not well retained after sonication without previous crosslinking. The 24-well plates or the 12-mm filters work best for the sonication procedure. Because the interactions to the substrate are more mature and more stable, membrane patches can be generated even at early time points after plating without cross-linking, but cross-linking of cells to the substrate greatly increases the number of membrane patches obtained.

7. Any hypotonic buffer should work. Both 5-min and 10-min swelling times give nice patches on cover slips. Filters may need a longer swelling time, so we swell cells for 10 min to be safe.

8. The distance between microprobe and cover slip/filter is measured by eye. The distance does not have to be precise because after sonication there is a gradient of whole cells to patches, and since there are thousands of patches per cover slip or filter, there will usually be plenty even if many whole cells remain. Cells that are more resistant to rupture, for example polarized epithelial cells on Transwell filters, should be held closer to the microprobe (4 mm), whereas more delicate cells, for example fibroblasts on cover slips, should be held a few mm further from the microprobe (7 mm). Once one gets the feel for it, obtaining nice patches is very reproducible.

9. The duty cycle on the sonifier is set at 20, and the output varies according to the substrate (coated-glass or filter), the type of cells, and how long the cells have been plated. The output setting ranges from around 19% maximum power for fibroblasts to around 23% maximum power for MDCK cells polarized on filters.

10. Preparation-to-preparation variation (the size of the hole in the cell monolayer after sonication, the number of basal patches, the number of membranes with microtubules) depends on how long cells are plated, the distance of the microprobe from the cover slip/filter, and the intensity of the sonication burst. There will usually be whole cells present towards the edge of the cover slip, but these cells are not suitable for microscopy as they are damaged. The presence of whole cells means that these preparations are not suitable for biochemical iso-

lation of basal membranes. Do not attempt to deliver more than one sonication burst to a cover slip/filter to increase the area of basal patches, because the patches already present will be ruined.

11. Although the isolated membranes are thin, approx 1 µm, the best high-resolution images are obtained by confocal or deconvolution microscopy. For the images in **Fig. 5**, image stacks of 0.20 µm in thickness were taken on an Olympus IX-70 inverted microscope with a 100X oil-immersion objective 1.35 N.A. (Olympus Corp.) and captured by a cooled CCD camera (Photometrics Ltd.). Images are collected and processed using Delta Vision de-convolution software (Applied Precision, Seattle, WA) on a Silicon Graphics workstation (Silicon Graphics Corp.).

Acknowledgments

This work was supported by grants (GM35527, NS42735) from the NIH to W. J. Nelson, and by a predoctoral fellowship from the Boehringer Ingelheim Foundation, Germany to F. Drees and a postdoctoral fellowship (5T32CA09151) from the NCI to A. Reilein.

References

1. Yeaman, C., Grindstaff, K. K., and Nelson W. J. (1999) New perspectives on mechanisms involved in generating epithelial cell polarity. *Physiol. Rev.* **79,** 73–98.
2. Mostov, K., Su, T., and ter Beest, M. (2003) Polarized epithelial membrane traffic: conservation and plasticity. *Nat. Cell Biol.* **5,** 287–294.
3. Takeichi, M. (1990) Cadherins: a molecular family important in selective cell-cell adhesion. *Annu. Rev. Biochem.* **59,** 237–252.
4. Gumbiner, B. M. (1996) Cell adhesion: the molecular basis of tissue architecture and morphogenesis. *Cell* **84,** 345–357.
5. Jamora, C. and Fuchs E. (2002) Intercellular adhesion, signaling and the cytoskeleton. *Nat. Cell Biol.* **4,** 101–108.
6. Chen, Y-T. and Nelson, W. J. (1996) Continuous production of soluble extracellular domains of type-I transmembrane in mammalian cells using an Epstein-Barr Virus Ori P-based expression vector. *Anal. Biochem.* **242,** 276–278.
7. Heuser, J. (2000) The production of 'cell cortices' for light and electron microscopy. *Traffic* **1,** 545–552.
8. Chen, Y. T., Holcomb, C., and Moore, H. P. (1993) Expression and localization of two low molecular weight GTP-binding proteins, Rab8 and Rab10, by epitope tag. *Proc. Natl. Acad. Sci. USA* **90,** 6508–6512.
9. Angres, B., Barth, A. I. M., and Nelson, W. J. (1996) Mechanism for transition from initial to stable cell-cell adhesion: kinetic analysis of E-cadherin-mediated adhesion using a quantitative adhesion assay. *J. Cell Biol.* **134,** 549–557.

22

Application of Microscope-Based FRET to Study Molecular Interactions in Focal Adhesions of Live Cells

Christoph Ballestrem and Benjamin Geiger

Summary

This chapter describes the use of microscope-based fluorescence resonance energy transfer to follow dynamic interaction of molecules localized at focal adhesions. We first outline the significance of studying dynamic interactions in focal adhesions of living cells, and second provide an overview of the method itself. This is followed by a protocol for fluorescence resonance energy transfer measurements in live cells.

Key Words: FRET; fluorescence resonance energy transfer; molecular interaction; focal adhesion; cell migration; green fluorescence protein; GFP; CFP; YFP.

1. Introduction

For migrating over external surfaces, cells need to form adhesions with the extracellular matrix and to modulate these adhesions in a spatially and temporally coordinated manner. Cell adhesion occurs at specialized sites such as focal adhesions (FA), where transmembrane adhesion receptors (mostly members of the integrin family), together with regulatory molecules (e.g., kinases and phosphatases) and a large variety of adapter proteins, interact with the cytoskeleton. In recent years a vast amount of literature has accumulated, describing the localization, interactions and functions of many of these focal adhesion proteins. Recent review articles *(1,2)* outline the molecular complexity of focal adhesions, with more than 50 known associated proteins. Moreover, biochemical studies indirectly suggest a huge number of possible molecular interactions between these molecules *(1)*. However, it is apparent that only a fraction of these potential interactions can actually take place at any given moment because most of the associated proteins have several domains through which they might interact with

From: *Methods in Molecular Biology, vol. 294: Cell Migration: Developmental Methods and Protocols*
Edited by: J-L. Guan © Humana Press Inc., Totowa, NJ

1

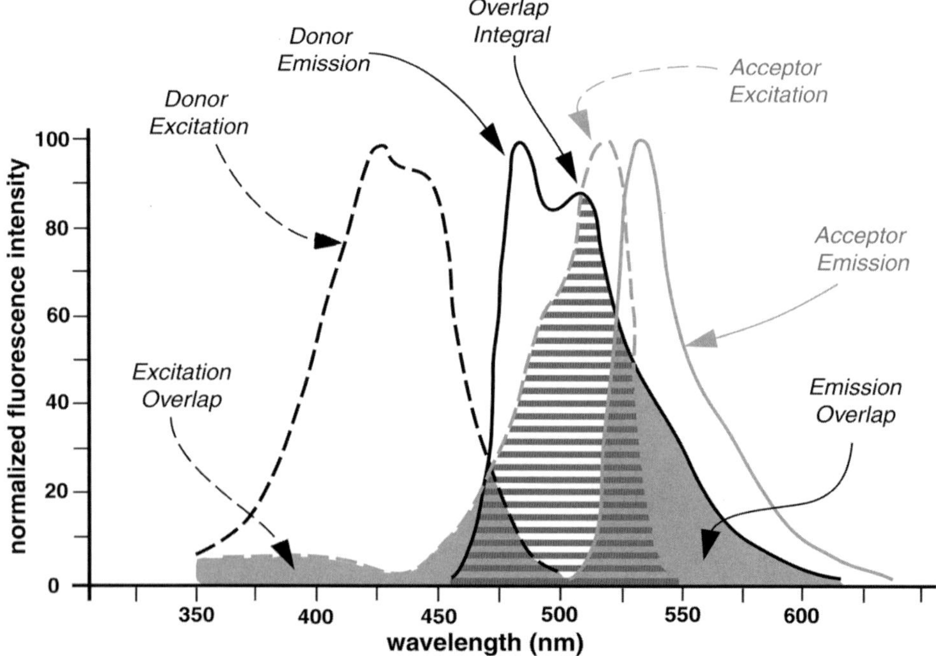

Fig. 1. Fluorescence intensity spectra of CFP (donor) and YFP (acceptor). The overlap area (striped) between CFP emission and YFP excitation is a prerequisite for FRET. Overlap areas (shaded) of CFP and YFP excitation spectra and of CFP and YFP emission spectra are the sources of bleed-through fluorescence in the FRET filters.

multiple (often more than 10) partners. In many instances several proteins might compete for binding to the same site. Therefore, it is important for the understanding of cell migration to measure interactions of focal adhesion proteins that actually occur in living cells. In this chapter we describe the application of microscope-based fluorescence resonance energy transfer (FRET) for visualizing such molecular interactions.

1.1. The Principle of FRET

FRET is based on energy transfer, via dipole–dipole interaction, from a "donor" fluorophore to an "acceptor" fluorophore, leading to an increase in the acceptor's fluorescence emission and quenching the donor's emission fluorescence *(3)*. A prerequisite for an efficient FRET is an overlap between the donor's emission and the acceptor's excitation spectra (for example, cyan fluorescent protein [CFP]/yellow fluorescent protein [YFP] in **Fig. 1**).

According to the Foerster equation:

$$E = R_0^6/(R_0^6 + r^6)$$

the efficiency (E) of energy transfer depends on the inverse sixth power of the distance between the donor and the acceptor *(4)*. Indeed, efficient FRET is only detectable when donor and acceptor molecules are in proximity of about 10 nm or closer, rendering this method suitable for measuring protein–protein interactions, intramolecular conformational changes and changes in intermolecular distances. The Foerster radius R_0 is defined as the distance at which the efficiency of energy transfer between the chromophores is 50% of the maximum and it depends on the fluorophores' spectral properties, the quantum yield of the donor, and the orientation factor k2 expressing the relative angular orientation of the chromophore's transition dipoles *(5)*. k2 can range from 0 to 4 according to the freedom of rotation of the fluorophores. It is thought that k2 in biological systems is around 2/3, though it can vary when fluorophores are fused to specific proteins of interest.

In the early seventies, FRET was extensively used as a spectroscopic ruler to measure proximities between molecules *(6)*. These experiments were usually performed in vitro with defined and selected partner molecules, and using fluorometers to measure FRET values.

More recently, with the introduction of fluorescently labeled proteins into live cells via microinjection, and especially since the wide application of green fluorescent protein (GFP) and its various spectral variants, cell biologists became interested in the use of live cell FRET in order to dissect signaling pathways and to map molecular interactions in the natural cytoplasmic environment. Obviously, to obtain FRET data from live cells it is necessary to use a fluorescence microscope as the main analytical tool, rather than a fluorometer. The primary advantage of such FRET measurements is the high spatial and temporal resolution of the light microscope. With a resolution of approx 0.2 µm along the focal plane of the microscope, and 0.5 µm in the perpendicular direction, molecular components of a wide variety of subcellular structures can be resolved and examined. Apparently, the same resolution obtained for regular imaging of cellular structures (2-D or 3-D) can be obtained for FRET. It is important to clarify that whereas FRET can report on molecular proximities of 1 to 10 nm, the actual spatial resolution is defined by the optical resolution of the microscope. The advantage and different readouts of microscope-based FRET measurements are outlined in **Fig. 2**. The temporal resolution of microscope-based FRET ranges from a few seconds to a few tens of seconds, depending on the exposure time, which, in turn, is determined by the intensity of the FRET signal and the sensitivity of the camera.

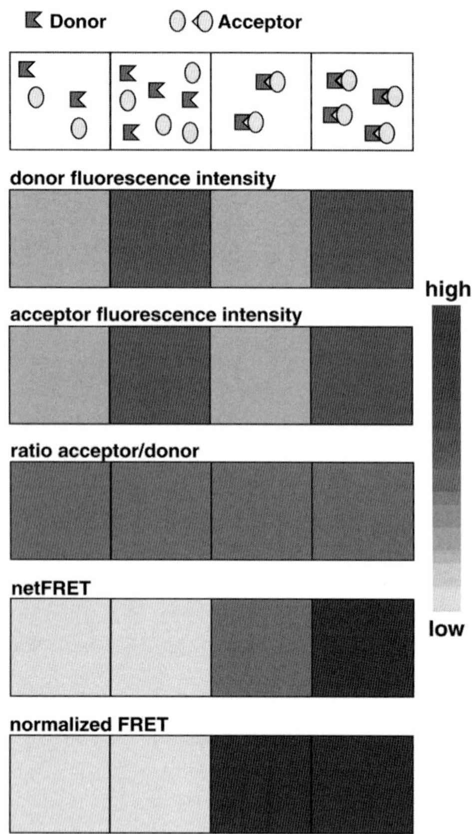

Fig. 2. FRET provides new information about molecular organization and interactions. The first row shows hypothetical distributions and interactions of two types of molecules, which may be present at high or low concentrations and may or may not interact with each other. The upper "boxes" represent one pixel each, and the fluorescence readout from that particular pixel is shown under each of the boxes. These rows display in a spectrum scale fluorescence readouts reporting on molecular distributions of the donor and acceptor molecules, the ratio between the two, netFRET and FRET normalized by the intensity of the donor. Note the effects of concentrations and interactions on the fluorescence intensity readout provided by the examples.

It is beyond the scope of this article to review the diverse possible applications of microscope-based FRET in modern biological research. A few examples are the use of this approach for studying specific molecular interactions within signaling networks, the assembly of cytoskeletal fibers and the formation of multiprotein complexes, such as the cell-matrix adhesions described here.

1.2. Alternative Approaches for Measuring FRET Values

Several methods were developed for measuring FRET through a microscope *(7)*. Here we will briefly mention the three main methods and highlight their advantages and limitations. In principle all three methods are suitable for mapping molecular interactions within cells, and the choice of one particular approach depends on the aim of study and the equipment available.

1.2.1. Fluorescence Lifetime Imaging Microscopy (FLIM)

FLIM is a powerful method for measuring FRET *(8)*, where the characteristic time that the molecule remains in the excited state before returning to its basal energy level is measured. Upon FRET to the acceptor fluorophore a decrease of the measured fluorescence lifetime of the donor molecule is detected. In contrast to other FRET approaches, this method is independent of fluorophore concentrations.

1.2.2. Donor Emission-Based FRET

Because efficient FRET between donor and acceptor leads to quenching of the donor a possible method to determine FRET efficiency is to measure the increase of the donor fluorescence after photobleaching of the acceptor (which eliminates its capacity to quench the donor molecules). FRET efficiency (E) would then be deduced from the ratio between donor intensity before and after acceptor photobleach *(9,10)*. FRET measurements, based on acceptor bleaching, can be performed with a confocal or a wide field microscope.

1.2.3. Acceptor Emission-Based FRET

This method is based on the measurement of increased emission fluorescence of the acceptor fluorophore after the excitation of the donor owing to FRET. This method is particularly useful to study spatial and temporal changes in protein localization, protein–protein interactions, and conformational changes in live cells. Because of the bleed-through of donor and acceptor fluorescence through the FRET filter sets (namely, direct fluorescence of the acceptor and donor fluorophores obtained with the FRET excitation and emission filter sets), the apparent FRET values need to be corrected for this "crosstalk" *(3)*.

1.2.4. Advantages and Limitations of the Different Approaches for Microscope-Based FRET Measurement

The three approaches for microscope-based FRET measurement are all suitable, in principle, for studying protein interactions in live cells. However, each approach has some unique advantages and limitations.

FLIM appears to offer some significant advantages for live cell studies. Its main drawback is the high cost and complexity of the equipment. Because this method allows FRET measurements independent of donor/acceptor concentrations, expression levels can be kept low and cells are less sensitive to photodamage.

Donor emission-based FRET provides reliable data using simple calculations and a rather standard microscope setup. Because of its laser power and rather fast bleaching time (a few seconds) a confocal microscope is more suitable for such measurements than a wide field microscope (bleaching time of 2–3 min). However, this method is destructive and is, therefore, not suitable for dynamic FRET analysis of developing and changing structures over time.

To track spatial and temporal changes of proteins, acceptor emission-based FRET measurements are the method of choice. These experiments can be performed with widefield and confocal (two- or multiphoton) microscopes. The main drawback of this approach is that the FRET values must be carefully corrected for "bleed-through" and background contributions. These fluorescence values are, in some cases, comparable to the positive signal.

However, in view of its compatibility with dynamic FRET recording, we chose to present here the acceptor emission-based approach as our method of choice, and describe this method in some more detail.

1.3. The Physical Basis for Acceptor Emission-Based FRET

An efficient FRET can occur only when the donor and acceptor fluorophores have overlapping emission and excitation wavelength, respectively (**Fig. 1**). In an ideal situation the donor fluorophore should have a maximum of spectral overlap of its emission wavelength with the acceptor's excitation wavelength. At the same time the donor's and the acceptor's excitation spectra and the donor's and the acceptor's emission spectra should not overlap. However, since in most cases some overlap exists (*see* **Fig. 1** for CFP/YFP), which cannot be completely eliminated by the FRET filters, corrections for the "bleed-through" fluorescence are essential. To minimize these cross-talks it is advisable to select narrow band-pass filters (*see* **Table 1** for CFP/YFP filter combinations). The amount of the donor and acceptor bleed-through in the FRET filter setting can be readily measured in cells expressing either donor or acceptor fluorescence. The ratio between the values measured in the FRET filters and the fluorescence of the donor or acceptor, measured with the regular absorbance filters is the "bleed-through coefficient" for each donor and acceptor fluorophore. Thus, to obtain the true FRET value (netFRET) of a cell transfected with both, donor and acceptor fluorophores, the bleed-through values have to be subtracted from the total fluorescence intensity measured in the FRET filter. Bleed-through values have to be precisely measured in each experiment otherwise this method is prone to major artifacts.

Table 1
Filter Sets for FRET Measurements Using CFP Fusion
Proteins as Donor and YFP Fusion Proteins as Acceptor

Filterset	Excitation (nm)	Dichroic mirror	Emission (nm)
CFP	436/10	455	480/40
FRET	436/10	455	535/25
YFP	495/10	525	535/25

Numbers for excitation and emission filters indicate the peak transmission wavelength and the band-width. For example, the 480/40 emission filter passes light from 460 to 500.

To obtain optimal FRET, using the acceptor emission-based measurement, the following points should be taken into consideration:

1. Use the right filter combinations to minimize fluorescence cross-talk (*see* **Table 1** for CFP and YFP).
2. Avoid autofluorescence in the specimen (including in the culture medium).
3. Avoid high donor and acceptor concentrations that can cause self-quenching.
4. Avoid too low expression levels of either donor and or acceptor fluorophores to improve signal to noise ratio.
5. Use shortest exposure times possible to prevent photobleaching and photodamage of the cell.
6. Calculate precisely the correction of the fluorescence bleed-through.

1.4. Fluorophores

The donor-acceptor combinations of blue fluoresecent protein (BFP)-GFP, CFP-yellow fluorescent protein (YFP), and GFP-DsRed of the living color panel were reported to be suitable for FRET experiments *(11,12)*. In our experience the CFP-YFP variant has some significant advantages. It appears to be superior to the BFP/GFP pair, because of the larger Foerster radius ($R_0 = 49$–52 vs $R_0 = 40$–43), the higher quantum yield, and the higher photostability of CFP compared to BFP *(13)*. It is also superior to the GFP/DsRed pair, since DsRed fusion proteins often miss-localize, most likely because of the strong oligomerization tendency of DsRed. It should be mentioned that there are new CFP, YFP, and DsRed variants (*see* **refs.** *14* and *15*) with improved spectral properties and reduced potential to oligomerize, which might offer some advantages for FRET.

There are additional pairs of FRET-compatible fluorophores, which consist of chemical dyes that can be conjugated to proteins. Such donor acceptor pairs can be introduced to cells by microinjection of the labeled fluorescent proteins. Dyes suitable to use as donor-acceptor pairs with its R_0 values can be found in the molecular probes catalogue (*see* www.probes.com/handbook/boxes/ 0422.html). The combination of living colors and labeled proteins offers further possibilities as it has been successfully shown by Kraynov and colleagues using a GFP-Rac construct together with an Alexa 546 labeled Rac binding protein domain of p21 Pak *(16)*.

1.5. Recent Application of Microscope-Based FRET

We would like to mention here just a few examples for recent live cell FRET studies that might be of special interest in the context of cell migration. One of the most prominent groups of regulators of cell adhesion and migration are members of the Rho family of GTPases. Recent studies have addressed the compartmentalization of the active forms of these Rho GTPases using microscope-based FRET. The group of Hahn performed "intermolecular FRET" experiments using a Rac1-GFP construct together with a Alexa 546 coupled PBD (p21-activated kinase-(PAK)-Binding Domain) to follow Rac activation *(16)*. It was found that active Rac localizes to the cell periphery whereas the total Rac1 pool was enriched around the cell nucleus and in ruffles. Similar results were obtained by the group of Matsuda, using a construct that consists of the Rac-1 fused to PBD and bridged by a flexible linker *(17)*. This fusion protein is further flanked by CFP and YFP *(17)*. In the active state Rac-1 binds to the PBD, thereby bringing the donor (CPF) and acceptor (YFP) to a "FRET range." This "intramolecular FRET" method has been successfully used to show the localized activation of a number of other GTPases, including *Ras*, Rap1, Cdc42, and Rho *(17–19)*. One of the advantages of this approach is that the ratio of the donor and acceptor is always 1:1 and therefore variations caused by different levels of expression of the two partner molecules do not exist *(20)*. The method of intramolecular FRET is also suitable to visualize conformational changes of proteins as it has been done with the CrkII adapter molecule *(21)*. Since a number of proteins localizing to focal adhesions (e.g., vinculin, Src), are known to undergo conformational changes, this type of constructs could be used to simultaneously monitor the localization and activation state of these proteins.

2. Materials

1. ECFP and EYFP expression vectors were from Clontech (Palo Alto, CA; www.clontech.com; *see* **Note 1**).
2. NIH 3T3 mouse fibroblasts (*see* **Note 2**) were from the American Type Culture Collection (Rockville, MD; www.atcc.org/SearchCatalogs/CellBiology.cfm).
3. Best transfection results were obtained using Lipofectamine Plus reagent from Invitrogen (Invitrogen Corporation, Carlsbad, CA; www.invitrogen.com/).
4. 35-mm Glass-bottom dishes (MatTek Corporation, Ashland, MA; www.mattek. com/).

5. Fluorescent (488 nm) 2.5-μm beads (Molecular Probes, Eugene, OR; www. probes.com/).
6. Fibronectin was from bovine plasma (Sigma, St. Louis, MO; www.sigmaaldrich. com/).
7. Dulbecco's modified Eagle's medium (DMEM) supplemented with 10% fetal calf serum and antibiotics for cell culture (complete medium). For FRET, DMEM (pH 7.4) medium supplemented with antibiotics and HEPES (25 mM), without sodium bicarbonate (for use in a CO_2-free environment), without phenol red, and without riboflavin (FRET medium; *see* **Note 3**).
8. Inverted, wide field microscope attached to a CCD camera and equipped with a 100W mercury lamp, appropriate filter sets (*see* **Table 1**) and analysis software. A sensitive CCD camera is important to prevent photodamage and bleaching usually occurring when exposure times are high. Filters appropriate for FRET can be purchased from Omega (www.omegafilters.com/) or Chroma (www.chroma.com/). To maintain stable temperature, the microscope was set under a box and temperature control system from Life Imaging Services (Switzerland; www.lis.ch/; *see* **Note 4**). Several companies offer image capturing and analysis software, some of them offer FRET analysis modules (Openlab, Improvision, UK, www.improvision.com; Intelligent Imaging Innovations, Denver, CO, www.intelligent-imaging.com). We are running our system by a DeltaVision system from Applied Precision (Applied Precision Inc., Issaquah, WA; www.api.com/) and analyze the images using Priism software (used in D. Agard's and J. Sedat's laboratories at UCSF, San Francisco; downloadable at http://msg.ucsf.edu/IVE/Download/index.html).

3. Methods

1. Prepare donor (CFP) and acceptor (YFP) fusion proteins according to standard molecular biology techniques. Proteins can be fused to the C- or N-terminus of the fluorophore (*see* **Note 5**).
2. Transfection: co-transfect cells with cDNA constructs encoding donor and acceptor fusion proteins using Lipofectamine Plus reagent according to the manufacturers instructions in 35-mm dishes. For reliable results the acceptor should exceed the donor protein when localized to FA.

   The following additional transfections are essential for every experiment:

   a. CFP-(donor)-fusion protein alone. This will be used to determine the bleed-through of CFP fluorescence through the "FRET filter."
   b. YFP-(acceptor)-fusion protein alone. This will be used to determine the bleed-through of YFP fluorescence through the "FRET filters."
   c. As a negative control, select two noninteracting CFP-donor and YFP-acceptor fusion proteins for co-transfection.

   After 3 h of transfection, trypsinize the cells and wash them once with complete DMEM. Resuspend the cells in DMEM culture medium and plate 50,000 cells per one 35-mm MatTek glass-bottom dish. Before plating cells, MatTek chambers were coated for 1 h at room temperature with 10 μg/mL FN.

3. Twenty-four hours after transfection, replace the culture medium with pre-warmed (37°C) serum-free "FRET medium" and place the dish in the warm microscope chamber. Wait 30 min before starting the experiment to allow equilibration of the cells to the new environment.

4. Image acquisition: acquire three images of the transfected cells, showing prominent focal adhesions: 1) an image with a YFP filter settings, 2) an image of the FRET filter setting, and 3) an image with CFP filter setting (**Table 1**). Use 63X or 100X magnification oil immersion objective with highest possible numerical aperture without phase ring and apply the same exposure times for all three images. Exposure time should be sufficient to obtain a clear image without causing photobleaching and photodamage to the cell.

5. Image processing: image filtration.

   a. Use an appropriate image filtration program that subtracts background fluorescence. We have used an image filtration method, which is described in detail *(22)* and is based on the Applied Precision software. It involves an algorithm that creates boxes around each pixel, which are considerably larger than typical focal adhesions, and subtracts the average intensity of such boxes from intensity values of the particular pixel. This process leads to an efficient "flattening" of the background and helps in image thresholding.

   b. Set threshold to include only pixels with high fluorescence signals of both donor and acceptor into calculations for FRET values.

   c. Correct for lateral shifts or magnification.

   Chromatic, and other aberrations can cause lateral or focus shifts, which have to be corrected. For correction of lateral shifts measure the location of 2.5-μm beads seeded in the glass-bottom dishes. The beads should be visible in all three channels. Lateral pixel shifts should be measured and corrected for all the captured images. Correction of focus shifts: test whether there is any focus shift using beads seeded on glass-bottom dishes. In case focus shifts are present, correct the Z-focus for the "out of focus image" during your time-lapse recording (*see* **Note 6**).

6. Determine the bleed-through coefficients of α and β for CFP and YFP. Take processed images of cells transfected with donor and acceptor only and calculate pixel-by-pixel ratios of I_{FRET}/I_{CFP} and I_{FRET}/I_{YFP}, respectively. You will get numbers around $\alpha = 0.7$ (70%) for CFP and $\beta = 0.2$ (20%) for YFP (*see* **Note 7**).

7. After subtracting the bleed-through fluorescence and getting "netFRET" there are various ways to normalize FRET values *(3,23)*. Our calculation defines the fraction of molecules that transfer energy to the acceptor fluorophore at a specific ($netFRET/I_{CFP}$).

   a. $netFRET = I_{FRET} - I_{CFP} \times \alpha - I_{YFP} \times \beta$

   b. $netFRET/I_{CFP} = (I_{FRET} - I_{CFP} \times \alpha \times I_{YFP} \times \beta) / I_{CFP}$

   where α = bleed-through coefficient for CFP.

   β = bleed-through coefficient for YFP.

   I_{FRET} = total fluorescence intensity measured in the FRET filter set.

   I_{CFP} = fluorescence intensity measured in the CFP filter set.

   I_{YFP} = fluorescence intensity measured in the YFP filter set.

8. FRET imaging: FRET value is obtained for each pixel in focal adhesions, which displays both donor and acceptor fluorescence above threshold. In order to visualize and analyze FRET values it is useful to display them in a spectrum color scale along with acquired fluorescence intensity and ratio images for the donor and acceptor (**Fig. 3**). Compare positive FRET values with those of negative controls that were scaled in the same range (*see* **Note 8**).

   Sequence of images will show dynamics of molecular interactions (*see* **Note 9**).

4. Notes

1. It has been reported that these EGFP variants tend to dimerize and it is preferable to use non-dimerizing EGFP variants such as those described by Tsien *(14,15)*. In our experience dimerization does not pose a serious problem in the case of focal adhesion proteins where protein localization is dependent on targeting domains in contrast to proteins that are free to move in the membrane or cytoplasm.
2. In terms of signaling events during cell adhesion, NIH 3T3 cells are a well-characterized cell line, which has reasonable-sized focal adhesions. However the choice of the cell line usually depends on the aim of the investigation. Rat embryonic fibroblasts are cells with very prominent focal adhesions, however they are slow in reorganizing them. B16 mouse melanoma cells are highly motile, however their focal adhesions are smaller than those of fibroblasts.
3. Riboflavin has high autofluorescence in the excitation range of 450–490 nm and emission range of 500–560 nm *(24)*. Special media without riboflavin can be ordered upon request from the supplier Gibco-BRL. Hams F12 medium can also be used; it contains only a very low amount of riboflavin. Because of some autofluorescence of FCS we performed our experiments in serum-free medium.
4. Precise temperature control is of great importance; small focus drifts resulting from slight temperature shifts increase the noise of calculated FRET values.
5. It is recommended, whenever possible to try both, N- and C-terminal CFP or YFP fusions to the proteins of interest. The location of the fluorophore might affect the FRET values.
6. Lateral and magnification shifts can also be visualized by co-expression of the same FA protein as donor fused to CFP and as acceptor fused to YFP.
7. For bleed-through and FRET measurements only focal adhesions with high signal to noise ratio will give correct values.
8. Since background noise and autofluorescence of substances in the cells cannot be entirely eliminated, FRET values in negative controls are not necessarily zero. It is important to improve signal to noise ratios for getting reliable results.
9. FRET movies can appear noisy. This might be caused by real changes in FRET, however, it is often caused by slight temperature shifts, unstable light sources, and/or low signal to noise ratios.

References

1. Zamir, E. and Geiger, B. (2001) Molecular complexity and dynamics of cell-matrix adhesions. J. Cell Sci. **114,** 3583–3590.

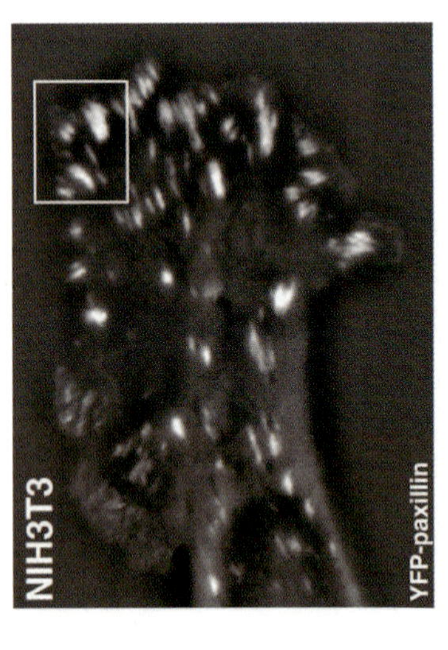

NIH3T3
YFP-paxillin

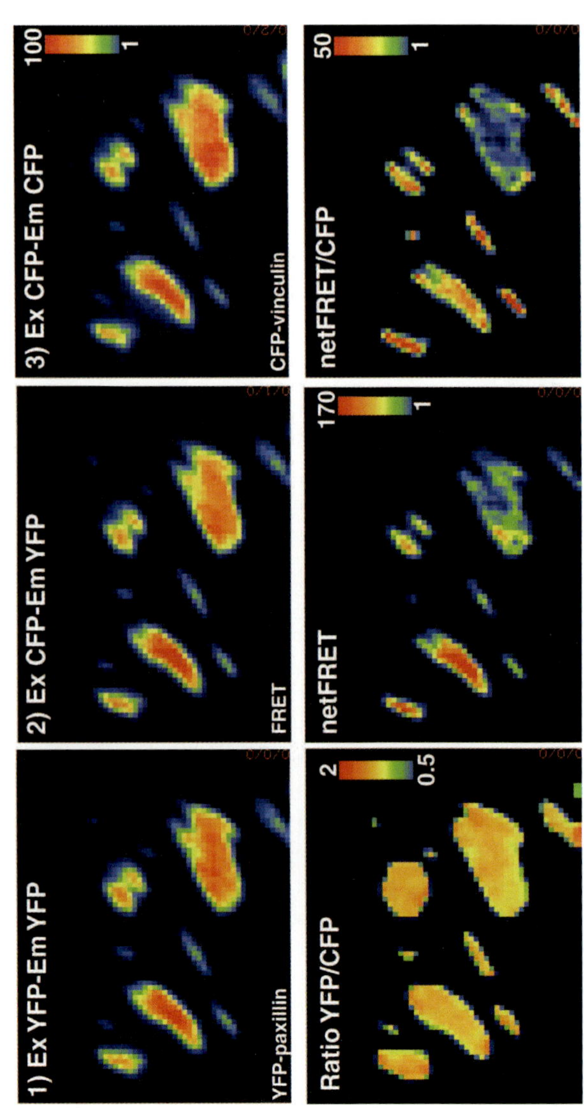

1) Ex YFP-Em YFP
YFP-paxillin
2) Ex CFP-Em YFP
FRET
3) Ex CFP-Em CFP
CFP-vinculin
Ratio YFP/CFP
2
0.5
netFRET
170
1
netFRET/CFP
50
1
100
1

2. Geiger, B., Bershadsky, A., Pankov, R., and Yamada, K. M. (2001) Transmembrane crosstalk between the extracellular matrix—cytoskeleton crosstalk. *Nat. Rev. Mol. Cell. Biol.* **2,** 793–805.

3. Gordon, G. W., Berry, G., Liang, X. H., Levine, B., and Herman, B. (1998) Quantitative fluorescence resonance energy transfer measurements using fluorescence microscopy. *Biophys. J.* **74,** 2702–2713.

4. Foerster, T. (1948) Zwischenmoleculare Energiewanderung und Fluoreszenz. *Ann. Physik. (Leipzig)* **2,** 55–75.

5. Patterson, G. H., Piston, D. W., and Barisas, B. G. (2000) Forster distances between green fluorescent protein pairs. *Anal. Biochem.* **284,** 438–440.

6. Stryer, L. (1978) Fluorescence energy transfer as a spectroscopic ruler. *Annu. Rev. Biochem.* **47,** 819–846.

7. Sekar, R. B. and Periasamy, A. (2003) Fluorescence resonance energy transfer (FRET) microscopy imaging of live cell protein localizations. *J. Cell Biol.* **160,** 629–633.

8. Bastiaens, P. I. and Squire, A. (1999) Fluorescence lifetime imaging microscopy: spatial resolution of biochemical processes in the cell. *Trends Cell Biol.* **9,** 48–52.

9. Kenworthy, A. K. and Edidin, M. (1998) Distribution of a glycosylphosphatidyl-inositol-anchored protein at the apical surface of MDCK cells examined at a resolution of <100 A using imaging fluorescence resonance energy transfer. *J. Cell Biol.* **142,** 69–84.

10. Siegel, R. M., Chan, F. K., Zacharias, D. A., Swofford, R., Holmes, K. L., Tsien, R. Y., et al. (2000) Measurement of molecular interactions in living cells by fluorescence resonance energy transfer between variants of the green fluorescent protein. *Sci. STKE* **2000,** PL1.

11. Miyawaki, A. and Tsien, R. Y. (2000) Monitoring protein conformations and interactions by fluorescence resonance energy transfer between mutants of green fluorescent protein. *Methods Enzymol.* **327,** 472–500.

Fig. 3. *(opposite page)* FRET for paxillin and vinculin in focal adhesions. NIH 3T3 cells were co-transfected with cDNA encoding CFP-vinculin and YFP-paxillin. A sequence of three images was required, starting with the YFP, then the FRET, and finally the CFP channel. Captured images were filtered and processed as described. The first row of images shows a cell displaying YFP-paxillin, total FRET measured, and CFP-vinculin fluorescence in an intensity spectrum scale. Second row shows the calculated YFP/CFP ratio, netFRET, and netFRET normalized by the CFP fluorescence intensity. The homogenous color distribution in the ratio image indicates an overall similar distribution of the donor and the acceptor in focal adhesions. The FRET images are considerably less homogenous, suggesting that there are hot spots where vinculin-paxillin is more prominent within focal adhesions. Note the difference between netFRET and normalized FRET. High netFRET occurs in areas with high donor and acceptor presence. Normalized FRET for the vinculin-paxillin pair is high also in smaller focal adhesions with lower concentrations of the two molecules (*see also* **Fig. 2**).

12. Mizuno, H., Sawano, A., Eli, P., Hama, H., and Miyawaki, A. (2001) Red fluorescent protein from Discosoma as a fusion tag and a partner for fluorescence resonance energy transfer. *Biochemistry* **40,** 2502–2510.

13. Tsien, R. Y. (1998) The green fluorescent protein. *Annu. Rev. Biochem.* **67,** 509–544.

14. Campbell, R. E., Tour, O., Palmer, A. E., Steinbach, P. A., Baird, G. S., Zacharias, D. A., et al. (2002) A monomeric red fluorescent protein. *Proc. Natl. Acad. Sci. USA* **99,** 7877–7882.

15. Zacharias, D. A., Violin, J. D., Newton, A. C., and Tsien, R. Y. (2002) Partitioning of lipid-modified monomeric GFPs into membrane microdomains of live cells. *Science* **296,** 913–916.

16. Kraynov, V. S., Chamberlain, C., Bokoch, G. M., Schwartz, M. A., Slabaugh, S., and Hahn, K. M. (2000) Localized Rac activation dynamics visualized in living cells. *Science* **290,** 333–337.

17. Itoh, R. E., Kurokawa, K., Ohba, Y., Yoshizaki, H., Mochizuki, N., and Matsuda, M. (2002) Activation of rac and cdc42 video imaged by fluorescent resonance energy transfer-based single-molecule probes in the membrane of living cells. *Mol. Cell. Biol.* **22,** 6582–6591.

18. Mochizuki, N., Yamashita, S., Kurokawa, K., Ohba, Y., Nagai, T., Miyawaki, A., and Matsuda, M. (2001) Spatio-temporal images of growth-factor-induced activation of *Ras* and *Rap*1. *Nature* **411,** 1065–1068.

19. Yoshizaki, H., Ohba, Y., Kurokawa, K., Itoh, R. E., Nakamura, T., Mochizuki, N., et al. (2003) Activity of Rho-family GTPases during cell division as visualized with FRET-based probes. *J. Cell Biol.* **162,** 223–232.

20. Miyawaki, A. (2003) Visualization of the spatial and temporal dynamics of intracellular signaling. *Dev. Cell* **4,** 295–305.

21. Kurokawa, K., Mochizuki, N., Ohba, Y., Mizuno, H., Miyawaki, A., and Matsuda, M. (2001) A pair of fluorescent resonance energy transfer-based probes for tyrosine phosphorylation of the CrkII adaptor protein in vivo. *J. Biol. Chem.* **276,** 31,305–31,310.

22. Zamir, E., Katz, B. Z., Aota, S., Yamada, K. M., Geiger, B., and Kam, Z. (1999) Molecular diversity of cell-matrix adhesions. *J. Cell. Sci.* **112,** 1655–1669.

23. Xia, Z. and Liu, Y. (2001) Reliable and global measurement of fluorescence resonance energy transfer using fluorescence microscopes. *Biophys. J.* **81,** 2395–2402.

24. Billinton, N. and Knight, A. W. (2001) Seeing the wood through the trees: a review of techniques for distinguishing green fluorescent protein from endogenous autofluorescence. *Anal. Biochem.* **291,** 175–197.

23

Use of Multiphoton Imaging
for Studying Cell Migration in the Mouse

Andrea Flesken-Nikitin, Rebecca M. Williams,
Warren R. Zipfel, Watt W. Webb, and Alexander Yu. Nikitin

Summary

We describe a method for studying cell motility in the living mouse using multiphoton microscopy. The procedure consists of mouse anesthesia, labeling of target cells with enhanced green fluorescent protein by infection with recombinant adenovirus, implantation of beads carrying chemoattractant, preparation of the mouse for imaging, and imaging of individual cell motions via multiphoton microscopy. Two-photon fluorescence excitation of enhanced green fluorescent protein allows visualization of cells within the dermis, whereas second harmonic generation (a non-linear scattering process) allows a simultaneous detailed definition of the dermis structure.

Key Words: Cell motility; live imaging; multiphoton microscopy; mouse; second harmonic generation; stromal cells; intradermal tissues.

1. Introduction

The recent deciphering of human and mouse genomes, together with technical advances in mouse genetic engineering, have resulted in the continuously increasing use of mice for modeling of human diseases, including cancer, neurological, cardiovascular and immune disorders. Since individual cell tracing provides important clues for understanding pathogenesis, the ability to detect single cells and monitor their behavior in the living mouse is particularly important. Multiphoton microscopy (MPM; **refs.** *1*,*2*), a type of laser scanning microscopy, has become a preferred fluorescence imaging technique for in vivo studies owing to its ability to image deeply into living tissue and its absence of out-of-focal plane excitation *(3–8)*. A type of MPM, two-photon microscopy

From: *Methods in Molecular Biology, vol. 294: Cell Migration: Developmental Methods and Protocols*
Edited by: J-L. Guan © Humana Press Inc., Totowa, NJ

uses the simultaneous absorption of two infrared photons to excite an electronic transition equivalent to the absorption of a single higher energy (bluer) photon. In the absence of endogenously added (or transfected) fluorophores, MPM can also be used to image morphology via detection of intrinsic tissue emissions of such compounds as NADH and retinoids *(9)*. However, genetically encoded green fluorescent proteins are significantly brighter *(10)*, and, if incorporated into the system of interest, they enable imaging with lower power and less biologically damaging wavelengths *(9)*. Using the same apparatus, second harmonic generation (SHG), a non-linear scattering emission, can be collected from fibrillar collagen to obtain simultaneous images of structural features within many types of tissue. Here we describe how to use MPM for non-invasive imaging of the derma and its motile cells at subcellular resolution. The procedure consists of mouse anesthesia, labeling of target cells with enhanced green fluorescent protein (EGFP) by infection with recombinant adenovirus, implantation of beads carrying chemoattractant, preparation of the mouse for imaging, and imaging of individual cells with multiphoton microscopy.

2. Materials

2.1. Mouse Anesthesia

1. Medical oxygen tank with pressure sensor (Airgas, Radnor, PA).
2. Oxygen flow regulator (Airgas).
3. Isoflurane vaporizer (Harvard Apparatus, Holliston, MA).
4. Isoflurane (Halocarbon Laboratories, River Edge, NJ).
5. Tubing manifold for directing the gas to either the induction chamber or the imaging platform (Harvard Apparatus).
6. Port to building exhaust.
7. Tygon tubing (5/16-in. inner diameter).
8. Anesthesia induction chamber (Harvard Apparatus).
9. Small animal heating pad (Fine Science Tools, Foster City, CA).

2.2. Labeling Target Cells With EGFP

1. FVB/N mice (Taconic, Germantown, NY).
2. Recombinant adenovirus Ad5*CMVEGFP*, purified and tittered at 10^{11}–10^{12} infectious particles/ml as described (**ref. *11***; Gene Transfer Vector Core; University of Iowa College of Medicine, Iowa City, IA). Albeit this virus is replication deficient, all NIH Recombinant DNA Guidelines must be closely followed.
3. Microscope Nikon SMZ 645 (Nikon, Melville, NY).
4. Fiber optic illuminator ACE (Schott-Fostec LLC, Auburn, NY).
5. Pet trimmer shaver (Wahl Clipper Corporation, Sterling, IL).
6. Isotonic saline (0.85% NaCl) sterile.
7. 70% Ethanol.
8. Tissue marking dye (Triangle Biomedical Sciences, Durham, NC).

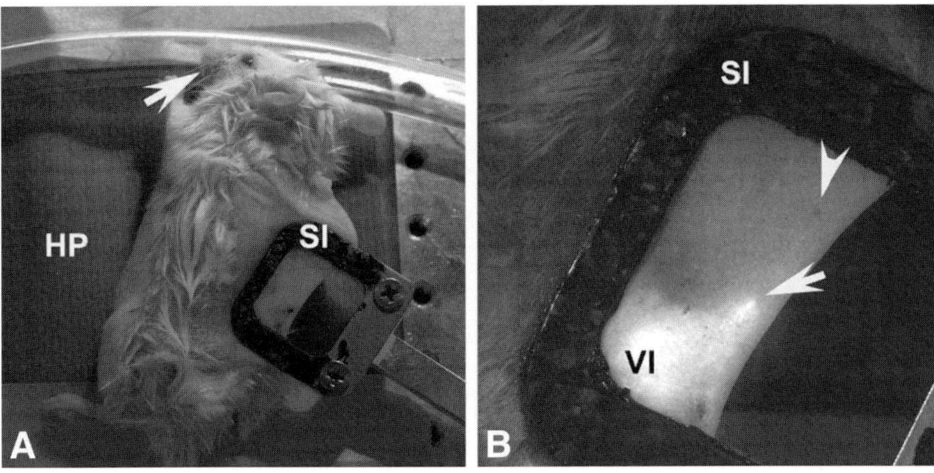

Fig. 1. Preparing mouse for imaging. (**A**), Mouse is placed on imaging platform consisting of an anesthesia line (arrow), a heating pad (HP), and a skin immobilizer (SI). (**B**), Wide-field fluorescence image of the EGFP distribution in skin. Note that leading edge of infected cells (arrow) is directed toward the location of heparin beads containing MCP1 (arrowhead). VI, place of Ad5*CMVEGFP* injection; SI, skin immobilizer. **A** and **B** are taken with white light and fluorescence microscopies, respectively.

   9. Hypodermic needles 21 and 30 gage.
 10. Sterile disposable syringes; 1 mL.
 11. Micro dissecting forceps serrated 3.5-in. long, 0.6-mm tip (Roboz, Rockville, MD).
 12. Pipetman for 1–20 µL, 10–200 µL, and 100–1000 µL.
 13. Sterile filter tips 1–20 µL and 10–200 µL.

2.3. Implanting Beads With Chemoattractant

1. Monocyte chemotactic protein 1 (MCP1), recombinant, mouse (derived from *Escherichia coli*), rmJE/MCP-1 (R&D Systems, Minneapolis, MN).
2. Heparin-acrylic beads (Sigma, St. Louis, MO).
3. PBS, Ca^{2+}/Mg^{2+}-free phosphate-buffered saline, sterile (Cellgro, Herndon, VA).

2.4. Preparation of Mouse for Imaging

1. Pet trimmer shaver (Wahl Clipper Corporation, Sterling, IL).
2. Nair® lotion hair remover (Carter Products, New York, NY).
3. Skin immobilizer (custom built; *see* **Fig. 1**).

2.5. Multiphoton Microscopy

1. Ti:Sapphire laser (Spectra Physics, Mountain View, CA). The Ti:Sapphire laser is a Class 4 laser. A specular reflection directed into the eye could result in retina

damage. A dark object in the beam path could result in a fire hazard. Appropriate cautions should be taken. The beam should be covered and stray reflections blocked from imaging personal.
2. Laser spectral analyzer (IST/Rees, Horseheads, NY).
3. Pockels cell for beam modulation (ConOptics, Danbury, CT).
4. Beam scanner with optics modified for IR transmission and associated image acquisition electronics (Bio-Rad, Hercules, CA).
5. Microscope with laser port (Olympus, Melville, NY).
6. Non-descanned fluorescence and SHG detection unit integrated with beam scanner electronics (Bio-Rad).
7. Appropriately selected microscope objective (Olympus).

3. Methods

The subsequent methods outline 1) mouse anesthesia, 2) labeling target cells with EGFP, 3) implanting beads with chemoattractant, 4) preparing mouse for imaging, and 5) multiphoton microscopy.

3.1. Mouse Anesthesia

Gas anesthesia (**Fig. 2**) enables the extended experimental durations that are necessary for these experiments.
1. Set the flow rate to 1 L/min.
2. Anesthetize the mouse in an induction chamber at 3.5% Isoflurane for several min (*see* **Note 1**).
3. Transfer the mouse to the heated imaging/surgery platform and maintain them at approx 1.5% Isoflurane. Anesthesia masks small enough to fit under the microscope are not yet commercially available. Holes cut in the anesthesia tubing are an operable solution (**Fig. 1A**).

3.2. Labeling Target Cells With EGFP

Connective tissue cells are labeled by infection with recombinant adenovirus Ad5*CMVEGFP*. Ad5*CMVEGFP* is a modification of the adenovirus-5 genome, from which the *e1a* and *e1b* regions required for viral replication have been deleted and replaced with *EGFP* encoding sequence driven by the CMV immediate early regulatory sequence (**ref.** *11*; Gene Transfer Vector Core, University of Iowa College of Medicine).
1. Freeze Ad5*CMVEGFP* in small aliquots and store at −80°C.
2. Thaw frozen Ad5*CMVEGFP* aliquots, dilute in isotonic saline to 5×10^8 pfu/mL and keep on ice until needed (*see* **Note 2**).
3. Anesthetize 2-mo-old FVB/N mice with Isoflurane (*see* **Subheading 3.1.**), place them on the heating pad, and shave on the left and right side of the upper and lower back.
4. Transfer the mice to a stereomicroscope with fiber optic illumination and position for the virus injection.

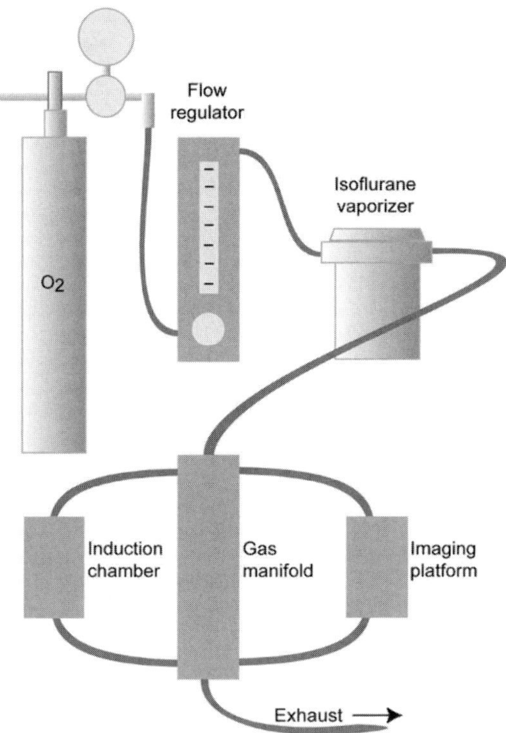

Fig. 2. Diagram of anesthesia apparatus. Oxygen flow from a pressurized cylinder is directed through an Isoflurane vaporizer. A gas manifold directs the gas through either an induction chamber or an imaging/surgery platform. The output of the system is funneled directly to the building exhaust.

5. Inject 50 µL of virus with a 30-gage hypodermic intradermally in both lower dorsal quadrants.
6. After injection, mark the areas of the visible virus bubble with four blue tissue dye dots (*see* **Note 2**).

3.3. Implanting Beads With Chemoattractant

MCP1 is a chemoattractant known to induce positive chemotaxis of mononuclear phagocytic cells *(12–14)*. It is used to direct positive motility of macrophages infected with Ad5*CMVEGFP*.

1. Reconstitute a lyophilized sample of MCP1 in sterile PBS containing 0.1% bovine serum albumin to a concentration of 100 µg/mL (*see* **Note 3**).
2. Suspend heparin beads in storage solution evenly by flipping the tube and add 90 µL of suspension to 1 µg/10 µL MCP1 in a 1.7-mL sterile micro centrifuge tube.

3. After incubation for 1 h at room temperature, centrifuge the beads for 5 min at 600*g*, take out the supernatant, add 1 mL of sterile PBS, suspend gently, and centrifuge as before.
4. After the washing, suspend the beads in sterile PBS to a final concentration of 10 µg/mL MCP1 and store at room temperature until injection.
5. Incubate and wash control beads in PBS as described in **step 3**.
6. Immediately after the virus administration, inject 100 µL of MCP1 chemoattractant releasing beads intradermally with a 21-gage hypodermic needle one centimeter above the area injected with adenovirus.
7. Control beads are injected in the opposite dorsal quadrant above the area injected with adenovirus.

3.4. Preparing Mouse for Imaging

1. Remove hair in region of interest with a small animal trimmer.
2. Apply Nair® lotion hair remover for 4 min (*see* **Note 4**).
3. Wash area with warm water and finish with 70% ethanol to remove optical interference caused by mineral oil component of the Nair®.
4. General motion of the skin as the result of breathing would make impossible the task of following individual cells with micron resolution. Thus, during imaging, the skin must be stretched into a skin immobilizer (**Fig. 1A**) to isolate the imaging area from the general body motion.
5. At this point the generalized distribution of fluorescent cells can be imaged using a low NA objective and fluorescent illumination (**Fig. 1B**).

3.5. Multiphoton Microscopy

Several commercial multiphoton microscopes now exist (such as the Bio-Rad Radiance 2100 MP, the Zeiss 510 NLO, and the Leica TCS MP) and should be capable of imaging into the dermis. Our setup consists of a Ti:Sapphire laser (Milenia/Tsunami combination, Spectra Physics), Bio-Rad 600 laser scanner and modified Olympus AX-70 upright microscope (**Fig. 3**).

The excitation light is focused into the mouse dermis with a large-barrel Olympus 20X/0.95NA water objective, which provides a large field-of-view, a relatively high NA, good IR transmission, and the few mm of working distance.

The multiphoton excitation wavelength peak (λex) is tuned to 900 nm because EGFP absorbs well at this wavelength *(10)*, whereas water and intrinsic cellular absorbers do not *(9)*. The EGFP emission spectral profile is independent of the excitation wavelength; it peaks at 515 nm with a spectral width of approx 40 nm. The collagen SHG emission wavelength, however, will tune with the excitation wavelength; it peaks at λex/2 (and possesses a spectral width equal to that) of the exciting beam reduced by $\sqrt{2}$ *(9)*. Non-linear emissions are collected in epi mode and immediately separated from the excitation beam directly after the objective with a 670DCXXRU long-pass dichroic filter. Emission filters are cho-

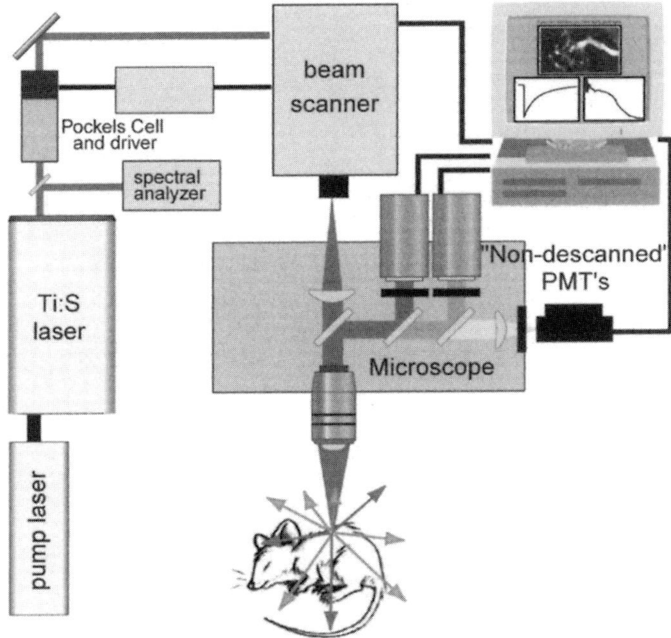

Fig. 3. Multiphoton microscope. Ti:Sapphire laser (Milenia/Tsunami combination, Spectra Physics) is directed into a Bio-Rad 600 beam scanner interfaced with a modified Olympus AX-70 upright microscope. A Conoptics 350–80 BKLA Pockel's Cell provides beam intensity modulation and blanking during scanner flyback when data is not being collected. A frequency spread in the output beam is necessary for supporting mode-locked (pulsed) operation and is monitored by a laser spectral analyzer (IST/Rees E201). The excitation light is focused into the mouse dermis with a large-barrel Olympus 20X/0.95NA water objective, which provides a large field-of-view, a relatively high NA, good IR transmission, and a few mm of working distance, which is usually necessary for maneuvering in live animals. The non-linear emissions are not directed back through the scanning mirrors like they are in confocal microscopy because background rejection is unnecessary. The emission beam is collected in epi mode, spectrally separated from the excitation beam and immediately detected within the microscope housing (termed "non-descanned" detection).

sen for a clean blue, green (450 and 515 nm) separation (BGG22 and 580/150 filters with a separating 500DCXR dichroic filter, Chroma Technology) and a 10^7 rejection ratio of the exciting to emitting wavelengths. The resulting two emission beams are collected by Hamamatsu HC125-02 bialkali photomultiplier tube assemblies (*see* **Note 5**).

1. Tune laser wavelength to 900 nm.
2. Peak laser power.
3. Tune laser prisms for a stable, Gaussian-like frequency spread.
4. Image test slide to ensure system is operating properly.
5. Image mouse skin. One of the benefits of multiphoton microscopy is its ability to collect clean optical sections from live tissue. The thickness (full-width at half maximum) of each section is given by *(15)*:

$$\Delta z = \frac{0.31\,\lambda\,\text{ex}}{n\,\sin^2\,(\theta/2)}$$

where $n\sin\theta$ is the NA of microscope objective and n is the refractive index of the immersion fluid. For the 20X/0.95NA water objective used in our experiments, this thickness is 1.4 µm. To follow cells, one can either acquire a time series at one plane or a time series of stacks at multiple planes. An example of time series at 8 min intervals shows cell movement with an average speed of 1.4 µm/min (**Fig. 4**).

4. Notes

1. The amount of sufficient Isoflurane varies from mouse to mouse and especially with age. Breathing rate is the easiest way to monitor the health of the mouse. Higher Isoflurane levels correspond to slower breathing rates. We try to maintain the mouse at an "ideal" breathing rate of about 0.3 Hz, or one to two breaths per image. A "typical" 8 h-day of surgery and imaging will consume about one "medical-sized" oxygen tank and about one 200-mL bottle of Isoflurane. Heating is critical to long-term maintenance of the mouse under anesthesia. We use a Fine Science Tools heating pad calibrated to a surface temperature (without the mouse) of 30–35°C. While under anesthesia, the mouse is unable to regulate its own temperature well. Too much or too little heat can result in the death of the patient.
2. All aliquots are only thawed once and the rest is discarded after use. The use of sterile filter tips to aliquot and dilute the virus is recommended. All plasticware and instruments that had contact with the virus are soaked in 70% ethanol for decontamination. To perform the intradermal injection, a skin fold is held up with a pair of micro dissection forceps. After penetrating the skin with a 30-gage needle, the forceps are loosened and the fluid is allowed to evenly disperse. The needle is slowly withdrawn so that no liquid is able to leak out.
3. Upon reconstitution MCP1 can be stored at –20°C no more than 3 mo. *See also* **Note 2** regarding intradermal injection.
4. For repeated imaging, the Nair® hair removal solution must be totally removed because it is quite caustic (pH > 11.0), and thus expected to be toxic to the mice when ingested during grooming.
5. Of utmost importance in imaging live specimens are the efficiency and placement of the emission detectors. Because non-linear emissions are only excited

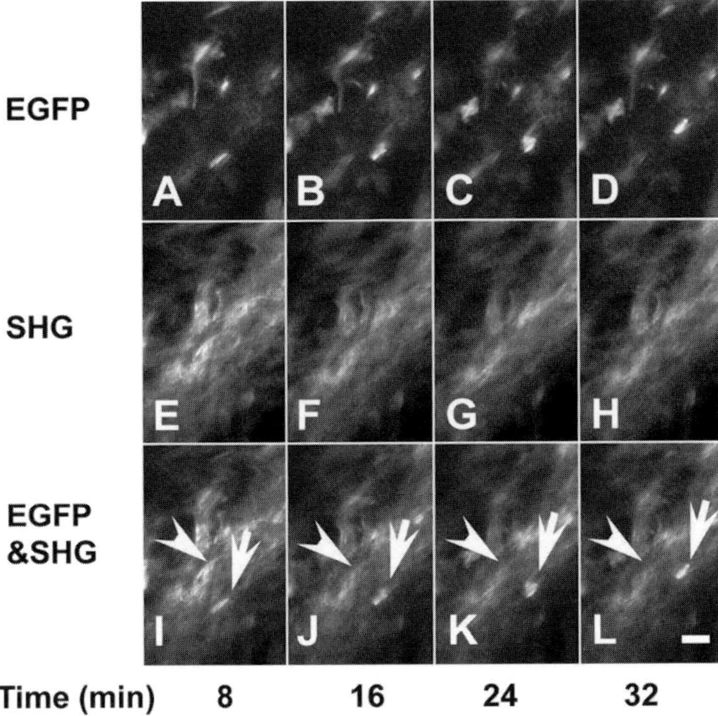

Fig. 4. Multiphoton images of cell migration. Detection of fluorescence of EGFP-expressing cells (EGFP, **A–D**) and SHG-producing collagen (SHG, **E-H**) in the dermis. **I–L,** Photon emissions from both channels added together in order to show cell motion with respect to the stationary dermal structure. Images are acquired each 30 s, and displayed at 8-min intervals. Note that EGFP-labeled cell (arrows) moves in relation to stationary collagen fibers (arrowheads). Calibration bar, 20 μm.

in the focal plane, they need not be descanned and focused through a confocal pinhole. Removing extraneous optics from the detection pathway and locating detectors close to the collecting objective increases the detector efficiency and enables the additional collection of photons deflected by the highly scattering skin on the way out.

Imaging a constant test slide, consisting of a fixed biological specimen or a fluorescent polymer or beads, before every microscopy session is an extremely valuable practice. Though the multiphoton microscope is a relatively complicated instrument, it should return consistent imaging results from day to day. When imaging highly variable specimens such as live mice, one should be sure that the instrument itself is not providing any of the experimental variability.

One irritating problem encountered in these experiments is that the immersion saline often wicks around the mouse and away from the objective, resulting in the loss of the image. A thin layer of bathroom caulk over the skin immobilizer eliminates this problem. Another difficulty encountered in the acquisition of time series images is focal plane drift over time. This problem is best solved by identifying and alleviating any slight temperature variations around the microscope or specimen. Because skin pigmentation significantly impairs photon penetration, all experiments should be performed on white albino mice, such as FVB/N.

Acknowledgments

This work was supported by NIH grants R01 CA96823 (to A. Y. N.), R33 CA094311 (to W. R. Z.) and 9-P41-EB0011976-16 (to W. W. W.). A. Y. N. is a recipient of the NCRR, NIH Midcareer Award in Mouse Pathobiology (RR017595).

References

1. Denk, W., Strickler, J. H., and Webb, W. W. (1990) Two-photon laser scanning fluorescence microscopy. *Science* **248,** 73–76.
2. Williams, R. M., Zipfel, W. R., and Webb, W. W. (2001) Multiphoton microscopy in biological research. *Curr. Opin. Chem. Biol.* **5,** 603–608.
3. Christie, R. H., Bacskai, B. J., Zipfel, W. R., Williams, R. M., Kajdasz, S. T., Webb, W. W., et al. (2001) Growth arrest of individual senile plaques in a model of Alzheimer's disease observed by in vivo multiphoton microscopy. *J. Neurosci.* **21,** 858–864.
4. Wang, W., Wyckoff, J. B., Frohlich, V. C., Oleynikov, Y., Huttelmaier, S., Zavadil, J., et al. (2002) Single cell behavior in metastatic primary mammary tumors correlated with gene expression patterns revealed by molecular profiling. *Cancer Res.* **62,** 6278–6288.
5. Brown, E. B., Campbell, R. B., Tsuzuki, Y., Xu, L., Carmeliet, P., Fukumura, D., et al. (2001) In vivo measurement of gene expression, angiogenesis and physiological function in tumors using multiphoton laser scanning microscopy. *Nat. Med.* **7,** 864–868.
6. Lendvai, B., Stern, E. A., Chen, B., and Svoboda, K. (2000) Experience-dependent plasticity of dendritic spines in the developing rat barrel cortex in vivo. *Nature* **404,** 876–881.
7. Svoboda, K., Denk, W., Kleinfeld, D., and Tank, D. W. (1997) In vivo dendritic calcium dynamics in neocortical pyramidal neurons. *Nature* **385,** 161–165.
8. Charpak, S., Mertz, J., Beaurepaire, E., Moreaux, L., and Delaney, K. (2001) Odor-evoked calcium signals in dendrites of rat mitral cells. *Proc. Natl. Acad. Sci. USA* **98,** 1230–1234.
9. Zipfel, W. R., Williams, R. M., Christie, R., Nikitin, A. Y., Hyman, B. T., and Webb, W. W. (2003) Live tissue intrinsic emission microscopy using multi-

photon-excited native fluorescence and second harmonic generation. *Proc. Natl. Acad. Sci. USA* **100,** 7075–7080.

10. Xu, C., Zipfel, W., Shear, J. B., Williams, R. M., and Webb, W. W. (1996) Multiphoton fluorescence excitation: New spectral windows for biological nonlinear microscopy. *Proc. Natl. Acad. Sci. USA* **93,** 10,763–10,768.

11. Anderson, R. D., Haskell, R. E., Xia, H., Roessler, B. J., and Davidson, B. L. (2000) A simple method for the rapid generation of recombinant adenovirus vectors. *Gene Ther.* **7,** 1034–1038.

12. Rollins, B. J., Morrison, E. D., and Stiles, C. D. (1988) Cloning and expression of JE, a gene inducible by platelet-derived growth factor and whose product has cytokine-like properties. *Proc. Natl. Acad. Sci. USA* **85,** 3738–3742.

13. Gu, L., Tseng, S. C., and Rollins, B. J. (1999) Monocyte chemoattractant protein-1. *Chem. Immunol.* **72,** 7–29.

14. Luini, W., Sozzani, S., Van Damme, J., and Mantovani, A. (1994) Species-specificity of monocyte chemotactic protein-1 and -3. *Cytokine* **6,** 28–31.

15. Williams, R. M., Piston, D.W., and Webb, W.W. (1994) Two-photon molecular excitation provides intrinsic 3-dimensional resolution for laser-based microscopy and microphotochemistry. *FASEB J.* **8,** 804–813.

24

Using Microfluidic Channel Networks
to Generate Gradients for Studying Cell Migration

Daniel S. Rhoads, Sharvari M. Nadkarni, Loling Song,
Camilla Voeltz, Eberhard Bodenschatz, and Jun-Lin Guan

Summary

In this chapter, we will discuss a method for the generation of gradients that can be
quantitatively used for studying directional cell migration. Microfluidic networks, which
serially split and remix small volumes of solutions under laminar flow conditions to
generate a series of microchannels of increasing protein concentration. At a juncture of
these microchannels, where a single broad channel is formed, a protein concentration
gradient can be easily achieved. This method is highly useful because of the ability with
which we can control, manipulate and analyze chemical gradients and cells' chemotactic
behavior in a quantitative manner.

Key Words: Gradient; microfluidics; migration; chemoattractant; microenvironment;
nanobiotechnology.

1. Introduction

The merger of physics, engineering, and biological disciplines into what is
currently termed nanobiotechnology has allowed for the development of a number of novel techniques that can be used to study biological activities on very
small scales. This area of research has sprung from an outgrowth of the micro-
processor industry, which for years has sought to fabricate smaller and smaller
structures using photolithography. Photolithography uses epoxy resins that when
illuminated by specific wavelengths of light cross-link to form insoluble micro-
structures, to pattern a series of physical features onto a silicon substrate (1). In
patterning microfluidic networks, a negative acting epoxy resin with broadband
sensitivity in the near-UV called SU-8 is the standard resist, which is ideal for
thick features having high aspect ratios (2–500 µm high off the substrate). The

From: *Methods in Molecular Biology, vol. 294: Cell Migration: Developmental Methods and Protocols*
Edited by: J-L. Guan © Humana Press Inc., Totowa, NJ

raised features formed by the patterned SU-8 then act as a negative mold for rapid prototyping, a process which uses a silicone elastomer, poly(dimethylsiloxane) (PDMS), to form an optically transparent network of microchannels. Bonding the PDMS to a glass slide then seals these microchannels *(2)*.

The microfluidic device is then used to make a biochemical gradient. Generation of the gradient is based upon the controlled diffusive mixing of laminar flow fluids by repeated splitting, mixing, and recombination of fluid streams within the network of microchannels. This gradient can be composed of virtually any biologically relevant molecule, either in solution or deposited on a surface depending upon the characteristics of the molecule in question. Here, in studying chemotactic cell migration we can generate a soluble gradient of growth factors or chemokines, or in studying haptotaxis, deposit a surface gradient of extracellular matrix (ECM) proteins.

The gradients of these and other classes of molecules play important roles in a variety of biological processes such as morphogenesis, angiogenesis, axon pathfinding, immunological response, and tumor metastasis, to name a few. Although other techniques are available for studying cell migration, such as the Boyden chamber or micropipets to release diffusible molecules, few are capable of generating and controlling spatially and temporally well-defined gradients and allowing data analysis in a quantitative manner. These characteristics make migration studies on concentration gradients a useful tool in studying dynamic processes that enable cells to sense and react to chemical changes in their local environment.

A variety of methods have been described for generating concentration gradients of biomolecules, including diffusion of alkanethiols through porous matrices *(3,4)*, photochemical activation of self-assembled monolayers (SAMs) for peptide coupling *(5,6)*, electrochemical desorption of SAMs *(7)*, depletion of protein inside microfluidic channels by adsorption *(8)*, and the use of pipets for forming transient gradients in solution *(9–12)*, in addition to the use of pyramidal microfluidic networks *(13,14)*. Although each of these can be useful in particular studies, pyramidal networks offer quantifiable gradients that make possible the quantitative study of cellular responses under a variety of spatially and temporally tailored chemical gradients.

2. Materials

1. CAD software, such as L-edit or Symbad pattern editor.
2. PDMS: Sylgard® 184 silicone elastomer kit (Dow Corning).
3. Nano™ SU-8 50 (MicroChem Corp.).
4. SU-8 developer (MicroChem Corp.).
5. Silicon wafer(s).
6. Resist spinner and hotplates.

7. Optical pattern generator (e.g., GCA PG3600F).
8. Resist-coated chrome mask.
9. Gastight® 1750TLL 500-µL syringes with a 3.26-mm inner diameter (cat. no. 81220; Hamilton Company).
10. Teflon tubing with hubs (24 gage; cat. no. 90624; Hamilton Company).
11. PHD2000 dual-syringe pump (cat. no. 70-2000; Harvard Apparatus).
12. Plasma cleaner.

3. Methods

The methods described below outlines the design of a microfluidic network, microfabrication of a device in PDMS, and the generation of a concentration gradient for use in quantitative analysis of directional cell responses.

3.1. Microfluidic Network Design

A number of considerations must be taken into account in designing a microfluidic network for the purposes of generating a gradient, and these are described as follows:

1. As shown in **Fig. 1**, the microfluidic network is an array of microchannels arranged in a pyramid-like structure. This design serially splits and recombines two fluids of different concentrations to give a continuous linear concentration gradient, and a useful means of arranging these mixing microchannels in a serpentine pattern, to conserve space and maximize diffusive mixing. The necessary lengths of serpentine mixing channels for complete diffusive mixing before branching and re-mixing for each level of the network are given by the formula:

   Mixing length = (flow rate) × [(channel width)^2/(diffusion constant)], where

   $$\text{diffusion constant} = \sqrt{(k^3/(m^3)} \times [T^{3/2}/ (a^2 P)].$$

   T, Temperature (°K); P, Pressure; k, Boltzmann constant; m, mass (molecular, of dissolving substance); a, Stokes diameter.

2. Microchannel lengths may be varied in successive levels in the pyramid, since flow velocity decreases, requiring less distance for complete diffusive mixing to occur. Also, in determining mixing microchannel lengths, a rule of thumb is to aim for flow rates of between 0.1 and 1.0 mm/s in the main channel. Appropriate network design and flow rate conditions are described more thoroughly by N. L. Jeon and S. K. W. Dertinger *(13,14)*. In addition, diffusive mixing may be facilitated by other methods *(15)* (*see* **Note 1**).

3. After an arbitrary number of levels in the pyramidal network (e.g., **ref. 8**), the microchannels must be brought back together to generate the gradient in a broad channel. For generation of simple linear gradients in the broad channel, widths of 0.5 to 1.0 mm allow for a sufficient cross-sectional distance for gradient formation and establishing convenient fields of view for observing cell migration by standard microscopy techniques using a 10X objective (*see* **Note 2**).

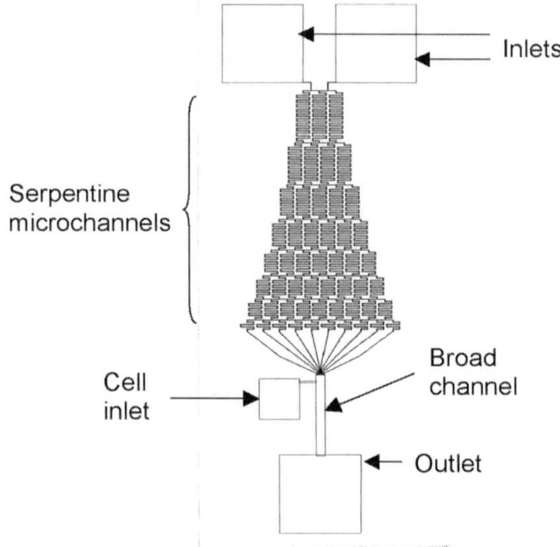

Fig. 1. CAD printout of a pyramidal microfluidic network design. Important features include the serpentine microchannels that maximize channel length while keeping network size compact, a broad channel for formation of the desired gradient, and inlet/outlet ports. The port to the left of the broad channel may be found useful for introducing cells for migration experiments under many conditions, especially for soluble gradients that require continuous flow during the experiment. Microchannels are 50-μm wide.

4. A final consideration for microchannel design is the fluid access ports; inlets and outlets for materials to flow into and out of the network. Two or more inlets are necessary for passage of buffered solute(s) to generate the gradient, and an inlet for introduction of cells is useful, as is an outlet connected to the opposite end of the gradient-forming broad channel of the network (*see* **Note 3**).

3.2. Device Microfabrication

Making the microfluidic devices involves a series of steps, including utilizing photolithography, and PDMS molding (*see* **Fig. 2**).

1. Microfluidic networks can easily be designed using computer-assisted design (CAD) software, such as L-Edit or SYMBAD/PED.
2. CAD patterns must be transferred to a mask for fabrication of silicon wafer masters. This is done with an optical pattern generator, such as a GCA/Mann 3600F PG, that exposes variably sized rotated rectangles onto the mask blank. The mask is a glass plate coated with a layer of about 80 nm of sputtered Cr that is coated

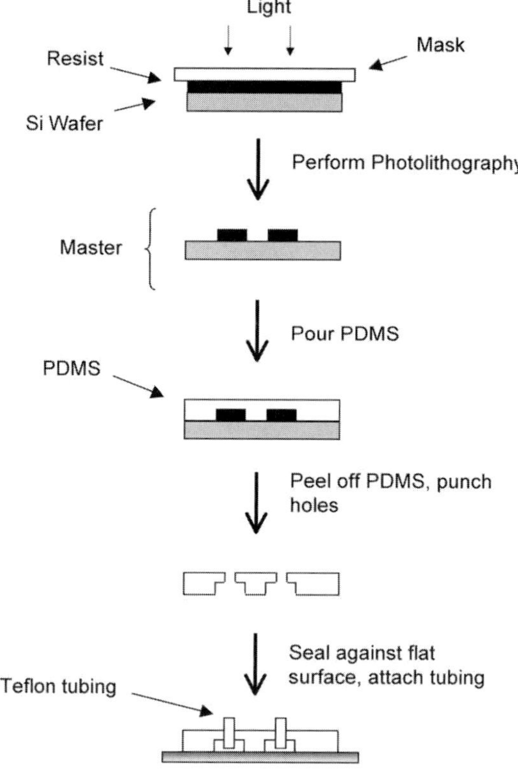

Fig. 2. From wafer to microfluidic device. A chrome mask etched with a microfluidic network pattern is exposed through to crosslink SU-8 resist, transferring the pattern by photolithography. After development, the wafer retains the SU-8 pattern and can function as a master for molding PDMS stamps. PDMS is poured over the pattern and cured, then peeled off, and holes are punched to provide access to fluid ports. The PDMS stamp is then sealed to a flat surface and tubing is attached via punched holes.

overtop with a photoresist. After exposure and development, the Cr is removed from the unprotected areas with an acid etch, and an image of the pattern is left in the Cr. The resulting mask is used to limit crosslinking of a negative photoresist (SU-8 50; MicroChem Corp.), which is spin-coated onto silicon wafers to create masters (**refs. *2,16*; *see* Note 4**).

3. Spin-coating of silicon wafers with SU-8 50 photoresist should be performed according to MicroChem Corp. specifications. However, a guideline for coating SU-8 50 (and therefore channels) 100 µm deep on 4-inch wafers is to spin for 30 s at 1500 rpm (*see* **Note 5**).

4. After coating, network patterns are crosslinked into the SU-8 photoresist by UV illumination through mask transparencies. Uncross-linked photoresist is removed by treatment with SU-8 Developer (MicroChem Corp.).
5. The raised pattern of SU-8 photoresist on the silicon wafer serves as a mold to imprint the microfluidic network into a silicone elastomer, PDMS. PDMS has a variety of characteristics that make it an excellent material for microfluidics, including durability, biological inertness, flexibility, optical transparency and the capability for strongly induced adhesion to glass and other PDMS-coated surfaces.
6. PDMS is poured over the patterned wafer in mixture with its curing agent in a 10:1 ratio, and air bubbles in the mixture are removed by applying a vacuum, and baked at 70°C for 60 min. Enough PDMS should be used to cover the wafer at least 5-mm deep. In this stage, the removal of air bubbles can also be done in open air, but this takes longer: the use of a vacuum should be observed closely as the expanding bubbles can cause the mixture to boil over (*see* **Note 6**).
7. The solid PDMS can now be cut into cover slip-sized pieces, each containing a patterned network, and peeled off the silicon wafer. In preparing each stamp for assembly into a microfluidic device, holes providing connection to the access ports must be punched through the PDMS, using a 19-gage needle or capillary, for instance (*see* **Note 7**).
8. Assemble each microfluidic device, by bonding the PDMS replica to either a glass cover slip or flat PDMS. This is done by plasma cleaning of the silicon-based materials *(2,16)*.
9. Within about half an hour of assembling the device, it is good to immerse and store the PDMS stamp in buffered water (A. Stroock, personal communication), so as to avoid drying of the PDMS, which easily becomes extremely hydrophobic (*see* **Note 8**).
10. Connections between the access ports in the PDMS pattern and tubing that carries fluid into the network can vary in form, but in this case, fewer inter-connections are preferred. Holes can be punched through the PDMS into the access ports using a 19-gage needle or capillary, and then inserting a 24-gage piece of Teflon tubing, which forms a tight connection which can be sealed with a small amount of epoxy.

3.3. Gradient Generation

Important considerations in generating the gradient include proper control of the syringe pump, calculating the appropriate pump rate to generate desired flow velocity, state of the fluids used with the network, and duration of the pump action.

1. A recommended setup for syringes and pumps to control flow rates includes using two 1750TLL series 500 µL of Gastight® syringes with hub-connected Teflon tubing from the Hamilton Company, and a PHD2000 dual syringe pump from Harvard Apparatus, shown in **Fig. 3** (*see* **Notes 9** and **10**).

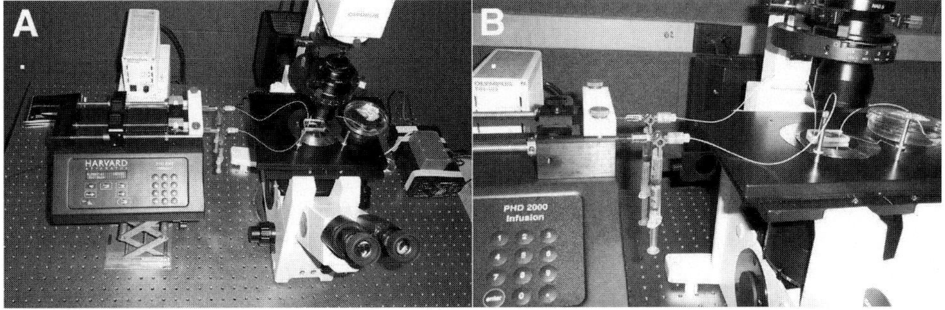

Fig. 3. Microfluidics experimental setup. **(A)**, View of entire microfluidic setup with syringe pump, connected syringes, and PDMS stamp on microscope. **(B)**, Closer image of syringes, tubing and PDMS stamp.

2. With the fluid mixtures prepared (i.e., 0% and 100% mixtures of your molecule of interest for a gradient) and in the syringes, the motor must be run at a given rate for an established length of time. The length of time depends upon the conditions of the experiment and the nature of the molecule. For soluble gradients, this must be for the entire duration of the migration experiment *(18)*, while for molecule deposition of insoluble gradients, it varies with the adsorption properties of the molecule, and should be countered with an inactive deposited molecule, for example, serum albumin or poly-L-lysine (PLL) *(19)*.

3. Desired syringe pump rate is established by two criteria: the resulting flow rate must be slow enough to allow time for diffusive mixing within the microchannels, but fast enough to limit diffusion for a large portion of the broad channel and keep a constant gradient profile for cell migration experiments to take place. For both criteria, the diffusion rate of your solute is critical, and is based upon the same formula described in **Subheading 3.1.1.** Of the mentioned criteria, the requirement for diffusive mixing within the microchannels is the limiting variable determining flow rate, and thus syringe pump rate. This can be exceeded slightly if necessary to achieve some degree of constancy for the gradient profile in the broad channel (*see* **Note 11**).

4. Detection and analysis of the resulting gradient can be done by fluorescence microscopy (*see* **Fig. 4**). In the case of soluble gradients, the molecule of which the gradient is being generated may be tagged previously with a fluorescent probe such as fluorescein isothiocyanate (FITC), or a separate fluorescent probe, such as dextran-conjugated FITC may be added to the original 100% mixture of your gradient molecule. For deposited gradients, again, the molecule of interest may be tagged previously with a fluorescent probe or an immunofluorescence assay of the gradient may be performed.

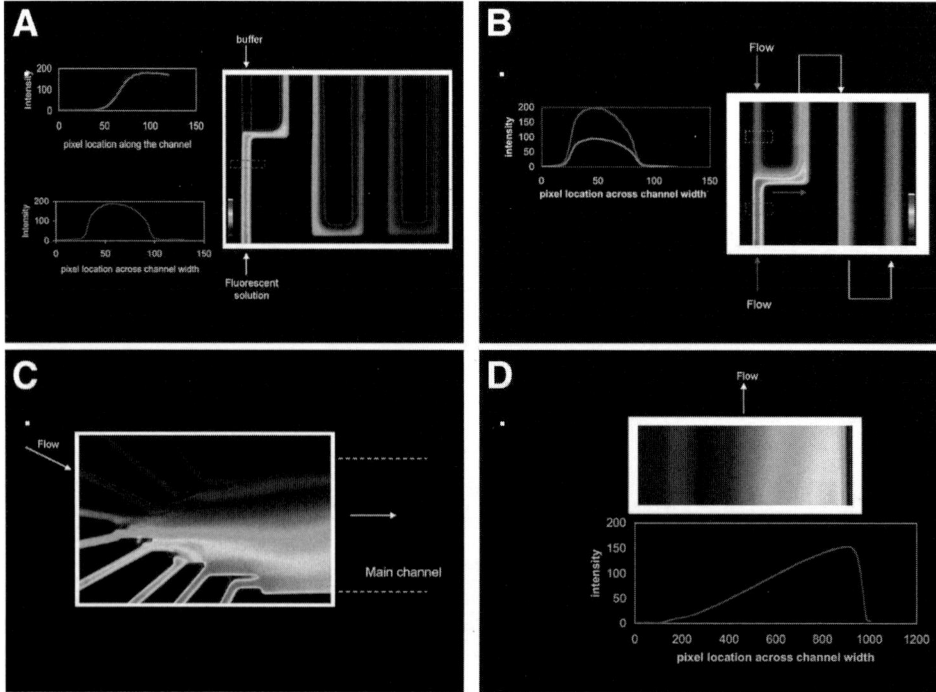

Fig. 4. Diffusive mixing and gradient generation with a pyramidal microfluidic network. **(A,B)**, Colorimetric images of diffusive mixing between 0% and 100%, and 25% and 75%, concentrations of Dextran–FITC within a serpentine microchannel, respectively. **(C,D)**, Colorimetric images of gradient formation as the microchannels rejoin into the broad channel, and an intensity profile analysis of the resulting gradient, respectively. Arrows indicate direction of flow within the networks.

4. Notes

1. An alternative to using serpentine microchannels in a network pattern, which provide enough time for diffusive mixing to occur, is to incorporate Staggered Herringbone Mixers (SHM), which facilitate chaotic mixing in laminar flow conditions *(14)*.

2. View of the entire range of the gradient may not be necessary, and indeed, it may be useful to focus in on small regions or individual cells during a migration assay. However, for cases where view of the entire gradient at a higher magnification is required, the broad channel may easily be scaled down, with fewer levels of mixing microchannels, and thus fewer microchannels to recombine into the broad channel.

3. In our experience, the cell inlet and the outlet must be comparable in size to the area of the broad channel, or many of the introduced cells will settle and attach in

these reservoirs and not in the broad channel where your biochemical gradient is generated. The size of the two top inlets is of lesser importance, as this does not interfere with cellular behavior. Also, we have found that making reservoirs conical can facilitate fluid transport by minimizing fluid tension.

4. Photolithography is the process of transferring geometric shapes into a photoresist (photo-reactive epoxy resin) pattern. The Cornell Nano-scale Facility and the National Nanofabrication Users Network have broad resources for learning more about and performing photolithography processes.

5. More detailed information on photolithography and photoresists can be found through MicroChem Corp., with specific information at http://www.microchem. com/products/pdf/SU8_50-100.pdf.

6. The baking time is adjustable; higher temperatures such as 95°C require approx 60 min of baking time, while lower temperatures such as 37°C require baking times of several hours, and overnight baking at low temperatures is acceptable as well.

7. Caution: as with all microscale devices, dust can always be a problem, contaminating or clogging surfaces and channels. All steps should be done under clean, dust-free conditions to the greatest extent possible.

8. Poly (dimethylsiloxane) is a very hydrophobic elastomeric substrate that strongly repels water, preventing steady-state flow conditions required for reproducible gradient generation. Plasma cleaning and forced exposure to water "wet" the material, allowing proper flow through the microchannel network.

9. A variety of other mechanical and non-mechanical types of motors are possible, and listings of these are provided in Whitesides et al., 2001 *(17)*. In our setup, the mechanical stability of the pushing block as well as the mechanism to stabilize the syringe on the syringe pump are critical in generating temporally stable gradients. We made appropriate modifications to our syringe pump (Harvard Apparatus, PHD2000) to meet these criteria.

10. Syringe and motor characteristics determine the range of motor speeds, and therefore the range in flow rates, that are useful in experiments. Syringes must be large enough to contain enough fluid for an experiment but must have relatively small inner diameter for an adequately slow flow rate in the broad channel, for example, a 6-cm long syringe with a 3.26 mm inner diameter that holds up to 500 µL is useful with a precision motor, because the network only holds a fraction of that amount of fluid at a time and flow constriction can be avoided. Refilling or changing solutes in the syringes can be done conveniently and with minimal risk of introducing air bubbles by connecting a three-way Leur cock stop as shown in the setup (as shown in **Fig. 3**).

11. A good starting point for estimating how long to allow gradient deposition to take place is to use the recommended coating density (typically given by manufacturers in $\mu g/cm^2$). This value can be used to calculate how much material to use, and thus how long to allow deposition to take place. However, if deposition is antagonized (e.g., by bovine serum albumin or poly-L-lysine), then the gradient generation could be allowed to progress for longer periods of time.

Acknowledgments

This work was supported by the STC Program of the National Science Foundation under Agreement No. ECS-9876771 and NIH grant GM48050.

References

1. MicroChem Corporation: Website: http://www.microchem.com.
2. Duffy, D. C., McDonald, J. C., Schueller, O. J., and Whitesides, G. M. (1998) Rapid prototyping of microfluidic systems in poly(dimethylsiloxane). *Anal. Chem.* **70,** 4974–4984.
3. Liedberg, B. and Tengvall, P. (1995) Molecular gradients of omega-substituted alkanethiols on gold—preparation and characterization. *Langmuir* **11,** 3821–3827.
4. Liedberg, B., Wirde, M., Tao, Y. T., Tengvall, P., and Gelius, U. (1997) Molecular gradients of omega-substituted alkanethiols on gold studied by X-ray photoelectron spectroscopy. *Langmuir* **13,** 5329–5334.
5. Herbert, C. B., McLernon, T. L., Hypolite, C. L., Adams, D. N., Pikus, L., Huang, C. C., et al. (1997) Micropatterning gradients and controlling surface densities of photoactivatable biomolecules on self-assembled monolayers of oligo (ethylene glycol) alkanethiolates. *Chem. Biol.* **4,** 731–737.
6. Hypolite, C. L., McLernon, T. L., Adams, D. N., Chapman, K. E., Herbert, C. B., Huang, C. C., et al. (1997) Formation of microscale gradients of protein using heterobifunctional photolinkers. *Biconjugate Chem.* **8,** 658–663.
7. Terrill, R. H., Balss K. M., Zhang, Y., and Bohn P. W. (2000) Dynamic monolayer gradients: active spatiotemporal control of alkanethiol coatings on thin gold films. *J. Am. Chem. Soc.* **122,** 988–989.
8. Caelen, I., Bernard, A., Juncker, D., Michel, B., Heinzelmann, H., and Delamarche, E. (2000) Formation of gradients of proteins on surfaces using microfluidic networks. *Langmuir* **16,** 9125–9130.
9. Bradke, F. and Dotti, C. G. (1999) The role of local actin instability in axon formation. *Science* **283,** 1931–1934.
10. Gallardo, B. S., Gupta, V. K., Eagerton, F. D., Jong, L. I., Craig, V. S., Shah, R. R., et al. (1999) Electrochemical principles for active control of liquids on submilliliter scales. *Science* **283,** 57–60.
11. Weiner, O. D., Servant, G., Welch, M. D., Mitchison, T. J., Sedat, J. W., and Bourne, H. R. (1999) Spatial control of actin polymerization during neutrophil chemotaxis. *Nat. Cell Biol.* **1,** 75–81.
12. Wilkinson, P. C. (1998) Assays of leukocyte locomotion and chemotaxis. *J. Immunol. Methods* **216,** 139–153.
13. Jeon, N. L., Dertinger, S. K., Chui, D. T., Choi, I. S., Stroock, A. D., and Whitesides, G. M. (2000) Generation of solution and surface gradients using microfluidic systems. *Langmuir* **16,** 8311–8316.
14. Dertinger, S. K., Chui, D. T., Jeon, N. L., and Whitesides, G. M. (2001) Generation of gradients having complex shapes using microfluidic networks. *Anal. Chem.* **73,** 1240–1246.

15. Stroock, A. D., Dertinger, S. K., Ajdari, A., Mezic, I., Stone, H. A., and Whitesides, G. M. (2002) Chaotic mixer for microchannels. *Science* **295,** 647–651.
16. McDonald, J. C. and Whitesides, G. M. (2002) Poly (dimethylsiloxane) as a material for fabricating microfluidic devices. *Acc. Chem. Res.* **35,** 491–499.
17. Whitesides, G. M., Ostuni, E., Takayama, S., Jiang, X., and Ingber, D. E. (2001) Soft lithography in biology and biochemistry. *Annu. Rev. Biomed. Eng.* **3,** 335–373.
18. Jeon, N. L., Baskaran, H., Dertinger, S. K., Whitesides, G. M., Van de Water, L., and Toner, M. (2002) Neutrophil chemotaxis in linear and complex gradients of interleukin-8 formed in a microfluidic device. *Nat. Biotechnol.* **20,** 826–830.
19. Dertinger, S. K., Jiang, X., Li, Z., Murthy, V. N., and Whitesides, G. M. (2002) Gradients of substrate-bound laminin orient axonal specification of neurons. *Proc. Natl. Acad. Sci. USA* **99,** 12,542–12,547.

W

Whole embryo culture, 253
    Leibovitz-15 media, 253
Wound assay, 255, 263
Wound-healing assay, 23, 24
    cell–cell interactions, 24
    cell interaction with extracellular
        matrix (ECM), 23, 24
    directional cell migration, 23
    image analysis software, 24
    time-lapse microscopy, 24
Wnt signaling non-canonical, 219

X

Xenopus, 9, 235
    cranial neural crest (CNC), 240
        migration assays, 242
        processing and imaging, 243
        removal of CNS, microsurgical, 240
        substrate preparation, 242
    embryo preparation, 237
        labeling, cell outline, 237
        labeling, intracellular, 237
    explant, mesendoderm, 238
        image analysis, 239
        microsurgery and preparation, 239
        migration assays, leading edge, 239
        substrates, preparation, 238
        viewing chamber, preparation, 238

Z

Zebrafish, 8, 212
    care of embryos, 229
    cell movements, 212
    Chordin, 223
    chordino, 219, 223
    convergence, 214
    convergence extension, 215
    extension, 214
    fate or movement, 223
    knypek, 219
    measuring embryo shape, 225
    morphogenetic defects, 216
    morphology, 216, 219
        somitabun (sbn), 222
        spadetail (spt), 222
    morphometric analysis, 216, 225
    movement of cell groups, 226
    photo-activated dye, 226
    preparation of photoactivatible dye,
        229
    silberblick, 219
    somitabun, 219
    stages of development, 212
    STAT3, 221
    trilobite/strabismus, 219
    Wnt 5, 221
    Wnt 11, 221